Soil Stabilization: Principles and Practice

Soil Stabilization: Principles and Practice

Contributors

Suzanne Simard, and Mary Austin et al.

AURIS
Reference

www.aurisreference.com

Soil Stabilization: Principles and Practice

Contributors: Suzanne Simard, and Mary Austin et al.

Published by Auris Reference Limited

www.aurisreference.com

United Kingdom

Soil Stabilization: Principles and Practice

ISBN: 978-1-78154-972-8

British Library Cataloguing in Publication Data

A CIP record for this book is available from the British Library

Printed in the United Kingdom
Exclusively distributed by CBS Publishers & Distributors Pvt. Ltd.
Sales & Distribution Rights only for India, Pakistan, Bangladesh, Sri Lanka, Nepal and Bhutan.This book is not to be sold outside these territories.

Contents

List of Abbreviations

ANN	Artificial Neural Networks
CART	Classification and Regression Trees
CBR	California Bearing Ratio
CCE	Calcium Carbonate Equivalent
CEC	Cation Exchange Capacity
CT	Classification Trees
DEM	Digital Elevation Model
DPM	Decomposable Plant Materials
EC	Electrical Conductivity
ECAF	European Conservation Agriculture Federation
ESCSI	Expanded Shale, Clay, and Slate Institute
FAC	Fulvic Acid Carbon
FFB	Fresh Fruit Bunches
FOM	Floatable Organic Matter
GAM	Generalized Additive Model
GE	Grazing Exclusion
GHG	Green House Gases
GLM	Generalized Linear Model
HAC	Humic Acid Carbon
IMGERS	Inner Mongolia Grassland Ecosystem Research Station
IOM	Inert Organic Matter
LSD	Least Significant Difference
MDD	Maximum Dry Density
NFOM	Non-water-Floatable Organic Matter
NPP	Net Primary Productivity
OM	Organic Matter
OMC	Optimum Moisture Content
PM	Plasma Membrane
RHA	Rice Husk Ash
ROC	Readily-Oxidizable Carbon
RPM	Resistant Plant Materials
RSG	Reference Soil Grous
SOC	Soil Organic Carbon
SOM	Soil Organic Matter
SWC	Soil-Water Content
TC	Total Carbon
TKN	Total Kjeldahl Nitrogen
TKP	Total Kjeldahl Phosphorus
TN	Nitrogen Content
UCS	Unconfined Compressive Strength
WRB	World Reference Base

List of Contributors

Suzanne W. Simard
University of British Columbia, Vancouver, Canada, and Corvallis, USA

Mary E. Austin
University of British Columbia, Vancouver, Canada, and Corvallis, USA

Shih-Hao Jien
Department of Soil and Water Conservation, National Pingtung University of Science and Technology, Pingtung 91201, Taiwan

Chung-Chi Wang
Department of Soil and Water Conservation, National Pingtung University of Science and Technology, Pingtung 91201, Taiwan

Chia-Hsing Lee
Department of Agricultural Chemistry, National Taiwan University, Taipei 10617, Taiwan

Tsung-Yu Lee
Department of Geography, National Taiwan Normal University, Taipei 10610, Taiwan

Chunyan Wang
Key Laboratory of Ecosystem Network Observation and Modeling, Institute of Geographic Sciences and Natural Resources Research, Chinese Academy of Sciences, Beijing, 100101, China
College of Geographical Science, Southwest University, Chongqing, 400715, China

Nianpeng He
Key Laboratory of Ecosystem Network Observation and Modeling, Institute of Geographic Sciences and Natural Resources Research, Chinese Academy of Sciences, Beijing, 100101, China

Jinjing Zhang
College of Resource and Environmental Science, Jilin Agricultural University, Changchun, 130118, China

Yuliang Lv
College of Geographical Science, Southwest University, Chongqing, 400715, China

Li Wang
College of Resource and Environmental Science, Jilin Agricultural University, Changchun, 130118, China

John J. Sloan
Texas AgriLife Research; Dallas, TX, USA

Peter A.Y. Ampim
Texas AgriLife Research; Dallas, TX, USA

Raul I. Cabrera
Texas AgriLife Research; Dallas, TX, USA

Wayne A. Mackay
Mid-Florida Research & Education Center, Apopka, FL, USA

Steve W. George
Texas AgriLife Extension Service; Dallas, TX, USA

B. Suneel Kumar
Graduate Student, Department of Civil Engineering, Geotechnical Engineering, SRM University, Kattankulathur-60320, Tamil Nadu, India

T. V. Preethi
Assistant professor, Department of Civil Engineering, SRM University, Kattanku-lathur-603203, Tamil Nadu, India

T. Watanabe
Graduate School of Agriculture, Hokkaido University, Kita-ku, Sapporo, Japan

M. S. H. Khan
Department of Soil Science, HMD Science and Technology University, Dinajipur, Bangladesh

I. M. Rao
Centro Internacional de Agricultura Tropical (CIAT), A.A.6713, Cali, Colombia

J. Wasaki
Graduate School of Biosphere Science, Hiroshima University, Higashi-Hiroshima, Japan

T. Shinano
National Agricultural Research Center for Hokkaido Region, Sapporo, Japan

M. Ishitani
Centro Internacional de Agricultura Tropical (CIAT), A.A.6713, Cali, Colombia

H. Koyama
Faculty of Applied Biological Sciences, Gifu University, Gifu, Japan

S. Ishikawa
National Institute for Agro-Environmental Science, Tsukuba, Japan

K. Tawaraya
Faculty of Agriculture, Yamagata University, Tsuruoka, Japan

M. Nanamori
Graduate School of Agriculture, Hokkaido University, Kita-ku, Sapporo, Japan

N. Ueki
Faculty of Agriculture, Yamagata University, Tsuruoka, Japan

T. Wagatsuma
Faculty of Agriculture, Yamagata University, Tsuruoka, Japan

O.M. Nieto
IFAPA Centro Camino de Purchil, Junta de Andalucía, Granada, Spain
Dpto. Edafología y Química Agrícola, Facultad de Ciencias, Universidad de Granada, Granada, Spain

J. Castro
IFAPA Centro Camino de Purchil, Junta de Andalucía, Granada, Spain

E. Fernández
Dpto. Edafología y Química Agrícola, Facultad de Ciencias, Universidad de Granada, Granada, Spain

Magdalena Borzecka-Walker
Institute of Soil Science and Plant Cultivation-State Research Institute Poland

Antoni Faber
Institute of Soil Science and Plant Cultivation-State Research Institute Poland

Katarzyna Mizak
Institute of Soil Science and Plant Cultivation-State Research Institute Poland

Rafal Pudelko
Institute of Soil Science and Plant Cultivation-State Research Institute Poland

Alina Syp
Institute of Soil Science and Plant Cultivation-State Research Institute Poland

Mareike Ließ
University of Bayreuth, Department of Geosciences/ Soil Physics, Germany

Bruno Glaser
Martin-Luther University Halle Wittenberg, Soil Biogeochemistry, Germany

Bernd Huwe
University of Bayreuth, Department of Geosciences/ Soil Physics, Germany

J. Shamshuddin
Department of Land Management, Faculty of Agriculture, Universiti Putra Malaysia, 43400 Serdang, Selangor, Malaysia

Noordin Wan Daud
Department of Crop Science, Faculty of Agriculture, Universiti Putra Malaysia, 43400 Serdang, Selangor, Malaysia

Michael A. Blazier
Louisiana State University Agricultural Center, USA

Hal O. Liechty
University of Arkansas Monticello, USA

Lewis A. Gaston
Louisiana State University Agricultural Center, USA

Keith Ellum
University of Arkansas Monticello, USA

Preface

Soil stabilization a general term for any physical, chemical, biological, or combined method of changing a natural soil to meet an engineering purpose. There are many techniques for soil stabilization, including compaction, dewatering and by adding material to the soil. The text *Soil Stabilization: Principles and Practice* focuses on the techniques of stabilizing soils. The objective of first chapter is to review the role of mycorrhizas and mycorrhizal networks in the stability of forest ecosystems and forest soils as climate changes. The stabilization of organic matter by biochar application in compost-amended soils with contrasting pH values and textures has been focused in second chapter. The objectives of third chapter are to investigate the influences of long-term grazing exclusion (GE) on soil organic matter (SOM) composition in semiarid grassland soils, and explore changes in SOM stability with long-term GE. The aim of fourth chapter is to evaluate the dynamics of water and nutrient adsorption by expanded shale (EXSH). Behavior of clayey soil stabilized with rice husk ash and lime has been discussed in fifth chapter. Physiological and biochemical mechanisms of plant adaptation to low-fertility acid soils of the tropics have been presented in sixth chapter. Seventh chapter describes the effect on the soil after the spreading of olive-pruning debris together with the residues of the olive-fruit cleaning in two predominant soils in Andalusian olive orchards. The purpose of eighth chapter is to evaluate carbon sequestration and to present potential bioenergy crops for carbon sequestration in Poland. Ninth chapter focuses on soil-landscape modelling and spatial prediction of soil attributes. Tenth chapter classifies the highly weathered soils in Malaysia and discusses the management of the soils for sustainable production of oil palm, rubber and cocoa. Poultry litter fertilization impacts on soil, plant, and water characteristics in loblolly pine plantations and silvopastures in the mid-South USA have been investigated in last chapter.

Chapter 1

THE ROLE OF MYCORRHIZAS IN FOREST SOIL STABILITY WITH CLIMATE CHANGE

Suzanne W. Simard and Mary E. Austin

University of British Columbia, Vancouver, Canada, and Corvallis, USA

INTRODUCTION

Global change and the related loss of biodiversity as a result of explosive human population growth and consumption are the most important issues of our time. Global change, including climate change, nitrogen deposition, land-use change and species invasions, are altering the function, structure and stability of the Earth's ecosystems (Vitousek, 1994; Lovelock, 2009). Climate change specifically has been marked by an 80% increase in atmospheric CO_2 levels and a 0.74 °C increase in average global near-surface temperature over the period 1906–2005, with average temperature projected to increase by an additional 1 to 6°C by 2100 (IPCC, 2007). Warming is expected to continue for centuries, even if greenhouse gas emission are stabilized, owing to time lags associated with climate processes and feedbacks (IPCC, 2007). Precipitation patterns have changed along with temperature, with average annual increases up to 20% in high-latitude regions but decreases up to 20% in mid- and low-latitudinal regions. The changes in temperature and precipitation patterns have resulted in higher sea levels, decreases in the extent of snow and ice, earlier timing of species spring events, upward and poleward shifts in species ranges, increases and earlier spring run-offs, and increases in forest disturbances by fires, insects and diseases. Of critical importance are the effects of global change on soils. Soils store one-third of the Earth's carbon and, therefore, small shifts in soil biogeochemistry could affect the global carbon balance (Schlesinger & Andrews, 2004). The effects of global change on soils are complex, however, with multiple feedbacks across broad spatiotemporal scales that have the potential to further amplify climate change effects on the ecology of the Earth. Changes in soils are already occuring as a result of climate change, and include increased soil temperatures, increased nutrient availability, melting of permafrost, increased ground instability in mountainous regions, and increased erosion from floods (IPCC, 2007).

Forests are especially important in the carbon balance of the Earth. Even though forests comprise only 30% of the terrestrial ecosystems, they store 86% of the above-ground carbon and 73% of the world's soil carbon (Sedjo, 1993). On average, forests store two-thirds of their carbon in soils, where much of it is protected against turnover in soil aggregates or in chemical complexes (FAO, 2006). Forest soils not only absorb and store large quantities of carbon, they also release greenhouse gases such as CO_2, CH_4 and N_2O. The carbon sink and source strengths of soils have been considered relatively stable globally, with the strong sink strength of northern-mid latitudes roughly balanced by the strong source strength of the tropics (Houghton et al., 2000). However, climate change can upset the soil carbon balance, or its functional stability, by reducing carbon storage and causing a large positive feedback to atmospheric CO_2 levels. Indeed, the amount of CO_2 emissions being sequestered by terrestrial ecosystems is declining and they may become a source by the middle of the 21st century (Cox et al., 2000; Kurz et al., 2008a). When this happens, the atmospheric carbon trajectory will become less dependent on human activities and more so on the much larger carbon pools in terrestrial ecosystems and oceans (Cox et al., 2000). To underscore the gravity of this shift, the magnitude of total belowground respiration is already approximately 10 times greater than fossil fuel emissions annually (Lal, 2004). The effect of climate change on soil functional stability is particularly concerning in high latitude ecosystems (boreal forests, taiga, tundra and polar regions) because these systems store 30% of the Earth's carbon, and are currently warming at the fastest rates globally (IPCC, 2007; Schuur et al., 2009). The tundra-polar regions recently became a net source of atmospheric CO_2 (Apps et al., 2005). The functional stability of soils or ecosystems is defined in this paper as the maintenance of soil or ecological complexity within certain bounds so that key processes (e.g., carbon cycling, productivity) are protected and maintained (Levin, 2005).

Although the climate change forecasts by the IPCC (2007) have the illusion of predictable and steady change over the next century, the real changes in climate will likely be sudden and unexpected (Lovelock, 2009). Indeed, non-linearity, unpredictability and disequilibrium characterize the Earth and its ecosystems as complex systems (Levin, 2005). Congruently, the IPCC (2007) is predicting an increase in the frequency of climatic extremes, such as heavy rains, heat waves and hot days/nights. These will affect disturbances caused by fire, drought, hurricanes, windstorms, icestorms, insect and disease outbreaks, and invasion by exotic species, and these are projected to increase in frequency, extent, severity and intensity as climate changes (Dale et al., 2001). Changes in natural disturbance regimes have the potential to increase the uncertainty in climate change projections because of their large effects on terrestrial carbon pools (Houghton et al., 2000; Kurz et al., 2008b). Disturbances could greatly overshadow the direct incremental effects of climate change on forest soil and ecosystem stabililty, or the effect of mitigation efforts (Kurz et al., 2008b). Large increases in forest fire and insect disturbances in Canada since 1980 have already reduced ecosystem carbon storage (Kurz & Apps, 1999). Disturbances not only kill plants and affect soil carbon storage, but they also accelerate nutrient cycling, alter mycorrhizal communities, and change soil foodweb dynamics.

Carbon storage in soils involves complex feedbacks between plants and soil organisms. Carbon storage depends on the balance between carbon inputs through photosynthesis and outputs through autotrophic (root and mycorrhiza) and heterotrophic (soil microbial) respiration (Bardgett et al., 2008). Both photosynthesis and respiration are directly affected by climate change factors; including atmpospheric CO_2 level, soil nutrient availability, and temperature and precipitation patterns. They are also clearly affected by tree mortality. The direct effects of these climate change factors on plants then feed back to indirectly affect the structure and activity of soil microbial communities, which drive nutrient cycling, soil carbon storage, and soil stability (Bever et al., 2002a). The intimate cascading interaction between plants and soil microbes in their response to climate change factors is likely of critical importance in predicting the consequences of climate change to ecosystem stability and the carbon balance. Although the feedbacks are complex and poorly understood, we are already measuring climate change effects on soil carbon in high latitude ecosystems (Apps et al., 2005; Schuur et al., 2009) as well as on the composition and activity of soil communities involved in soil nutrient cycling in northern forests (Treseder, 2008).

Of the soil microbes, mycorrhizal fungi are likely the most intimately involved and responsive to carbon fluxes between plants, soils and the atmosphere, and hence are important to consider in climate change impacts on terrestrial ecosystems. This is because of their pivotal position at the root-soil interface, where they link the aboveground and belowground components of biogeochemical cycles. Mycorrhizal fungi are obligate symbionts with all forest tree species, where they scavenge soil nutrients and water from the soil in exchange for photosynthate from the tree. Without their fungal symbionts, most trees cannot acquire enough soil resources to grow or reproduce; without the trees, the fungi have insufficient energy to carry out their life cycle. Because of this obligatory exchange, mycorrhizal fungi are considered the primary vectors for plant carbon to soils (Talbot et al., 2008) and, conversely, the primary vectors of soil nutrients to plants (Hobbie & Hobbie, 2006). The fungal partner plays a role in other essential services as well, such as increasing soil structure, protecting soil carbon against mineralization, and protecting tree roots against disease or drought. A single mycorrhizal fungus can also link different plants together, thus forming mycorrhizal networks. These networks have been shown to facilitate regeneration of new seedlings, alter species interactions, and change the dynamics of plant communities (Selosse et al., 2006). As such, mycorrhizas are considered key players in the organization and stability of terrestrial ecosystems (Smith & Read, 1997; Simard, 2009).

The objective of this synthesis paper is to review the role of mycorrhizas and mycorrhizal networks in the stability of forest ecosystems and forest soils as climate changes. We start by reviewing the role mycorrhizal fungi play in soil carbon flux dynamics. We then review some of the direct effects of climate change factors (specifically increased CO_2, nutrient availability, temperature and drought) on plants and mycorrhizal fungi. Next, we briefly review the current and potential effects of climate change on forests in North America. The crux of our review, however, is on the role of mycorrhizas and mycorrhizal networks in helping to mitigate the effects of climate change through their stabilizing effects on forest ecosystems. We use our own research in the interior Douglas-fir forests of western North America to illustrate these stabilizing effects, including the role of mycorrhizal networks in forest recovery following disturbance and in soil carbon flux dynamics. We then discuss the potential roles that management can play in helping maintain forest stability as climate changes. The body of studies suggests that mycorrhizal fungi, and their capacity to stabilize forests, will have a significant impact on the terrestrial portion of the global carbon budget.

THE ROLE OF MYCORRHIZAL FUNGI IN SOIL CARBON FLUXES

The soil carbon pool is 3.3 times larger than the atmospheric carbon pool and 4.5 times larger than the biological carbon pool (Lal, 2004). As a result, the global carbon balance is strongly influenced by soil carbon flux dynamics. The global soil carbon pool is 2500 Gt, and is comprised of 1500 Gt organic carbon (70%) and 950 Gt inorganic carbon (30%) (Schlesinger & Andrews, 2004; Lal, 2004). The organic portion of the soil pool is comprised of plant roots, fungal biomass, microbial biomass, and decaying residues. It includes fastcycling sugars, amino acids and proteins, and slow-cycling cellulose, hemicellulose and lignin. The soil organic pool is highly dynamic, variable, and greatly influenced by land use practices (Rice et al., 2004).

There are three functional groups of fungi in soils: mycorrhizal (i.e., mutalists), saprotrophic (i.e., decomposers) and pathogenic (i.e., parasitic) fungi. Of these, the mycorrhizal (plantfungal) symbiosis is ancient, having evolved over 4.5 million years into a tight mutualism (generally speaking, but there is a continuum in the symbiosis between mutualism and parasitism; Jones & Smith, 2004). The mycorrhizal symbiosis involves thousands of fungal species world-wide (Molina et al., 1992). Mycorrhizas are universally present in all terrestrial biomes, including native forests, woodlands, savannas, grasslands and tundra (Smith & Read, 1997). The three dominant groups are ectomycorrhizas (ECM, primary on trees and shrubs in boreal and temperate ecosystems), ericoid mycorrhizas (ERM, primarily on Ericaceae species in high latitude and high altitude ecosystems), and arbuscular mycorrhizas (AM, primarily on grasses, herbs and tropical tree species).

Many mycorrhizal taxa associate with a broad range of plant species, and thus are considered host generalists. Most fungi in the AM group are considered host generalists, whereas fungi in the ECM and ERM groups include both host generalists and host specialists (i.e., that associate with a narrow group of plant species) (Molina et al., 1992). The low host specificity of many mycorrhizal taxa allows a single mycorrhizal fungal mycelium to link the roots of two or more plants of one or more species in a mycorrhizal network. Increasingly, mycorrhizal networks are recognized as ubiquitous in terrestrial ecosystems, including tropical, temperate and boreal forests (van der Heijden & Horton, 2009). Mycorrhizal networks can function in the mycorrhizal colonization of new seedlings, spread of fungal mycelia, or transfer of carbon, nutrients or water between plants (Simard et al., 2002), thus affecting plant and fungal community dynamics. The architecture of mycorrhizal networks can follow regular, random or scale-free models. In both regular and random networks, links (e.g., fungi) tend to be distributed equally among nodes (e.g., trees). In scalefree models, however, some nodes (e.g., hub trees) are highly linked (Bray, 2003). The architecture of the network reflects its resilience against disturbance (e.g., removal of trees).

All mycorrhizas take up nutrients and water from the soil in exchange for photosynthate carbon from host trees. Photosynthate carbon has been shown to transfer from host plants to mycorrhizal hyphae within hours (Johnson et al., 2002) and this drives half of the belowground microbial activity, with the rest fueled by heterotrophic metabolism of dead organic matter (Högberg & Högberg, 2002). Plants invest photosynthate carbon in mycorrhizas (instead of building their own roots) because the small and profuse hyphae have 60 times more absorptive area than fine roots (Simard et al., 2002). Generally, as nutrient and water limitations increase, plants allocate more photosynthate to mycorrhizal hyphae to increase soil resource uptake. This explains their increasing dominance (relative to bacteria) in high latitude, high altitude or upslope ecosystems (Hobbie, 2006; Högberg et al., 2007). In turn, colonization by mycorrhizal fungi has been shown to up-regulate photosynthesis (Rygiewicz & Anderson, 1994; Miller et al., 2002).

A large portion of photosynthate carbon is allocated belowground and metabolized by roots, mycorrhizal fungi and heterotrophic organisms. The proportion of carbon that is allocated belowground to roots and mycorrhizas has been shown to range from 27-68% of net primary productivity (NPP) in ECM culture studies (Hobbie, 2006). The proportion of carbon allocated directly to mycorrhizal fungi ranges from 1-21% of total NPP (Hobbie, 2006). The amount allocated to root exudation represents 1-10% of NPP, and is important in fueling soil foodwebs and soil organic matter formation (Cardon & Gage, 2006).

Mycorrhizal fungi have a diversity of functions in carbon metabolism. They not only directly access mineral nutrients and water in soil in exchange for photosynthate, they can also decompose soil carbon for energy and nutrient uptake. For example, mycorrhizal fungi have been shown to assimilate simple organic compounds (e.g., amino acids) from the soil solution while in symbiosis (Näsholm et al., 1998). Recently, ECM fungi, ERM fungi and, to a lesser extent AM fungi, have also been found to act as decomposers of larger organic molecules (e.g., proteins, chitin, pectin, hemicellulose, cellulose, polyphenols) by producing extracellular enzymes (e.g., proteases, polyphenol oxidases) (Read & Perez-Moreno, 2003; Tu et al., 2006; Talbot et al., 2008). Talbot et al. (2008) proposed three conditions under which mycorrhizas act as decomposers: (1) when plant photosynthate is low (e.g., in shade, winterearly spring, or when plants are declining) and mycorrhizas require an alternative energy source, (2) when soils are highly organic (e.g., at high latitude or high altitude) and mycorrhizas are required to mine organic nutrients, or (3) when plant productivity is high (especially in crop plants), and mycorrhizal decomposition is primed by large belowground photosynthate carbon fluxes. The model of Talbot et al. (2008) differs from the traditional decomposition model where saprotrophic fungi were considered exclusively responsible for all soil organic matter decomposition. In addition to saproptrophs, however, different taxa of mycorrhiza fungi are now recognized as targeting different carbon sources, implying niche partitioning. This niche partitioning can help explain why such a dazzling diversity of fungi are involved in carbon and nutrient metabolism in soils (Hansen et al., 2008). It also points to the importance of understanding the diverse roles of plants and fungi in global carbon flux dynamics.

Arbuscular mycorrhizal fungi generally do not break down soil organic matter, but they do play important roles in promoting soil aggregation and soil carbon storage. Soil aggregation occurs when hyphae pervade soil pores and entwine soil particles. Mycorrhizal hyphal growth in soils is extensive, with mycelial lengths reaching 111 m cm^{-3} (0.5 mg g^{-1}, or up to 900 kg ha^{-1}) in a prairie soil (Miller et al., 1995). Though AM hyphae turn over quickly (in days to a few months), they also deposit significant quantities of relatively recalcitrant carbon compounds such as chitin and glomalin. Glomalin is a carbon-, nitrogen- and ironrich glycoprotein produced in fungal cell walls (Treseder & Turner, 2007). When it is deposited during decomposition, glomalin joins hyphae in binding small soil particles, thus promoting aggregation and soil stability. Although it constitutes only 0.4-6% of hyphal biomass, glomalin accumulates in soil macro-aggregates at much higher masses (e.g., >100 mg g^{-1}) than does hyphae. In soil aggregates, glomalin carbon is protected from decomposition by chemicals and soil organisms, allowing it to remain in soils for decades and accumulate over time (Rillig et al., 2001; Zhu & Miller, 2004). Carbon in bulk soil, by contrast, is more vulnerable to decomposition. Hence, AM glomalin represents a large pathway for storage of stable carbon in soils. Glomalin content of soils generally increases with the abundance of AM plants and carbon allocation to AM hyphae, and has been shown to represent 3-8% of soil carbon in undisturbed AM grassland and chaparral communities (Rillig et al., 2001).

The composition of mycorrhizal communities shifts with changes in the balance of carbon and nutrients in soils because of fungal species variation in demands for carbon, nitrogen and phosphorus. For example, increases in carbon allocated belowground with CO_2 enrichment or warming may shift the mycorrhizal community toward dominance by high biomass fungi with proteolytic or long-distance exploration capabilities that enable them to compete for scarce nutrients or contribute to soil carbon storage (Treseder, 2005; Hobbie & Hobbie, 2006). These fungi are also considered important in forming mycorrhizal networks with high transfer capacity (Simard & Durall, 2004). In the next section we discuss how climate change can trigger such shifts in the mycorrhizal fungal community.

EFFECTS OF CLIMATE CHANGE FACTORS ON MYCORRHIZAL FUNGI

Climate change is resulting in increasing atmospheric CO_2 concentrations, increasing soil nutrient availability, regional warming and regional drying as a result of fossil fuel burning, land use change, and nutrient pollution. These changes are having multi-faceted effects on plants, mycorrhizal fungi and ecosystems. In this section, we review the key climate factors individually and their potential effects on mycorrhizal fungi.

Atmospheric CO_2 Enrichment

Carbon as atmospheric CO_2 has increased from a pre-industrial level of 280 ppm to 392 ppm in 2010 (Keeling, 1998; IPCC, 2007; http://co2now.org/). The most important effects of atmospheric CO_2 enrichment on mycorrhizal fungi are expected to be indirect through their impacts on plants (Staddon & Fitter, 1998). Plants generally respond to CO_2 enrichment with increased photosynthesis, decreased stomatal conductance, and increased net primary productivity (Poorter, 1993). They also distribute greater amounts of carbon belowground to roots, mycorrhizas, soil foodwebs and exudates (Pritchard et al., 2008; Drigo et al., 2008), due either to greater productivity or shifts in allocation patterns (Zak et al., 2000). Increased availability of carbon to mycorrhizas belowground is considered an important strategy for plants to meet their increasing needs for nutrients and water (Bazazz, 1990; Rogers et al., 1994). In addition to these predicted shifts, mycorrhizal function may also change with increasing CO_2, resulting in lower net carbon costs or increased nutrient-uptake benefits for host plants (Johnson et al., 2005). In addition to acquiring carbon, mycorrhizal fungi also mediate the return of CO_2 to the environment through metabolism and decomposition. The degree to which the increased carbon allocated belowground is rapidly released as CO_2 or allocated to a more recalcitrant soil pool is not well understood.

In keeping with the above predictions, Treseder (2004) found in a meta-analysis of field studies that mycorrhizal abundance increased on average by 47% (84% for AM fungi; 19% for ECM fungi) with increased atmospheric CO_2 concentration; these increases occurred irrespective of biome, level of CO_2 enrichment or measurement method. Meta-analyses are powerful tools that can be used to detect general responses in ecosystems that are often difficult to isolate in individual studies. Individual studies are still critical, however, in uncovering sources of variation and response mechanisms. In long-term CO_2 enrichment experiments, for example, Allen et al. (2005) and Treseder et al. (2003) were able to determine that AM fungal abundance response increased with CO_2 enrichment and peaked at 550-650 ppm. (Some caution is needed in interpreting such experiments because abrupt rises in CO_2 enrichment can over-estimate mycorrhizal responses (Klironomos et al., 2005)). Treseder et al. (2003) also determined that net ecosystem exchange to the atmosphere declined with increasing CO_2, where the extra carbon was added to bulk soil and, to a greater degree, soil macro-aggregates through increased AM hyphal growth and glomalin production. Staddon et al. (1999) also showed that AM fungi stimulated carbon flow belowground with elevated CO_2, but they estimated that most of this belowground carbon was respired. Allen et al. (2005) found that the standing crop of fungi, bacteria and soil organisms did not increase with elevated CO_2 in arid chaparral ecosystems, but they speculated that microbial turnover increased in response to increased carbon allocation belowground. A few other studies have found no effect or even reduced AM fungal colonization with increased CO_2 levels (Staddon & Fitter, 1998). Though the meta-analysis of Treseder (2004) showed strong trends, clearly there are multiple environmental and species influences on mycorrhizal responses to elevated CO_2 that remain to be explored.

Enrichment of CO_2 is expected to cause shifts in mycorrhizal community composition. These shifts will depend on the relative abilities of different fungal taxa to exploit carbon, nitrogen and phosphorus pools, or to acclimatize to the changing environment. Where elevated CO_2 increases belowground carbon allocation and stimulates nutrient deficiencies, "late stage" or medium or long distance "exploration types" of mycorrhizal fungi may be favoured because of their specific exploration strategies for accessing immobile or distant nutrient patches (Agerer, 2001; Hobbie & Agerer, 2010). Where phosphorus is limiting in particular, fungi that invest more carbon into hyphal branching should be favoured because of the relative immobility of this nutrient (Treseder, 2005). In environments where nitrogen is more limiting, however, fungal groups that invest in rhizomorphs that forage over long distances to nitrogen-rich patches should be favoured. Other mycorrhizal fungal taxa may also be favoured in these low nutrient environments because of their adaptations for producing extracellular enzymes to decompose soil organic complexes (see above), or for cultivating associative N-fixing bacteria in their hyphospheres in exchange for nitrogen (Agerer, 2001; Treseder, 2005). Studies examining ECM communities under the low nutrient conditions expected under CO_2 enrichment have found shifts toward morphotypes dominated by extraradical hyphae, rhizomorphs and thin fungal sheaths, and to communities dominated by Cortinarius, Suillus, Tricholoma or Cenococcum (Lilleskov et al., 2001 and 2002; Treseder, 2005). These exploration fungal types are also considered important in the formation of mycorrhizal networks and transfer of nutrients between plants, suggesting that elevated CO_2 may favour the development of more extensive networks that link plants over long distances.

In addition to its effect on the composition of mycorrhizal communities, elevated CO_2 has also been shown to alter the composition of the broader soil microbial community (Allen et al., 2005). Increased carbon allocation to roots and mycorrhizas stimulates soil foodweb activity, but variation in the amount and quality of carbon can favour specific members of the foodweb. For example, saprotrophic fungi have been shown to increase in abundance with rising CO_2 because of greater inputs of root and leaf litter to the soil (Parrent & Vilgalys, 2007). Modifications in litter chemistry, including increases in lignin concentrations with increasing CO_2 levels (Norby et al., 2001), should also have consequences for soil microbial communities (Bradley et al., 2007).

Soil Nutrient Enrichment

Nutrient availability is generally increasing in two ways with global change: through localized anthropogenic nutrient deposition via fertilization and pollution and, to a lesser extent, through increased microbial decomposition with soil warming. Although global change is having the strongest impact on the nitrogen cycle, soil warming has the potential to affect the availability of all soil nutrients. Nitrogen deposition specifically has increased by 3-5 times through industrial fixation and fossil fuel burning, and now exceeds levels of natural nitrogen fixation world-wide (Vitousek, 1994). Because plant productivity is nitrogen-limited globally, NPP has increased and plant distributions have shifted in response to nitrogen enrichment (Vitousek, 1994; Treseder et al., 2005). Currently most nitrogen deposition is in the temperate regions of the USA and Europe, where nitrogen is considered most limiting, but future nitrogen deposition is expected to increasingly occur in the tropics (Dentener et al., 2006). Nutrient enrichment through soil warming can result in increased NPP, but this may ultimately be limited by depletion of the soil nutrient capital (e.g., phosphorus).

In global change studies, scientists have investigated nitrogen enrichment effects on plants and mycorrhizas along nitrogen deposition gradients, in fertilization experiments, and in experiments that have artificially increased soil temperature. These studies have generally shown positive effects of nitrogen enrichment on aboveground plant productivity but negative effects on the belowground foodweb (Treseder et al., 2004). As soil nutrient availability increases, plants have less need for investing carbon into roots, mycorrhizas and microbial activity for nutrient uptake, and therefore they allocate more carbon to aboveground biomass. A recent meta-analysis has indeed shown that industrial nitrogen deposition not only stimulated aboveground forest growth (Thomas et al., 2010), but also reduced soil microbial activity, diversity and soil organic matter decomposition, thus stimulating carbon sequestration in temperate forests (Janssens et al., 2010). Congruently, in a meta-analysis of field fertilization studies, (Treseder, 2004) found that mycorrhizal biomass declined on average by 15% with soil nitrogen enrichment (25% decline in AM biomass versus 5% decline in EM biomass) and by 32% with soil phosphorus enrichment. Similar declines (15%) in total microbial biomass (fungi plus bacteria) with nitrogen additions were observed in a separate meta-analysis of 82 field studies, with greater declines where fertilizer was added over longer periods and at higher amounts (Treseder et al., 2008). Janssens et al. (2010) caution, however, that saturating levels of nitrogen deposition could lead to declines in forest productivity, both above- and belowground, because of soil acidification, leaching of ions and nitrogen, and increasing phosphorus deficiencies. These negative effects may overwhelm any positive effects of nitrogen deposition world-wide, particularly in tropical forests where phosphorus is the primary limitation to tree growth.

Fertilization studies suggest that smaller changes tend to occur in ECM than AM fungal communities and in deciduous than coniferous forests (Peter et al., 2001; Aber et al., 2003; Treseder et al., 2007; Vitousek et al., 2008). Correspondingly, ECM fungal diversity and richness declined in coniferous forests along a nitrogen deposition gradient (Lilleskov et al., 2002), but appeared to decline to a lesser degree in deciduous forests (Arnolds, 1991). The differences in these responses is likely related to the degree to which plant species are nitrogen or phosphorus limited, the diversity of associated fungal species, and the availability of soil mineral and organic nitrogen (Talbot et al., 2008). For example, many deciduous tree species are more nutrient-rich than coniferous species (Simard et al., 1997a; Jerabkova et al., 2006), and should therefore be less sensitive to nutrient additions. In addition, AM plants generally occur in more nutrient rich environments (Smith & Read, 1997), but the wider diversity of fungi that ECM plants host for accessing nutrients may provide a degree of functional similarity that buffers the community against increases in nutrient availability (Jones et al., 2010).

Where nutrients are elevated, "early stage", contact or short distance "exploration types" of mycorrhizal fungi may be favoured because of their ability to rapidly colonize new seedlings and exploit nutrient-rich environments (Deacon & Donaldson, 1983; Hobbie & Agerer, 2010). When plants are initially establishing on disturbed or enriched sites, carbon can be briefly limiting to mycorrhizal growth. Under these conditions, mycorrhizal taxa that allocate more biomass to exchanges sites, such as arbuscules in AM fungi, or the Hartig net in ECM fungi, or those taxa that can acquire carbon from alternate sources, may also be favoured (Treseder, 2005). The decline in ECM fungal diversity observed by Lilleskov et al., (2001, 2002) along a nitrogen deposition gradient corresponded with a shift toward early successional fungi such as Laccaria, Paxillus and Lactarius that posess these characteristics. Early successional fungi have been shown to form mycorrhizal networks in forests and facilitate carbon and nitrogen transfer over short distances (Simard et al., 1997a), but to a smaller degree than later successional fungi (Teste et al., 2009a). Reductions in mycorrhizal richness, whether involving early or later successional fungi, reduces the complexity of mycorrhizal networks, which has corresponded with lower rates of nutrient transfer and survival of establishing seedlings in temperate forests (Teste et al., 2009a).

Nitrogen deposition can not only reduce mycorrhizal activity and diversity, but it can also favour specific saprotrophic communities (Janssens et al., 2010). After 19 years of annual fertilization at Toolik Lake, Alaska, for example, Deslippe et al. (2010) found an increase in the abundance of saprotrophs and small changes in the ECM fungal community. The increasing group of saprotrophs (as discussed by Janssens et al. (2010)) can be superior at producing cellulose-decomposing and phosphate-acquiring enzymes, but not be very efficient at producing lignin-degrading enzymes. Ironically, the saprotrophs can therefore leave more recalcitrant organic matter, ultimately leading to greater accumulation of soil carbon and reducing respiration. Studies show that a large fraction of this soil organic matter is chemically or physically protected from further microbial decay, particularly where it is associated with clay particles. It is important to note that the more decay resistant carbon is the result of saprotrophic biochemical transformations rather than increased soil aggregation; this is because mycorrhizal abundance and rhizodeposition generally decline with increasing nitrogen availability. The long-term stability of these changes in soil carbon is therefore uncertain.

Soil Warming

Plant growth generally increases with soil temperature, but it can also decline where nutrient deficiencies are induced or soil water availability is reduced through increased rates of evapo-transpiration (Pendall et al., 2004). Where plant productivity increases with soil temperature, mycorrhizal and microbial activity are also predicted to increase to help meet increasing nutrient and water demands (Pendall et al., 2004). In keeping with these predictions, mycorrhizal fungal abundance has been shown to increase with soil warming. However, they have also declined initially where limiting thresholds of nutrient or water availability were exceeded (Rustad et al., 2001). Thus, temperature effects on mycorrhizal activity can be mediated through nutrient and water cycles.

Plants and mycorrhizas are not necessarily limited by the same resources at the same time, and feedbacks between climate change factors will mediate plant, mycorrhizal and soil responses to warming (Hobbie, 2000; Pendall et al., 2004). Moreover, plants and mycorrhizal fungi may acclimate to soil temperature changes (Allison et al., 2010). This suggests we should expect variable effects of soil warming on mycorrhizal fungi depending on the type of plant community, the length of time since warming, and feedbacks among different climate processes. For example, mycorrhizal fungal growth increased following 14 years of warming in the Arctic tundra at Toolik Lake, Alaska (Clemmenson et al., 2006) but declined in mature black spruce forests of Alaska (Allison and Treseder, 2008). After 19-years in the warming treatment at Toolik Lake, Deslippe et al. (2010) found that ECM colonization of the dominant tundra shrub, Betula nana, had returned to control levels, suggesting the ECM community had acclimatized to the new conditions. However, they also found an increase in high biomass mycorrhizal fungi with proteolytic capacity, especially Cortinarius, and a reduction in fungi with high affinities for nitrogen, especially Russula, supposedly reflecting Betula nana's increased demand for nutrients bound in soil organic matter with warming. In the black spruce forest, the decline in fungal biomass likely coincided with reductions in soil moisture and increases in nitrogen availability. Exceeding minimum thresholds in soil moisture due to evapotranspiration appears to constrain the predicted mycorrhizal increases with warming, and this ought to occur more commonly in dry than moist forests or than moist tundra underlain by permafrost.

The positive effects of soil warming on the abundance of mycorrhizal fungal taxa with high biomass and long distance exploration types found by Deslippe et al. (2010) suggests that soil warming should promote development of larger mycorrhizal networks. At Toolik Lake, Deslippe & Simard (2010) found that mycorrhizal networks transferred fixed carbon between Betula nana shrubs, but not to other plant species, and the amount was potentially sufficient to affect the performance of Betula nana. Development of larger mycorrhizal networks with warming should therefore favour community dominance by Betula nana and may help explain its current expansion on the Arctic tundra with warming. The increase in carbon uptake throught expansion of Betula nana in the tundra, however, will likely be exceeded by carbon release resulting from permafrost thawing effects (Schuur et al., 2009). In addition to general increases in mycorrhizal biomass with soil temperature, increases in microbial activity should lead to increases in soil organic matter decomposition. This is because microbes produce extracellular enzymes that catalyze the conversion of soil organic matter to dissolved organic carbon, which is the rate-limiting step in decomposition (Allison et al., 2010). The increase in decomposition with soil warming is speculated to offset the increases in carbon allocation belowground with increased atmospheric CO_2. In keeping with this expectation, soil CO_2 and CH_4 emissions have been found to initially increase in soil warming experiments. However, emission rates have been found to then decline back to control levels within a few years once microbes acclimate to the elevated temperature and allocate less carbon toward biomass growth (i.e., they reduce their carbon-use efficiency) (Allison et al., 2010). The short-term nature of respiration increases have also resulted partly from rapid depletion of labile carbon pools (Bradford et al., 2008).

Reduced Soil Water Availability

Reduced precipitation predicted for mid-latitude ecosystems will likely result in increasing water limitations to plant growth (IPCC, 2007). When soil water availability declines, plants should allocate more carbon to mycorrhizal fungi so that they can access scarce soil water (Augé, 2001). Conversely, in areas where precipitation increases, less plant carbon should be allocated to mycorrhizal growth. Studies show that drier conditions have tended to increase arbuscular mycorrhizas as predicted but have had variable effects on ectomycorrhizas (Allison & Treseder, 2008). While water limitations are generally expected to increase with climate change, increases in water use efficiency may buffer some of the negative effects of drought. For example, colonization with mycorrhizal fungi can increase plant water use efficiency due to improved phosphorus nutrition (Augé, 2001). Water-use efficiency of plants has also been shown to increase with atmospheric CO_2 (Bazazz, 1990).

Overall Climate Change Effects

The inter-related effects of climate change factors on forest ecosystems, plants and mycorrhizal fungi are complex and difficult to predict. The results of field studies generally suggest that increased CO_2, soil warming and soil drying should increase plant carbon allocation to mycorrhizas and shift the fungal community to species characterized by high biomass, long distance exporation strategies, and proteolytic capabilities for meeting nutrient demands. These mycorrhizal types should favour development of mycorrhizal networks, which would promote tree species establishment and survival, but they should also have greater decomposition capabilities. Conversely, nutrient enrichment should reduce mycorrhizal fungal abundance and favour early successional mycorrhizal species that are nitrophillic and with more limited networking capacity. This may limit growth of older trees or promote invasion of weedy plants that are less reliant on mycorrhizal fungi for meeting their resource needs. On balance, the present state of knowledge regarding climate change trajectories suggests that forest health will decline in the future and forest soils will become a net source of atmospheric CO_2 (Jones et al., 2004; Pendall et al., 2008; Kliejunas et al., 2009; Kurz et al., 2008a). Changes at high latitudes, including thawing and warming of Arctic and boreal soils are especially at risk of strong positive CO_2 and CH_4 feedbacks to the atmosphere (Schuur et al., 2009), as is evidenced by the recent shift in Arctic soils from being a net carbon sink to a net carbon source (Apps et al., 2005). This has the potential to greatly amplify climate change in the near future (Schuur et al., 2009).

Climate change factors could also alter the functional roles of mycorrhizal species in soil carbon dynamics (e.g., as vectors, scavengers or decomposers) (Talbot et al., 2008). Similarly, these changes could shift the compatibility and cooperation between hosts and fungi along the mutualism-parasitism continuum, and the relative fitness of various mycorrhizal fungi and other microbes that currently protect roots or suppress root disease (Kiers & van der Heijden, 2006; Hoeksema & Forde, 2008; Kliejunas et al., 2009). Specific changes in plant growth and physiology, population genetics, and interactions with changes in the mycorrhizal community, will also affect interplant interactions, plant community composition, and mycorrhizal fungal community composition. Therefore, the direct and indirect effects of climate change on both plants and mycorrhizas should have direct consequences for the global carbon balance.

EFFECTS OF CLIMATE CHANGE ON FORESTS AND THEIR MYCORRHIZAL COMMUNITIES

The effects of climate change on forests are expected to be profound (Aber et al., 2001; Dale et al., 2001). Climate change is expected to affect tree species and forest distributions, forest dynamics and succession, the interactions and co-evolution between trees, mycorrhizal fungi and other mutualists, and ecosystem function (Malcolm et al., 2006; Hamann & Wang, 2006; Whitham et al., 2006). Forest productivity can change slowly in response to the relatively slow and directional changes in mean CO_2 levels, temperature and precipitation, but it can also change rapidly in response to extreme events (e.g., drought, fire, insect outbreaks), which are occurring with greater frequency and severity world-wide (IPCC, 2007; Liu et al., 2010). There are currently 120 documented cases of forest dieback worldwide that are directly attributed to climate change (Allen et al., 2010). Forest decline resulting from climate change, whether due to slow increases in stress or sudden diebacks, has the potential to transform soil microbial communities and cause massive CO_2 feedbacks to the atmosphere. The dieback of 12 million hectares of lodgepole pine due to the mountain pine beetle epidemic in British Columbia, for example, has changed these forests from a net sink to a net source of carbon to the atmosphere (Kurz et al., 2008a).

In the Northern Hemisphere, the distribution of plant communities is expected to change dramatically and idiosyncratically over the next century. The IPCC (2007) generally predicts a northward migration of the boreal and temperate forests, an expansion of prairie and shrub-lands, and a dramatic reduction in the taiga and Arctic tundra. In British Columbia, climate models predict that the sub-boreal, montane and subalpine forests will almost disappear by 2100, while the grasslands and temperate forests will greatly expand to the north (Hamann & Wang, 2006; Spittlehouse, 2008). For each 1°C increase in temperature, forest zones will have to move 160 km (Petit et al., 2004; Hamann & Wang, 2006); for an increase in 4°C over the next century, species in the Northern Hemisphere may have to move northward by 500 km (or 500 m higher in altitude), or a few kilometers per year to find a suitable habitat (IPCC, 2007). This far outpaces the historical tree migration rate of 100-200 m per year estimated from pollen records and chloroplast DNA analyses (MacLachlan & Clark, 2004). If tree species are unable to migrate as predicted or adapt rapidly, they will face extirpation (Aitken et al., 2008). Gene flow of pre-adapted alleles from warmer climates will help tree species migrations at the leading edges of their ranges; however, populations at the rear edges will have greater chance of dieback due to lags in adaptation and migration ability (Aitken et al., 2008). Diebacks and declines are already evident in North American species such as paper birch, trembling aspen, ponderosa pine, pinyon pine and lodgepole pine (Hogg et al., 2002; Mueller et al., 2005; Bouchard et al., 2008; Heineman et al., 2010). Conversely, northward or upward migration is evident in lodgepole pine, white spruce and green alder (Johnstone & Chapin, 2003; Danby & Hik, 2007). Mycorrhizal fungi, through their obligate role in tree establishment, survival and growth, will play a key role in the conservation of core native forests, minimizing diebacks of forests at the trailing edges of tree species ranges, and facilitating migration of tree species at the leading edges of their ranges. Mycorrhizal networks, and their role in mycorrhization and mediation of resource distribution among trees according to need, will likely play a key role in maintaining both the integrity and reorganization of old and new forests. In the next section, we describe how mycorrhizal networks play an important role in the selforganization and stability of forests.

ROLE OF MYCORRHIZAL NETWORKS IN FOREST STABILITY WITH CLIMATE CHANGE

In this section, we argue that the most important role of mycorrhizas with climate change may be in their stabilizing effects on forests that are under increasing environmental stress. The functional significance of mycorrhizal fungi at these higher levels of ecosystem organization is increasingly recognized, including the role of mycorrhizal networks in forest regeneration, succession and resistance against exotic invasions (Nara & Hogetsu, 2004; Simard & Durall, 2004; Selosse et al., 2006; McGuire et al., 2007; Simard, 2009). Mycorrhizal networks may thus provide a community-based model for feedback pathways that promoteforest stability with climate change. We explore this concept with our research on mycorrhizal networks in the interior Douglas-fir forests of British Columbia.

Interior Douglas-fir forests vary widely in composition and structure, from predominantly single-species, uneven-aged forests in the arid and cool climatic regions, to multi-species, even-aged forests in the moist, warm climatic regions of British Columbia. Regardless of this variation in forest composition and structure, Douglas-fir is a dominant tree species. The composition of the fungal community changes with succession, where a few pioneering taxa such as Wilcoxina rehmii and Mycelia atrovirins radicans dominate the roots of Douglas-fir germinants in the first few years following wildfire or harvesting disturbances (Jones et al., 1997; Teste et al., 2009a; Barker et al., 2010). This is followed by rapid succession to a more diverse, late-stage ECM fungal community increasingly dominated by the Rhizopogon vinicolor/R. visiculosus complex (Twieg et al., 2007). The Rhizopogon complex joins up to 63 other ectomycorrhizal species in a complex fungal community colonizing interior Douglasfir (Twieg et al., 2007). Even with shifts in ECM fungal species composition with disturbance and succession, there is enough functional similarity among taxa that total enzyme production by the community remains unchanged (Twieg et al., 2009; Jones et al., 2010). Congruently, seedling nutrient uptake and growth remain stable over a wide range of disturbance severities (Barker & Simard, 2010).

Early and late successional ECM fungi form mycorrhizal networks linking together Douglas-fir trees of many ages in the arid temperate forests (Teste et al., 2009b, Beiler et al., 2010). Douglas-fir can also form linkages with several other tree and shrub species in these forests (Simard et al., 1997b; Hagerman et al., 2001; Twieg et al., 2007). In a dry, uneven-aged interior Douglas-fir forest, we used multi-locus, microsatellite DNA markers to determine that all Douglas-fir trees were interconnected and that the young trees had regenerated within the extensive Rhizopogon network of old veteran Douglas-fir trees (Beiler et al., 2010). Most of the young trees were linked to large, old hub (i.e., highly connected) trees, suggesting the network had scale-free properties; thus, the hub trees were important in selfregeneration of the old-growth forests. In similar forests nearby, we examined this experimentally and showed that seedling establishment success increased by four times where they had full access to the mycorrhizal network of older Douglas-fir trees (Teste & Simard, 2008; Teste et al., 2009b). Access to the network not only improved seedling survival and physiology, but seedlings were colonized by a more complex fungal community and received carbon, nitrogen and water transferred from the older trees (Schoonmaker et al., 2008; Teste et al., 2009a,b). The finding that the network had scale-free properties suggests they were robust against random removal or death of individual trees, which would have little effect on the connectivity of the network (Bray, 2003). By contrast, targeted removal of hub trees, such as through high-grade logging, or insects that selectively attact large trees (e.g., bark beetles), would have negative effects on the regeneration system. In fact, network models have shown that removal of highly connected nodes can cause the scale-free network to stop functioning (Bray, 2003). Random networks, where links are distributed equally among nodes, such as those found in the widely spaced Quercus garryana forests of California (Southworth et al., 2005) are conversely more likely to unravel from random tree removal.

The mixed Douglas-fir – paper birch stands in the moist, warm Interior Cedar-Hemlock forests are more productive and regenerate more readily after disturbance than do the pure Douglas-fir stands of the dry forests (Simard et al., 2005), but mycorrhizal networks alsoplay a role in Douglas-fir regeneration. In the understory of century-old paper birch and Douglas-fir mixtures, establishment success of Douglas-fir has increased where seedlings were linked into the mycorrhizal network of older trees (Simard et al., 1997b). There, greater regeneration success was associated with seedling colonization by a more complex mycorrhizal network associated with the mature trees (Simard et al., 1997b). In nearby clearcuts, Douglas-fir seedlings have also benefited from simple mycorrhizal networks by receiving carbon from neighbouring paper birch, particularly where Douglas-fir was shaded (Simard et al., 1997a). Net carbon transfer followed a source-sink photosynthate gradient, from carbon- and nutrient-rich paper birch source seedlings to increasingly light-stressed Douglas-fir sink seedlings. Traditional models of forest dynamics predict that regeneration patterns are controlled by competitive interactions with neighbours (Oliver & Larson, 1997), but this study showed that facilitation by networks increased regeneration performance and affected interspecific interactions between paper birch and Douglas-fir, encouraging a more diverse tree community. These tree-species-rich forests are also more resilient to insect attack and disease than pure Douglas-fir forests, as shown when deciduous species are removed by weeding or thinning (Morrison et al., 1988; Baleshta et al., 2004; Simard et al., 2005).

Forest ecosystems are dynamic, and this is illustrated by dynamic patterns and processes in mycorrhizal networks. Not only do the complexity and composition of mycorrhizal networks change over time (Twieg et al., 2007), but belowground fluxes of nutrients change over the growing season with shifts in source-sink gradients among networked plants (Lerat et al., 2002). Using dual $^{13}C/^{14}C$ labelling in the field, Philip (2006) found that the direction of net carbon transfer reversed twice over the growing season: (1) from shooting Douglas-fir to bud-bursting birch in spring; (2) then reversing, from nutrient and photosynthateenriched paper birch to stressed understory Douglas-fir in summer; and (3) reversing again, from still-photosynthesizing Douglas-fir to senescent paper birch in the fall. The carbon moved back-and-forth between birch and fir through multiple belowground pathways, including mycorrhizal networks, soils, and a non-networked mycorrhizal-soil pathway (Philip et al., 2010). Here, there appears to be a dynamic interplay between birch, fir and the interconnecting fungi, with carbon and nutrients moving in the direction of greater need over the growing season, resulting in an integrated, dynamic system.

Where severe disturbances remove forest floor and trees in dry climates, mycorrhizal networks are disrupted, resulting in greater reliance of new germinants on mycorrhizal colonization by spores or mycorrhizal fragments in the soil (Teste et al., 2009a; Barker et al., 2010). Although severe disturbances are part of the historic mixed fire regime in the interior Douglas-fir forests, they have usually been infrequent and restricted to small patches (Klenner et al., 2008). As the climate of these forests becomes warmer and drier (Hamann & Wang, 2006), severe disturbances are expected to increase in extent and frequency (Dale et al., 2001), raising concerns about mycorrhizal spore production and dispersal into the disturbed areas. Production of fruiting bodies declines in dry summers (Durall et al., 2006), and belowground dispersal of spores by truffle-forming species such as Rhizopogon may be limited over extensive openings. Douglas-fir seed rain and regeneration have also been sporadic in these forests (Vyse et al., 2006), and the resulting regeneration lags in dry summers could cause local extinction of Rhizopogon and other network-forming ECM fungal taxa. Nevertheless, early successional ECM fungi are host-generalists, and they will continue to play a critical role in seedling establishment following disturbance.

In spite of the greater risk of severe fires disrupting networks in arid climates, the stressgradient hypothesis suggests that biotic facilitation of Douglas-fir regeneration by mycorrhizal networks should be even greater in stressed environments. We tested this hypothesis along an environmental stress gradient caused by soil disturbance in the dry interior Douglas-fir forests. We found that naturally regenerated Douglas-fir seedlings received more transferred carbon through mycorrhizal networks from their neighbours where soils were disturbed by forest floor removal and compaction than where soils were undisturbed, but only where the seedlings were initially well colonized by EM fungi (Teste et al., 2010). Here, disturbance created a sufficient source-sink gradient between seedlings for carbon transfer to occur, but receiving seedlings also had to be healthy and colonized well enough to generate adequate sink strength. We are also testing network facilitation along a regional precipitation gradient across the interior Douglas-fir forests, from the very dry climate of the Interior Douglas-fir zone to the moist climate of the Interior CedarHemock zone (M. Bingham, unpublished data); this regional climate gradient is serving as a proxy for climate change. Early results suggest that, as expected, older Douglas-fir trees transferred more carbon through Rhizopogon-dominated mycorrhiza networks, and thus facilitated tree regeneration more strongly, in dry than in wet climates. A decade of drought combined with western spruce budworm and Douglas-fir bark beetle attack, however, has resulted in extensive dieback of these older trees (Campbell et al., 2003; Maclauchlan et al., 2007). Extensive hub tree mortality in some stands may be exceed thresholds where Rhizopogon networks are no longer sufficiently intact to facilitate regeneration (M. Bingham, unpublished data). Simpler networks comprised of early successional fungi, however, ought to continue to play a critical role in regeneration after disturbance (Barker et al., 2010).

FACILITATING THE STABILIZING EFFECTS OF MYCORRHIZAS THROUGH FOREST MANAGEMENT

Historical Management Practices

Forest management practices that sequester carbon include conservation of native forest, siliviculture practices that emulate natural processes, reforestation of crop-lands, manipulations of tree chemistry to favour lignin, and changes to the soil microbial community. Conversely, those forest management practices that result in net losses of soil organic matter include deforestation or conversion of native forests to plantations (Giller et al., 1997; Lal, 2004; Guo & Gifford, 2002). In an analysis of forest harvesting studies, Nave et al. (2009) showed that harvesting of native forests reduced soil carbon storage by an average of 8%, with considerably more lost from the forest floor (30%) than the mineral soil (no consistent change). In general, forest practices that result in carbon sequestration favour fungi over bacteria (e.g., since fungi have half the respiration rate of bacteria (30-40% versus 60%)). They should also favour fungal taxa that produce prodigious mycorrhizal networks for increased soil aggregation and connectivity, or that produce decay-resistant compounds. In the interior Douglas-fir forests of British Columbia, forest management practices have generally ignored the importance of ECM fungi or mycorrhizal networks in the natural forest regenerative capacity following disturbance. Over the past century, clearcut or highgrade harvesting along with severe insect attacks and severe fire have taken over mixed fire and insect attacks as the primary disturbance agents in these forests (Campbell et al., 2003; Maclauchlan et al., 2007; Klenner et al., 2008). The standard harvesting practice has been to remove the tallest, straightest, largest diameter stems (i.e., the hub trees) for their economic value, and leave patches of smaller residual trees and advance regeneration to grow and disperse seed into the harvested gaps (Vyse et al., 2006). These management practices are characterized by high mortality of establishing seedlings and patchy regeneration (40% survival of planted seedlings in the very dry forests; Simard, 2009). High-grading not only compromises mycorrhizal networks and regenerative capacity but probably also affects genotypic diversity of the trees comprising the forests. Indeed, the high-grading management approach, combined with summer drought (Hamann & Wang, 2006), episodic seed dispersal (Vyse et al., 2006) and gap-phase disturbance regime characteristic of interior Douglas-fir forests (Klenner et al., 2008), has lead to variable natural regeneration success across the dry climatic zones of interior Douglas-fir(Vyse et al., 2006; Stark et al., 2006).

In the moist, warm forests, or at the upper elevations of the dry forests, clearcutting followed by planting has become the most common practice. The interior Douglas-fir nursery stock that is planted on to these sites is grown under high watering and nutrient regimes and thus seedlings are non-mycorrhizal when lifted for planting (Kazantseva et al., 2009). Historically, regeneration of interior Douglas-fir under these conditions has been more or less successful provided site preparation is suitable, frost pockets recognized and weather conditions are favourable. However, the recent increase in extended summer droughts and more variable weather conditions in the wetter forest types, combined with the cumulative high-grading effects and increased severity of natural disturbances (Flannigan et al., 2005; Maclauchlan et al., 2007; Klenner et al., 2008; Kurz et al., 2008a), have changed the structure of these forests and lead to greater uncertainty in regeneration outcomes. Tree mortality is expected to increase even further at species trailing edges as summer drought and disturbance severity increase (Dale et al., 2001; Campbell et al., 2003; Parmesan, 2006). Mycorrhizal colonization and networks should become increasingly important to the recovery of these forests from disturbance as they become increasingly drought-stressed during the summer months as predicted by climate models (Hamann & Wang, 2006).

Forest practices, therefore, can play an important role in carbon sequestration and forest stability under climate stress. Conservation of whole intact forests should be a global priority given the alarming trends in climate change and loss of biodiversity. Where harvesting is necessary, however, retention of hub trees and their mycorrhizal networks should help maintain the strong carbon storage capacity of forests that is critical to the global carbon balance. Just as conserving living trees plays a critical role in conserving mycorrhizal diversity and function, mycorrhizas in turn play a critical role in the self organization and productivity of forests. By contrast, large-scale clearcutting not only increases greenhouse gas emissions (Kurz et al., 2008b), it also removes critical hub trees, threatens biodiversity (Jones et al., 2003; Martin et al., 2004) and could promote decline of nearby forests.

Future Management Practices

Forests may shift to new stability domains with climate change (Suding et al., 2004). As discussed in Section 4, climate models predict a dramatic shift in tree species ranges in North America (Parmesan, 2006; Hamann & Wang, 2006), typically with northward or eastward migration at the leading edges and extensive mortality at the trailing edges. To help mitigate lags in forest re-assembly and minimize the potential for large carbon pulses to the atmosphere, humans can play an important role in mitigating the decline of existing forests and in assisting tree migrations (Rehfeldt et al., 2001). At the trailing edges of tree species ranges, conservation of hub trees forming complex mycorrhizal networks should increase ecosystem stability by facilitating natural regeneration. Conserving forests in these warmer ecosystems will be very important for the source of pre-adapted alleles they provide for currently colder climates (Aitken et al., 2008), as well as for their strong carbon storage capacity. At the leading edges, an important potential barrier in tree migration may be in the colonization of non-local tree genotypes by weakly compatible local mycorrhizal fungi. Although most temperate trees are colonized by both host-specific and host-generalist ECM fungi (Molina et al., 1992), thus providing insurance against negative fungal community composition shifts, the symbiosis of specific plant and fungal pairings can range from mutualistic to parasitic (Bever, 2002a; Klironomos, 2003; Jones & Smith, 2004). Moreover, recent research shows that plants and fungi benefit more frequently with locally adapted associates (Johnson et al., 2010). Indeed, strong feedbacks between compatible symbionts have historically contributed to species coexistence and stability in plant and fungal communities (Bever, 2002b; Klironomos, 2003), whereas weak or antagonistic feedbacks have resulted in forest plantation failures. This research suggests trees that are migrated to new environments may suffer from poor matchings with local mycorrhizal fungi as well as intense competition with existing plants, leading to uncertain performance within local, existing mycorrhizal networks. Such poor marriages could limit the success of assisted tree migrations and contribute to the loss of forest stability (Suding et al., 2004). The loss could be magnified where management practices fail to conserve a diversity of tree and fungal genotypes (Levin, 2005). By contrast, conserving a genetically diverse community of mycorrhizal fungi at the leading edge of tree species ranges may reduce the risk of deleterious matchings and facilitate regeneration of genetically diverse forests with high adaptive capacity (sensu Whitham et al., 2006).

Even with good management and assisted migrations, mature and juvenile tree mortality is expected to increase from disturbances associated with climate change (IPCC, 2007). Mortality can be managed in a manner that eases the transition from one forest type to another, however, by conserving the structural and functional legacies of the original forest and establishing the new forest before the old trees are completely dead. It is well established that healthy plants can transfer nutrients to other healthy plants directly through mycorrhizal networks (Simard & Durall, 2004; Selosse et al., 2006), but even larger amounts may transfer from dying trees to healthy roots (Simard et al., 2002; Pietikäinen & Kytöviita, 2007). Where the dying native forest is protected (i.e., not salvage logged) until the new generation or community of trees is established, the new seedlings may be poised to capture nutrients released from the mycorrhizal network of the dying trees before they are acquired by soil microbes. Where new seedlings are not established during the dying process, the organic compounds exuded from senescing roots may be rapidly immobilized by the rhizosphere microbial community and, through turnover, the CO_2 respired back to the atmosphere and the inorganic nutrients released to the soil solution for microbial uptake, other plant uptake, or leaching. If germinants of native plants can avoid competition with soil microbes by acquiring carbon and nutrients directly from dying trees through a mycorrhizal network, they may establish more rapidly, thus increasing competiveness with non-networking invasive plants and reducing CO_2 feedback to the atmosphere. Mycorrhizal networks connecting new generations with old in forests under climate stress may thus be important in conserving existing forests, facilitating native plant establishment and migration, providing barriers to weed invasion, and mitigating large carbon losses.

CONCLUSIONS

Forests have been diminishing world-wide because of land-use changes and are experiencing additional stress from climate change. While CO_2 enrichment, warming and nutrient pollution are increasing forest productivity and belowground carbon sequestration in North America and Europe, increases in drought, extreme weather events and deforestation practices are also pushing disturbance regimes outside of their natural range. Wide-spread forest diebacks or decline are already occuring in response to increasing drought, wildfire, and insect and disease attacks with climate change. These have the potential to outweigh any positive effects of climate change factors on increased belowground carbon allocation to mycorrhizas, soil microbes or roots. Recovery of these forests is uncertain given the changing dynamics between climate, trees and their mutualists, as well the changing severity and extent of disturbances. The potential for positive feedbacks from dying forest respiration to atmospheric CO_2 levels is high. Humans can play an important role in mitigating forest mortality and assisting migration of species, thus dampening the impacts of climate change.

Mycorrhizas play an important role in the recovery and organization of forests, and it therefore follows that conservation of mycorrhizal fungal communities should help stabilize forests and soils with climate change. Mycorrhizal networks form rapidly following disturbance in the interior Douglas-fir forests of British Columbia, providing critical water and nutrients to establishing seedlings. In mature Douglas-fir forests, most trees, even those of different species, ages and sizes, are connected by a mycorrhizal network. The extensive networks of large hub trees facilitate regeneration of younger trees in the understory, helping them tolerate the stressful environmental conditions. Mycorrhizal networks and hub trees are foundational to the organization of forests because they create favorable local conditions for tree establishment and growth. Therefore, conserving hub trees and mycorrhizal networks appears important to the conservation, regeneration and restoration of forests. Conserving forests, mitigating or managing forest diebacks or declines, and assisting migration of tree species are all important strategies for adapting to the effects of climate change.

REFERENCES

1. Aber, J.D., Neilson, R.F., McNulty, S., Lenihan, J.M., Bachelet, D. & Drapek, R.J. 2001. Forest processes and global environmental change: predicting the effects of individual and multiple stressors. BioScience 51: 735–751.

2. Aber, J.D., Goodale, C.L., Ollinger, S.V., Smith, M.-L., Magill, A.H., Martin, M.E., Hallett, R.A. & Stoddard, J.L. 2003. Is nitrogen deposition altering the nitrogen status of north-eastern forests? BioScience 53: 375-389.

3. Agerer, R. 2001. Exploration types of ectomycorrhizae: A proposal to classify ectomycorrhizal mycelial systems according to their patterns of differentiation and putative ecological importance. Mycorrhiza 11: 107–114.

4.   Aitken, S.N., Yeaman, S., Holliday, J., Wang, T. & Curtis-McLane, S. 2008. Adaptation, migration or extirpation: climate change outcomes for tree populations. Evolutionary Applications 1: 95-111.

5.   Allen, C.D., Macalady, A.K., Chenchouni, H., Bachelet, D., McDowell, N., Vennetier, M., Kitzberger, T., Rigling, A., Breshears, D.D., Hogg, E.H., Gonzalez, P., Fensham, R., Zhang, Z., Castro, J., Demidova, N., Lim, J.-H., Allard, G., Running, S.W., Semerci, A., & Cobb., N. 2010. A global overview of drought and heat-induced tree mortality reveals emerging climate change risks for forests. Forest Ecology & Management 259: 660-684.

6.   Allen, M.F., Klironomos, J.N., Treseder, K.K. & Oechel, W.C. 2005. Responses of soil biota to elevated CO_2 in a chaparral ecosystem. Ecological Applications 15: 1701-1711.

7.   Allison, S.D. & Treseder, K.K. 2008. Warming and drying suppress microbial activity and carbon cycling in boreal forest soils. Global Change Biology 14: 2898-2909.

8.   Allison, S.D., Hanson, C.A. & Treseder, K.K. 2007. Nitrogen fertilization reduces diversity and alters community structure of active fungi in boreal ecosystems. Soil Biology & Biochemistry 39: 1878-1887.

9.   Allison, S.D., Czimczik, C.I. & Treseder, K.K. 2008. Microbial activity and soil respiration under nitrogen addition in Alaskan boreal forest. Global Change Biology 14: 1156- 1168.

10. Allison, S.D., Wallenstein, M.D. & Bradford, M.A. 2010. Soil-carbon response to warming dependent on microbial physiology. Nature Geoscience 3: 336-340.

11. Apps, M. J., Kurz, W.A., Luxmoore, R.J., Nilsson, L.O., Sedjo, R.A., Schmidt, R., Simpson, L.G. & Vinson, T.S. 2005. Boreal forests and tundra. Water, Air, & Soil Pollution 70: 39-53.

12. Arnolds, E. 1991. Decline of ectomycorrhizal fungi in Europe. Agr., Ecos. & Env. 35: 209-244.

13. Augé, R.M. 2001. Water relations, drought and vesicular-arbuscular mycorrhizal symbiosis. Mycorrhiza 11: 3–42.

14. Baleshta K., Simard, S.W., Guy, R.D. & Chanway, C. 2005. Reducing paper birch density increases Douglas-fir growth and Armillaria root disease incidence. Forest Ecology & Management 208: 1-13.

15. Bardgett, R.D., Freeman, C. & Ostle, N.J. 2008. Microbial contributions to climate change through carbon cycle feedbacks. ISME Journal 2: 805-814.

16. Barker, J.S. & Simard, S.W. 2010. Natural regeneration potential of Douglas-fir along wildfire and clearcut severity gradients. In preparation.

17. Barker, J.S., Simard, S.W., Jones, M.D. & Durall, D.M. 2010. The influence of wildfire and clear-cutting on ectomycorrhizas of naturally regenerating interior Douglas-fir. In preparation.

18. Bazazz, F.A. 1990. The response of natural ecosystems to rising global CO_2 levels. Annual Review of Ecology and Systematics 21: 167-196.

19. Beiler, K.J., Durall, D.M., Simard, S.W., Maxwell, S.A. & Kretzer, A.M. 2010. Mapping the wood-wide web: mycorrhizal networks link multiple Douglas-fir cohorts. New Phytologist 185: 543-553.

20. Bever, J.D. 2002a. Host-specificity of AM fungal population growth rates can generate feedback on plant growth. Plant & Soil 244: 281–290.

21. Bever, J.D. 2002b. Negative feedback within a mutualism: Host-specific growth of mycorrhizal fungi reduces plant benefit. Proc. Royal Soc. London B 269: 2595–2601.

22. Bouchard, M., Kneeshaw, D. & Bergeron, Y. 2008. Ecosystem management based on largescale disturbance pulses: A case study from sub-boreal forests of western Quebec (Canada). Forest Ecology & Management 256: 1734-1742.

23. Bradford, M.A., Davies, C.A., Frey, S.D., Maddox, T.R., Melillo, J.M., Mohan, J.E., Reynolds, J.F., Treseder, K.K. & Wallenstein, M.D. 2008. Thermal adaptation of soil microbial respiration to elevated temperature. Ecology Letters 11: 1316-1327.

24. Bradley, K.L., Hancock, J.E., Giardina, C.P., Pregitzer, K.S. 2007. Soil microbial community responses to altered lignin biosynthesis in Populus tremuloides vary among three distinct soils. Plant & Soil 294: 185-201.

25. Bray, D. 2003. Molecular networks: the top-down view. Science 301: 1864–1865.

26. Campbell, R., Smith, D.J. & Arsenault, A. 2003. Multicentury history of western spruce budworm outbreaks in interior Douglas-fir forests. Can. J. For. Res. 36: 1758–1769.

27. Cardon, Z.G. & Gage, D.J. 2006. Resource exchange in the rhizosphere: molecular tools and the microbial perspective. Ann. Rev. Ecol., Evol. Syst. 37: 459-488.

28. Clemmensen, K.E., Michelsen, A., Jonasson, S. & Shaver, G.R. 2006. Increased ectomycorrhizal fungal abundance after long-term fertilization and warming of two Arctic tundra ecosystems. New Phytologist 171: 391–404.

29. Cox P.M, Betts R.A, Jones C.D, Spall S.A, Totterdell I.J. 2000 Acceleration of global warming due to carbon-cycle feedbacks in a coupled climate model. Nature 408: 184–187.

30. Dale, V.H., Joyce, L.A., McNulty, S., Neilson, R.P., Ayres, M.P., Flannigan, M.D., Hanson, P.J., Irland, L.C., Lugo, A.E., Peterson, C.J., Simberloff, D., Swanson, F.J., Stocks, B.J & Wotton, B.W. 2001. Climate change can affect forests by altering the frequency, intensity, duration, and timing of fire, drought, introduced species, insect and pathogen outbreaks, hurricanes, windstorms, ice storms, or landslides. BioScience 51: 723-734.

31. Danby, R.K. & Hik, D.S. 2007. Variability, contingency and rapid change in recent subarctic alpine tree line dynamics. Journal of Ecology 95: 352-363.

32. Deacon, .JW. & Donaldson, S.J. 1983. Sequences and interactions of mycorrhizal fungi on birch. Plant & Soil 71: 257–262.

33. Dentener, F., Drevet, J., Lamarque, J.F., Bey, I., Eickhout, B., Fiore, A.M., Hauglustaine, D., Horowitz, L.W., Krol, M., Kulshrestha, U.C., Lawrence, M., Galy-Lacaux, C., Rast, S., Shindell, D., Stevenson, D., Van Noije, T., Atherton, C., Bell, N., Bergman, D., Butler, T., Cofala, J., Collins, B., Doherty, R., Ellingsen, K., Galloway, J., Gauss, M., Montanaro, V., Müller, J.F., Pitari, G., Rodriguez, J., Sanderson, M., Solmon, F., Strahan, S., Schultz, M., Sudo, K., Szopa, S. & Wild, O. 2006. Nitrogen and sulfur deposition on regional and global scales: A multimodel evaluation. Global Biogeochemical Cycles 20:

34. Deslippe, J.R. & Simard, S.W. 2010. Carbon transfer through mycorrhizal networks may facilitate shrub expansion in Low-Arctic tundra. Ecology Letters, submitted.

35. Deslippe, J.R., Haartman, M., Mohn, W.W. & Simard, S.W. 2010. Long-term experimental manipulation of climate alters the ectomycorrhizal community of Betula nana in Arctic tundra. Global Change Biology, in press.

36. Drigo, B., Kowalchuk, G.A. & van Veen, J.A. 2008. Climate change goes underground: effects of elevated atmospheric CO_2 on microbial community structure and activities in the rhizosphere Biology & Fertility of Soils 44: 667-679.

37. Durall, D.M., Gamiet, S., Simard, S.W., Kudrna, L. & Sakakibara, S.M. 2006. Effects of clearcutting and tree species composition on the diversity and community composition of epigeous fruit bodies formed by ectomycorrhizal fungi. Can. J. Bot. 84: 966-980.

38. FAO. 2006. Global forest resources assessment 2005. FAO Forestry Paper 147, Rome.

39. Flannigan, M.D., Logan, K.A., Amiro, B.D., Skinner, W.R. & Stocks, B.J. 2005. Future area burned in Canada. Climate Change 72: 1–16.

40. Giller, K.E., Beare, M.H., Lavelle, P., Izac, -M.N. & Swift, M.J. 1997. Agricultural intensification, soil biodiversity and agroecosystem function. Applied Soil Ecology 6: 3-16.

41. Guo, L.B. & Gifford, R.M. 2002. Soil carbon stocks and land use change: a meta analysis. Global Change Biology 8: 345-360.

42. Hagerman, S.M., Sakakibara, S.M. & Durall, D.M. 2001. The potential for woody understory plants to provide refuge for ectomycorrhizal inoculum at an interior Douglas-fir forest after clear-cut logging. Can. J. For. Res. 31: 711-721.

43. Hamann, A. & Wang, T. 2006. Potential effects of climate change on ecosystem and tree species distribution in British Columbia. Ecology 87: 2773-2786.

44. Hansen, C.A., Allison, S.D., Bradford, M.A., Wallenstein, M.D. & Treseder, K.K. 2008. Fungal taxa target different carbon sources in forest soil. Ecosystems 11: 1157-1167.

45. Heineman, J.E., Sachs, D.L., Mather, W.J. & Simard, S.W. 2010. Investigating the influence of climate, site, location and treatment factors on damage to young lodgepole pine in British Columbia. Can. J. For.Res. 40: 1109-1127.

46. Hobbie, E.A. 2006. Carbon allocation to ectomycorrhizal fungi correlates with belowground allocation in culture studies. Ecology 87: 563-569.

47. Hobbie, E.A. & Agerer, R. 2010. Nitrogen isotopes in ectomycorrhizal sporocarps correspond to belowground exploration types. Plant & Soil 327: 71-83.

48. Hobbie, J.E. & Hobbie, E.A. 2006 15N in symbiotic fungi and plants estimates nitrogen and carbon flux rates in Arctic. Ecology 87: 816-822.

49. Hobbie, S.E. 2000.Interactions between litter lignin and soil nitrogen availability during leaf litter decomposition in a Hawaiian montane forest. Ecosystems 3: 484-494.

50. Hoeksema, J.D. & Forde, S. E. 2008. A meta-analysis of factors affecting local adaptation between interacting species. Americal Naturalist 171: 275–290.

51. Högberg, P. & Högberg, M.N. 2002. Extramatrical ectomycorrhizal mycelium contributes one-third of microbial biomass and produces,

together with associated roots, half of the dissolved organic carbon in a forest soil. New Phytologist 154: 791-795.

52. Högberg, P., Högberg, M.N., Göttlicher, S.G., Berson, N.R., Keel, S.G., Metcalfe, D.B., Campbell, C., Schindlbacher, A., Hurry, V., Lundmark, T., Linder, S. & Näsholm, T. 2007. High temporal resolution tracing of photosynthate carbon from tree canopy to forest soil microorganisms. New Phytologist 177: 220-228.

53. Hogg, E.H., Brandt, J.P. & Krochtubajda, R. 2002. Growth and dieback of aspen forests in northwestern Alberta in relation to climate and insects. Can. J. For. Res. 32: 823- 832.

54. Houghton, R.A., Skole, D.L., Nobre, C.A., Hackler, J.L., Lawrence, K.T. & Chomentowski, W.H. 2000. Annual fluxes of carbon from deforestation and regrowth in the Brazilian Amazon. Nature 403: 301-304.

55. IPCC. 2007. Climate Change 2007: Synthesis Report. Contribution of Working Groups I, II and III to the Fourth Assessment Report of the Intergovernmental Panel on Climate Change [Core Writing Team, Pachauri, R.K. & Reisinger, A. (eds.)]. IPCC, Geneva, Switzerland, 104 pp.

56. Janssens, I.A., Dieleman, W., Luyssaert, S., subke, J.A., Reichstein, M., Ceulemans, R., Ciais, P., Dolman, A., Grace, J., Matteucci, G., Papale, D., Piao, S.I., schulze, E.D., Tang, J., & Law, B.E. 2010. Reduction of forest soil respiration in response to nitrogen deposition. Nature GeoScience 3: 315-322.

57. Jerabkova, L., Prescott, C.E. & Kishchuk, B.E. 2006. Nitrogen availability in soil and forest floor of contrasting types of boreal mixedwood forests. Can. J. For. Res. 36: 112–122.

58. Johnson, D., Leake, J.R., Ostle, N., Ineson, P. & Read, D.J. 2002. In situ 13CO$_2$ pulse-labelling of upland grassland demonstrates a rapid pathway of carbon flux from arbuscular mycorrhizal mycelia to the soil. New Phytologist 153: 327-334.

59. Johnson, N.C., Wolf, J., Reyes, M.A., Panter, A., Koch, G.W. & Redman, A. 2005. Species of plants and associated arbuscular mycorrhizal fungi mediate mycorrhizal responses to CO$_2$ enrichment. Global Change Biology 11: 1156–1166.

60. Johnson, N.C., Wilson, G.W.T., Bowker, M.A., Wilson, J.A. & Miller, R.M. 2010. Resource limitation is a driver of local adaptation in mycorrhizal symbioses. PNAS 107: 2093–2098.

61. Johnstone, J.F. & Chapin, F.S. 2003. Non-equilibrium succession dynamics indicate continued northern migration of lodgepole pine.

Global Change Biology 9: 1401- 1409.

62. Jones, C., McConnell, C., Coleman, K., Cox, P., Falloon, P., Jenkinson, D. & Powlson, D. 2004. Global climate change and soil carbon stocks; predictions from two contrasting models for the turnover of organic carbon in soil. Global Change Biology 11: 154- 166.

63. Jones, M.D. & Smith, S.E. 2004. Exploring functional definitions of mycorrhizas: Are mycorrhizas always mutualisms? Botany 82: 1089–1109.

64. Jones, M.D., Durall, D.M., Harniman, S.M.K., Classen, D.C. & Simard, S.W. 1997. Ectomycorrhizal diversity on Betula papyrifera and Pseudotsuga menziesii seedlings grown in the greenhouse or outplanted in single-species and mixed plots in southern British Columbia. Can. J. For. Res. 27: 1872-1889.

65. Jones, M.D., Durall, D.M. & Cairney, J.W.G. 2003. Ectomycorrhizal fungal communities in young forest stands regenerating after clearcut logging. New Phytologist 157: 399- 422.

66. Jones MD, Ward V, Twieg BD, Durall DM, & Simard SW. 2010. Functional diversity and redundancy for extracellular enzyme activity of Douglas-fir ectomycorrhizas after wildfire or clearcut logging. Functional Ecology, in press.

67. Kazantseva, O., Bingham, M.A., Simard, S.W. & Berch, S.M. 2009. Effects of growth medium, nutrients, water and aeration on mycorrhization and biomass allocation of greenhouse-grown interior Douglas-fir seedlings. Mycorrhiza 20: 51-66.

68. Keeling, C.D. 1998. Rewards and penalties of monitoring the Earth. Annual Review of Energy and the Environment 23: 25-82.

69. Kiers, E.T. & van der Heijden, M.A. 2006. Mutualistic stability in the arbuscular mycorrhizal symbiosis: exploring hypotheses of evolutionary cooperation. Ecology 87: 1627- 1636.

70. Klenner, W., Walton, R., Arsenault, A. & Kramseter, L. 2008. Dry forests in the southern interior of British Columbia: historic disturbances and implications for restoration and management. Forest Ecology & Management 206: 1711-1722.

71. Kliejunas, J.T., Geils, B.W., Glaeser, J.M., Goheen, E.M., Hennon, P., Kim, M.-S., Kope, H., Stone, J., Sturrock, R. & Frankel, S.J. 2009. Review of literature on climate change and forest diseases of western North America. Gen. Tech. Rep., PSW-GTR-225. Albany, CA: U.S. Department of Agriculture, Forest Service, Pacific Southwest Research

Station. 54 p.

72. Klironomos, J.N. 2003. Variation in plant response to native and exotic mycorrhizal fungi. Ecology 84: 2292–2301.

73. Klironomos, J.N., Allen, M.F., Rillig, M.C., Piotrowski, J., Makvandi-Nejad, S., Wolfe, B.E. & Powell, J.R. 2005. Abrupt rise in atmospheric CO_2 overestimates community response in a model plant–soil system. Nature 433: 621-624.

74. Kurz, W.A. & Apps, M.J. 1999. A 70-year retrospective analysis of carbon fluxes in the Canadian forest sector. Ecological Applications 9: 526-547.

75. Kurz, W.A., Stinson, G. & Rampley, G.J. 2008a. Could increased boreal forest ecosystem productivity offset carbon losses from increased disturbances? Phil. Trans. R. Soc. B 363: 2259-2268.

76. Kurz, W.A., Dymond, C.C., Stinson, G., Rampley, G.J., Neilson, E.T., Carroll, A.L., Ebata, T. & Safranyik, L. 2008b. Mountain pine beetle and forest carbon feedback to climate change. Nature 452: 987-990.

77. Kurz, W.A., Stinson, G., Rampley, G.J., Dymond, C.C. & Neilson, E.T. 2008c. Risk of natural disturbances makes future contribution of Canada's forests to the global carbon cycle highly uncertain. PNAS 105: 1551-1555.

78. Lal, R. 2004. Soil carbon sequestration impacts on global climate change and food security. Science 304, 1623-1627.

79. Lerat, S., Gauci, R., Catford, J.G., Vierheilig, H., Piché, Y. & Lapointe, L. 2002. 14C transfer between the spring ephemeral Erythronium americanum and sugar maple saplings via arbuscular mycorrhizal fungi in natural stands. Oecologia 132: 181–187.

80. Levin, S.A. 2005. Self-organization and the emergence of complexity in ecological systems. BioScience 55: 1075-1079.

81. Lilleskov, E.A., Fahey, T.J. & Lovett, G.M. 2001. Ectomycorrhizal fungal aboveground community change over a nitrogen deposition gradient. Ecological Applications 11: 397-410.

82. Lilleskov, E.A., Fahey, T.J., Horton, T.R. & Lovett, G.M. 2002. Belowground ectomycorrhizal fungal community change over a nitrogen deposition gradient in Alaska. Ecology 83: 104-115

83. Liu, Y., Stanturf, J. & Goodrick, S. 2010. Trends in global wildfire potential in a changing climate. Forest Ecology & Management 259: 685-697.

84. Lovelock, J. 2009. The vanishing face of Gaia. Basic Books, New York.

85. Maclauchlan, L., Cleary, M., Rankin, L., Stock, A. & Buxton, K. 2007.

2007 Overview of Forest Health in the Southern Interior Forest Region. BC Min. For., Kamloops, Canada.

86. McLachlan, J.S. & Clark, J.S. 2004. Reconstructing historical ranges with fossil data at continental scales. Forest Ecology & Management 197: 139-147.

87. Malcolm, J.R., Liu, C., Neilson, R.P., Hansen, L. & Hannah, L. 2006. Global warming and extinctions of endemic species from biodiversity hotspots. Conservation Biology, 20: 538–548.

88. Martin, K., Aitken, K.E.H. & Wiebe, K.L. 2004. Nest sites and nest webs for cavity-nesting communities in interior British Columbia, Canada: nest characteristics and niche partitioning. The Condor 106: 5-19.

89. McGuire, K.L. 2007. Common ectomycorrhizal networks may maintain monodominance in a tropical rain forest. Ecology 88: 567–574.

90. Miller, R.M., Jastrow, J.D. & Reinhardt, D.R. 1995. External hyphal production of vesiculararbuscular mycorrhizal fungi in pasture and tallgrass prairie. Oecologia 103: 17-23.

91. Miller, R.M., Miller, S.P., Jastrow, J.D. & Rivetta, C.B. 2002. Mycorrhizal mediated feedbacks influence net carbon gain & nutrient uptake in Andropogon gerardii. New Phytologist 155:149-162.

92. Molina, R., Massicotte, H. & Trappe, J.M. 1992. Specificity phenomenon in mycorrhizal symbiosis: community-ecological consequences and practical implications. Page 357-423 in Allen, M.F., ed. Mycorrhizal Functioning: An Integrative Plant-Fungal Process. Chapman Hall, New York, NY, USA.

93. Morrison, D.J., Wallis, G.W. & Weir, L.C. 1988. Control of Armillaria and Phellinus root diseases: 20-year results from the Skimikin stump removal experiment. Can. For. Serv., Pac. For. Cen., Victoria, B.C. Inf. Rep. BC-X-302.

94. Mueller, R.C., Scudder, C.M., Porter, M.E., Trotter III, R.T., Gehring, C.A. & Whitham, T.G. 2005. Differential tree mortality in response to severe drought: evidence for longterm vegetation shifts. Journal of Ecology 93: 1085-1093.

95. Näsholm, T., Ekblad, A., Nordin, A., Giesler, R, Högberg, M. & Högberg, P.1998. Boreal forest plants take up organic nitrogen. Nature 392: 914-916.

96. Nara, K. and Hogetsu, T. 2004. Ectomycorrhizal fungi on established shrubs facilitate subsequent seedling establishment of successional plant species. Ecology 85: 1700– 1707.

97. Nave, L.E., Vance, E.D., Swanston, C.W., & Curtis, P.S. 2010. Harvest impacts on soil carbon storage in temperate forests. Forest Ecology & Management 259: 857-866.

98. Norby, R.J., Cotrufo, M.F., Ineson, P., O'Neill, E.G., Canadell, J.G. 2001. Elevated CO_2, litter chemistry, and decomposition: a synthesis. Oecologia 127:153–165.

99. Oliver, C. & Larson, B.C. 1997. Forest stand dynamics: updated edition. Wiley, New York.

100. Parmesan, C. 2006. Ecological and evolutionary responses to recent climate change. Annual Rev. Ecol. Evol. Syst. 37: 637–669.

101. Parrent, J.L. & Vilgalys, R. 2007. Biomass and compositional responses of ectomycorrhizal fungal hyphae to elevated CO_2 and nitrogen fertilization. New Phytologist 176: 164 – 174.

102. Pendall, E., Bridgham, S., Hanson, P.J., Hungate, B., Klicklighter, D.W., Johnson, D.W., Law, B.E., Luo, Y., Megonigal, J.P., Olsrud, M., Ryan, M.G. & Wan, S. 2004. Belowground process responses to elevated CO_2 and temperature: A discussion of observations, measurement methods, and models. New Phytologist 162: 311-322.

103. Pendall, E., Rustad, L. & Schimel, J. 2008. Toward a predictive understanding of belowground process responses to climate change: have we moved any closer? Functional Ecology 22: 937-940.

104. Peter, M., Ayer F. & Egli, S. 2001. Nitrogen addition in a Norway spruce stand altered macromycete sporocarp production and below-ground ectomycorrhizal species composition. New Phytologist 149: 311-325.

105. Petit, R.J., Aguinagalde, I., de Beaulieu, L., Bittkau, C., Brewer, S., Cheddadi, R., Ennos, R., Fineschi, S., Grivet, D., Lascoux, M., Mohanty, A., Müller-Starck, G., DemesureMusch, B., Palmé, A., Martin, J.P., Rendell, S. & Vendramin, G.G. 2003. Glacial refugia: hotspots but not melting pots of genetic diversity. Science 300: 1563-1565.

106. Pietikäinen, A. & Kytöviita, M.-M. 2007. Defoliation changes mycorrhizal benefit and competitive interactions between seedlings and adult plants. Journal of Ecology 95: 639-647.

107. Philip, L.J. 2006. Carbon transfer between ectomycorrhizal paper birch (Betula papyrifera) and Douglas-fir (Pseudotsuga menziesii). PhD thesis, UBC, Vancouver, Canada.

108. Philip, L.J., Simard, S.W. & Jones, M.D. 2010. Pathways for belowground carbon transfer between paper birch and Douglas-fir seedlings. Plant Ecology & Diversity, in press.

109. Poorter, H. 1993. Interspecific variation in the growth response of plants to an elevated ambient CO_2 concentration. Vegetatio 104/105: 77-97.

110. Pritchard, E. T. G., Strand, A.E., McCormack, M.A., Davis, M.A., Finzi, A.C., Jackson, R.B., Roser, M., Rogers, H.H. & Oren, R. 2008. Fine root dynamics in a loblolly pine forest are influenced by free-air-CO_2-enrichment: a six-year-minirhizotron study. Global Change Biology 14: 1–15.

111. Read D. J. & Perez-Moreno, J. 2003. Mycorrhizas and nutrient cycling in ecosystems – a journey towards relevance? New Phytologist 157: 475–492.

112. Rehfeldt, G.E., Wykoff, W.R. & Ying, C.C. 2001. Physiological plasticity, evolution and impacts of a changing climate on Pinus contorta. Climatic Change 50: 355-37.

113. Rice, C.W., White, P.M., Fabrizzi, K.P. & Wilson, G.W.T. 2004. Managing the microbial community for soil carbon management. Supersoil 2004: 3rd Australian New Zealand Soils Conference 5-9 December 2004, University of Sydney, Australia.

114. Rillig, M.C., Wright, S.F., Nichols, K.A., Schmidt, W.F. & Torn, M.S. 2001. Large contribution of arbuscular mycorrhizas to soil C pools in tropical forest soils. Plant & Soil 233: 167-177.

115. Rogers, H.H., Runion, G.B., Krupa, S.V. 1994. Plant responses to atmospheric CO_2 enrichment with emphasis on roots and the rhizosphere. Environmental Pollution 83: 155-189.

116. Rustad, L.E., Campbell, J.L., Marion, G.M., Norby, R.J., Mitchell, M.J., Hartley, A.E., Cornelissen, J.H.C. & Gurevitch, J. 2001. A meta-analysis of the response of soil respiration, net nitrogen mineralization, and aboveground plant growth to experimental ecosystem warming. Oecologia 126: 542–562.

117. Rygiewicz, P.T. & Anderson, C.P. 1994. Mycorrhizae alter quality and quantity of carbon allocated below ground. Nature 369: 58-60.

118. Schlesinger, W.H. & Andrews, J.A. 2004. Soil respiration and the global carbon cycle. Biogeochemistry 48: 7-20.

119. Schoonmaker, A.L., Teste, F.P., Simard, S.W. & Guy, R.D. 2007. Tree proximity, soil pathways and common mycorrhizal networks: their influence on utilization of redistributed water by understory seedlings. Oecologia 154: 455-466.

120. Schuur, E.A.G., Vogel, J.G., Crummer, K.G., Lee, H., Sickman, J.O. & Osterkamp, T.E. 2009. The effect of permafrost thaw on old carbon

release and net carbon exchange from tundra. Nature 459: 556-559.

121. Sedjo, R. 1993. The carbon cycle & global forest ecosystem. Water, Air, Soil Poll. 70: 295-307

122. Selosse, M.-A., Richard, F., He, X. & Simard, S.W. 2006. Mycorrhizal networks: les liaisons dangeureuses? Trends in Ecology & Evolution 21: 621-628.

123. Simard, S.W. 2009. The foundational role of mycorrhizal networks in the self-organization if interior Douglas-fir forests. Forest Ecology & Management 258S: S95–S107.

124. Simard, S.W. & Durall, D.M. 2004. Mycorrhizal networks: a review of their extent, function, and importance. Botany 82: 1140–1165.

125. Simard, S.W., Perry, D.A., Jones, M.D., Myrold, D.D., Durall, D.M. & Molina, R. 1997a. Net transfer of carbon between ectomycorrhizal tree species in the field. Nature 388: 579-582.

126. Simard, S.W., Perry, D.A., Smith, J.W. & Molina, R. 1997b. Effects of soil trenching on occurrence of ectomycorrhizae on Psuedostuga menziesii seedlings grown in mature forests of Betula papyrifera and Psuedotsuga menziesii. New Phytologist 136: 327-340.

127. Simard, S.W., Jones, M.D. & Durall DM. 2002. Carbon and nutrient fluxes within and between mycorrhizal plants. Pages 33-61 in M. van der Heijden and I. Sanders, eds. Mycorrhizal Ecology. Springer-Verlag, Heidelberg. Ecological Studies, Vol. 157.

128. Simard, S.W., Hagerman, S.M., Sachs, D.L., Heineman, J.L. & Mather, W.J. 2005. Conifer growth, Armillaria ostoyae root disease and plant diversity responses to broadleaf competition reduction in temperate mixed forests of BC. Can. J. For. Res. 35: 843- 859.

129. Smith, S.E. & Read, D.J. 1997. Mycorrhizal symbiosis, 2nd ed. Academic Press, London, UK.

130. Southworth, D., He, X.-H., Swenson, W. & Bledsoe, C.S. 2005. Application of network theory to potential mycorrhizal networks. Mycorrhiza 15: 589–595.

131. Spittlehouse, D.L. 2008. Climate change, impacts, and adaptation secenarios: climate change and forest and range management in British Columbia. B.C. Min. For. Range, Res. Br., Victoria, BC. Tech. Rep. 045.

132. Staddon, P.L., & Fitter, A.H. 1998. Does elevated atmospheric carbon dioxide affect arbuscular mycorrhizas? Trends in Ecology & Evolution 13: 455-458.

133. Staddon, P.L., Fitter, A.H. & Robinson, D. 1999. Effects of mycorrhizal

colonization and elevated atmospheric carbon dioxide on carbon fixation and below-ground carbon partitioning in Plantago lanceolata. Journal of Experimental Botany 335: 853-860.

134. Stark, K.E., Arsenault, A. & Bradfield, G.E. 2006. Soil seed banks and plant community assembly following disturbance by fire and logging in the interior Douglas-fir forests of south-central British Columbia. Botany 84: 1548-1560.

135. Suding, K.N., Gross, K.L. & Houseman, G.R. 2004. Alternative states and positive feedbacks in restoration ecology. Trends in Ecology & Evolution 19: 46-53.

136. Talbot, J.M., Allison, S.D. & Treseder, K.K. 2008. Decomposers in disguise: mycorrhizal fungi as regulators of soil carbon dynamics in ecosystems under global change. Functional Ecology 22: 955-963.

137. Teste, F.P. & Simard, S.W. 2008. Mycorrhizal networks and distance from mature trees alter patterns of competition and facilitation in dry Douglas-fir forests. Oecologia 158: 193-203.

138. Teste, F.P., Simard, S.W. & Durall, D.M. 2009a. Role of mycorrhizal networks and tree proximity in ectomycorrhizal colonization of planted seedlings. Fungal Ecology 2: 21-30.

139. Teste, F.P., Simard, S.W., Durall, D.M., Guy, R.D., Jones, M.D. & Schoonmaker, A.L. 2009b. Access to mycorrhizal networks and tree roots: importance for seedling survival and resource transfer. Ecology 90: 2808–2822.

140. Teste FP, Simard SW, Durall DM, Guy RD, & Berch SM. 2010. Net carbon transfer under soil disturbance between Pseudostuga menziesii seedlings in the field. Journal of Ecology 98:429-439.

141. Thomas, C.D., Cameron, A., Green, R.E. Bakkenes, M., Beaumont, L.J., Collingham, Y.C., Erasmus, B.F.N., de Siqueira, M.F., Grainger, A., Hannah, L., Hughes, L., Huntley, B., van Jaarsveld, A.S., Midgley, G.F., Miles, L., Ortega-Huerta, M.A., Peterson, A.T., Philips, O.L. & Williams, S.E. 2004. Extinction risk from climate change. Nature 427: 145-148.

142. Thomas, R.Q., Canham, C.D., Weathers, K.C. & Goodale, C.L. 2010. Increased tree carbon storage in response to nitrogen deposition in the US. Nature Geoscience 3: 13-17.

143. Treseder, K.K. 2004. A meta-analysis of mycorrhizal responses to nitrogen, phosphorus, and atmospheric CO_2 in field studies. New Phytologist 164: 347-355.

144. Treseder, K.K. 2005. Nutrient acquisition strategies of fungi and their

relation to elevated atmospheric CO_2. In: The Fungal Community: Its Organization and Role in the Ecosystem, Third Edition. Edited by: J. Dighton, J.F. White & P. Oudemans. Taylor & Francis Group, LLC. ISBN: 978-1-4200-2789-1. Chapter 26, pp 713-731.

145. Treseder, K.K & Turner, K. 2007. Glomalin in ecosystems. Soil Sci. Soc. Am. J. 71: 1257-1266.

146. Treseder, K.K. 2008. Nitrogen additions and microbial biomass: a meta-analysis of ecosystem studies. Ecology Letters 11: 1111-1120.

147. Treseder, K.K., Egerton-Warburton, L.M., Allen, M.F., Cheng, Y. & Oechel, W.C. 2003. Alteration of soil carbon pools and communities of mycorrhizal fungi in chaparral exposed to elevated carbon dioxide. Ecosystems 6: 786-796.

148. Treseder, K.K., Czimczik, C.I., Trumbore, S.E. & Allison, S.D. 2008. Uptake of an amino acid by ectomycorrhizal fungi in a boreal forest. Soil Biology & Biochemistry 40: 1964- 1966.

149. Tu, C., Booker, F.L., Watson, D.M., Chen, X., Rufty, T.W., Shi, W. & Hu, S.J. 2006. Mycorrhizal mediation of plant N acquisition and residue decomposition: impact of mineral N inputs. Global Change Biology 12, 793–803.

150. Twieg, B., Durall, D.M. & Simard, S.W. 2007. Ectomycorrhizal fungal succession in mixed temperate forests. New Phytologist 176: 437-447.

151. Twieg, B, Durall DM, Simard SW, Jones MD. 2009. Influence of stand age and soil properties on ectomycorrhizal communities in mixed temperate forests. Mycorrhiza 19: 305– 316.

152. Van der Heijden, M.G.A. & Horton, T.R. 2009. Socialism in soil? The importance of mycorrhizal fungal networks for facilitation in natural ecosystems. Journal of Ecology 97, 1139–1150.

153. Vitousek, P.M. 1994. Beyond global warming: ecology and global change. Ecology 75: 1861- 1876.

154. Vitousek, P.M., Aber, J.D., Howarth, R.W., Likens, G.E., Matson, P.A., Schindler, D.W., Schlesinger, W.H. & Tilman, D.G. 1997. Human alteration of the global nitrogen cycle: sources and consequences. Ecological Applications 7: 737-750.

155. Vyse, A., Ferguson. C., Simard, S.W., Kano, T. & Puttonen, P. 2006. Growth of Douglas-fir, lodgepole pine, and ponderosa pine underplanted in a partially-cut, dry Douglasfir stand in south central British Columbia. Forestry Chronicle 82: 723-732.

156. Whitham, T.G., Bailey, J.K., Schweitzer, J.A., Shuster, S.M., Bangert,

R.K., LeRoy, C.J., Lonsdorf, E.V., Allan, G.J., DiFazio, S.P., Potts, P.M., Fischer, D.G., Gehring, K.A., Lindroth, R.L., Marks, J.C., Hart, S.C., Wimp, G.M. & Wooley, S.C. 2006. A framework for community and ecosystem genetics: from genes to ecosystems. Nature Review Genetics 7: 510-523.

157. Zak, D.R., Pregitzer, K.S., Curtis, P.S., Vogel, C.S., Holmes, W.E. & Lussenhop, J. 2000. Atmospheric CO_2, soil-N availability, and allocation of biomass and nitrogen by Populus tremuloides. Ecological Applications 10: 34-46.

158. Zhu, Y.-G. & Miller, R.M. 2004. Carbon cycling by arbuscular mycorrhizal fungi in soil-plant systems. Trends in Plant Science 8: 407-409.

Chapter 2

STABILIZATION OF ORGANIC MATTER BY BIOCHAR APPLICATION IN COMPOST-AMENDED SOILS WITH CONTRASTING PH VALUES AND TEXTURES

Shih-Hao Jien[1], Chung-Chi Wang[1], Chia-Hsing Lee[2] and Tsung-Yu Lee[3]

[1]Department of Soil and Water Conservation, National Pingtung University of Science and Technology, Pingtung 91201, Taiwan

[2]Department of Agricultural Chemistry, National Taiwan University, Taipei 10617, Taiwan

[3]Department of Geography, National Taiwan Normal University, Taipei 10610, Taiwan

ABSTRACT

Food demand and soil sustainability have become urgent concerns because of the impacts of global climate change. In subtropical and tropical regions, practical management that stabilizes and prevents organic fertilizers from rapid decomposition in soils is necessary. This study conducted a short-term (70 days) incubation experiment to assess the effects of biochar application on the decomposition of added bagasse compost in three rural soils with different pH values and textures. Two rice hull biochars, produced through slow pyrolization at 400 °C (RHB-400) and 700 °C (RHB-700), with application rates of 1%, 2%, and 4% (w/w), were separately incorporated into soils with and without compost (1% (w/w) application rate). Experimental results indicated that C mineralization rapidly increased at the beginning in all treatments, particularly in those involving 2% and 4% biochar. The biochar addition increased C mineralization by 7.9%–48% in the compost-amended soils after 70 days incubation while the fractions of mineralized C to applied C significantly decreased. Moreover, the estimated maximum of C mineralization amount in soils treated with both compost and biochar were obviously lower than expectation calculated by a double exponential model (two pool model). Based on the micromorphological observation, added compost was wrapped in the soil aggregates formed after biochar application and then may be protected from decomposing by microbes. Co-application of compost with biochar may be more efficient to stabilize and sequester C than individual application into the studied soils, especially for the biochar produced at high pyrolization temperature.

INTRODUCTION

Climate change and food demand are currently the two most crucial concerns for agricultural scientists throughout the world. How to maintain the soil organic matter (SOM) levels in soils is a key consideration in agricultural productivity and carbon sequestration, particularly in agricultural lands in subtropical and tropical regions. Mekuria *et al.* (2014) [1] mentioned that mulches, compost, or manure can be effective in enhancing soil organic carbon pool and agricultural productivity in the tropic regions, but these amendments were often short-lived. The added organic matters were usually mineralized to CO_2 rapidly leading to large-scale leakage in subtropical/tropical regions. Therefore, developing strategies for reducing the mineralization of added OM and increasing carbon sequestration in subtropical/tropical rural soils is necessary to facilitate land sustainability [2,3].

Biochar is the by-product of the pyrolysis of organic wastes and is regarded as a chemically- and biologically-stable C pool [4]. Applying biochar to agricultural soils is considered to improve soil quality effectively [5,6,7,8] while sequestering carbon and reducing greenhouse gas emission from soils [7,8]. Recently, co-application of biochars and other organic amendments has been determined to be an effective management practice for compensating for the limitations of applying biochar or an organic amendment alone [9,10,11,12]. Nevertheless, most of those studies were conducted at arid, Mediterranean, and temperate regions, and the biochars were incorporated into the soils for stimulating microbial activities to facilitate decomposition of added organic matters. Awad *et al.* (2013) [11] considered that rapid decomposition of plant residue was desired in double-cropping systems in temperate climates in order to maintain a proper supply of nutrients for crop growth and substrates for soil microorganisms. Several inconsistent results including facilitation and inhibition of SOM decomposition in soils co-amended with biochar and other organic amendments have been reported [2,12,13,14], which resulted from the original characteristics of biomass, differences in biochar properties (pyrolization processes and temperatures), and soil environments, such as soil pH value, native SOC contents, and soil texture. Van Veen and Kuikman (1990) [15] and Qayyum *et al.* (2012) [10] indicated that fine soil texture can physically protect SOM from decomposition caused by microbes through the adsorption of organic matter onto clay surfaces.

Regarding C mineralization in biochar-amended soils, Hamer *et al.* (2004) [16] and Kuzyakov *et al.* (2009) [14] reported that co-metabolism occurs in soils incorporated with a biochar and a fresh C source, thus facilitating cumulative CO_2 emission from the amended soils. Zimmerman *et al.* (2011) [17] observed both positive (< 90 days) and negative priming (>250 days) effects of incubating grass and wood biochars in sandy soils. This negative effect could be due to the adsorption of native SOM onto the surfaces and pore spaces of biochars, thus protecting the SOM from decomposition. Furthermore, Keith *et al.* (2011) [18] indicated that adding biochar can lead to a positive priming effect on native SOM but not on added organic matter (OM); however, the added OM can cause a positive priming effect on the biochar.

Many of the aforementioned studies clearly demonstrated the enhancing effects of biochar application on the physiochemical properties of soil. However, to further clarify the interaction between biochar and added organic fertilizers, we aimed to (1) determine the effects of rice husk biochar application on the C mineralization of compost-amended soils with various soil pH values and textures, and (2) examine the micro-structure by using a polarized microscope to determine the possible processes of interaction among the biochar, added organic fertilizer, and clay particles.

MATERIALS AND METHODS

Soil Collection and Biochar Preparation

Surface soil samples (0–15 cm) were collected from three agricultural slopelands in Taiwan. Laopi (Lo) soil is generated from quaternary-aged materials and is widely distributed in the terrace landscapes of Southern Taiwan. Shanhuipu (Sp) and Choutseunlun (Ct) soils are slate alluvial sediments along streams in Southern Taiwan. These three soils were selected because of their wide range of physical and chemical properties (Table 1). The soil samples were air-dried, sieved through a 2-mm screen, and stored in covered plastic containers at 25 °C.

Rice hull was used as a feedstock to separately produce two types of biochar through slow pyrolysis in a furnace equipped with an N_2-purged retort referred to Streubel *et al.* (2011) [19] at 400 °C (RHB-400) and 700 °C (RHB-700), respectively. The furnace was initially heated to 100 °C, and the temperature was then increased to 400 °C and 700 °C, respectively at a rate of 5 °C min^{-1} with a resident time of 30 min. The biochars were subsequently cooled overnight while the N_2 flush was maintained. The biochars were then gently crushed and ground to pass a 2-mm sieve before use and analysis.

Analytical Methods

The pH values of the soil samples and biochars were mixed with deionized water and determined using a glass electrode (1:1 *w/v* for soil; 1:10 *w/v* for biochar) [20]. The electrical conductivity (EC) of the saturated paste extracts of soils was measured using a conductivity meter [21]. The soil particle-size distribution was determined using a pipette method [22]. The cation exchange capacity (CEC) was determined using an ammonium acetate method (pH 7.0) [23]. The organic C content of the tested soils was determined using a wet oxidation method [24]. The total N was measured using the semi-micro-Kjeldahl procedure [25]. The inorganic N was extracted using 2 M KCl (1:10 *w/v*); concentrations of NH_4^+-N and NO_3^--N were estimated using steam distillation involving MgO and the Devarda alloy [26]. The calcium carbonate contents were determined by simple titrimetric method [27], which finely-ground soil and biochar samples (2.0 g) were reacted with 2 M HCl for 16 h. The emitted CO_2 in the reacted bottle was captured by NaOH, and then the base solution was titrated with 0.1 M HCl to calculate carbonate contents. All chemical analyses were conducted intriplicate. Table 1 and Table 2 present summaries of the relevant properties of the soils and biochars.

The readily-oxidizable carbon (ROC) was determined using a method proposed by Blair *et al.* (1995) [28]. Air-dried soil samples containing 15 mg of C were weighed into centrifuge tubes and reacted with333 mM $KMnO_4$ for 1 h at 25 °C. After centrifugation, the supernatants were diluted at a ratio of 1:250 with deionized water. The absorbance of the diluted samples and standards was recorded using a split-beam spectrophotometer at 565 nm. The change in the $KMnO_4$ concentration was used to estimate the amount of C oxidized, assuming that 1 mM $KMnO_4$wasconsumedin the oxidation of 0.75 mM or 9 mg of C. The $KMnO_4$-C fraction, suggested by Blair *et al.* (1995) [28], encompasses all the organic components that can be readily oxidized by $KMnO_4$, including labile humic material and polysaccharides [29], and accounts for 5%–30% of total organic carbon.

Incubation Experiment

A 70-days incubation experiment was conducted to investigate the effects of applying biochars and compost on CO_2emission in three agricultural slopeland soils. Twenty grams of each air-dried soil sample was placed in small plastic cups. A commercial bagasse compost was added as a substrate to the soils for each treatment at a rate of 20 t ha^{-1}. The biochars were then thoroughly mixed with the soilsat 0%, 2%, and 4% (w/w) (approximately 0, 40, and 80 t ha^{-1}, respectively). The experimental design consisted of 10 treatments for each soil in triplicate: (1) O (control; soil only), (2) O + C (soil with 1% compost), (3) O + C + 2% RHB-400, (4) O + C + 4% RHB-400, (5) O + C + 2% RHB-700, (6) O + C + 4% RHB-700, (7) O + 2% RHB-400, (8) O + 4% RHB-400, (9) O + 2% RHB-700, and (10) O + 4% RHB-700. Deionized water was added to the soils to reach 60% water-holding capacity. Each cup with the treated soil and a plastic vessel containing 10 mL of 1 N NaOH solution was placed in a wide-mouth plastic jar, which was subsequently sealed. Jars without treated soils were used as blanks. After 3, 7, 14, 21, 28, 42, and 70 days, the emitted CO_2 was measured in nondestructive determination by titrating the NaOH solution with 0.5 N HCl following addition of $BaCl_2$. The jars were then sealed again for incubation until the next measurement. The incubation experiment was performed in the dark at 25 ± 2 °C [30]. The 70 days of incubation duration was conducted based on Novak et al. (2010) [31] and Streubel et al. (2011) [19] who denoted that the CO_2 evolution rate may approach to a minimum value after 67–75 days.

Calculations and Statistical Analysis

The percentage of applied C mineralized (ACM) in the treatments involving compost and/or biochar was calculated according to Ribeiro et al. (2010) [32]:

$$\text{ACM, \%} = \frac{CMC_{treatment(70\ day)} - CMC_{control(70\ day)}}{organic\ C\ applied} \times 100 \tag{1}$$

where CMC is the cumulative C mineralized in the form of CO_2-C emitted during incubation.

The measured carbon emission results under the treatment of co-application of the biochar and the compost (O + C + RHB) indicate the final results of an overall interaction between the organic amendments. Expected results were calculated from the values of related treatments as follows: (O + C) + (O + RHB) – (O). The expected values indicate no interactions between the compost and biochars. Although the priming effects between compost and biochars were difficult to be determined, the difference between the expected and measured values could be explained as the interaction effect. We used the double exponential model to fit the expected and measured carbon emission of the incubation experiment. The two-pool model involved a labile fraction and a resistant fraction, which can well describe the decomposition of soil N and C [31]:

$$C_{min} = C_l \times \left(1 - \exp^{(-k_l \times t)} \right) + C_r \times \left(1 - \exp^{(-k_r \times t)} \right) \tag{2}$$

where C_{min} is the mineralized C amount at time t (day), C_l and C_r mean the amounts of potentially mineralizeable C (mg C/g C applied) of the labile and resistant fractions, respectively, and k_l and k_r are the respective mineralization rate constants (day^{-1}). The model fitting was carried out using the statistical program of Sigmaplot 8. The maximum values of the unstable C pool were calculated by C_l plus C_r which were derived from the model fitting.

Soil Micromorphology

Kubiena boxes were used to collect undisturbed soil blocks from the experimental pots (Stoops, 2003) [33]. The same solid mixture of each treatment was placed into a pot with a size larger than the Kubiena box followed by the same incubation process as described. After the 70-days incubation, the soil blocks were taken using Kubiena boxes. Thin sections of 30-μm thickness were then prepared following air drying using a microtome by Spectrum Petrographics Inc. (Washington, USA). The thin sections were then used for observing distribution and structure of organic matters among soil particles under a polarized microscope (Leica DM EP, TX, USA).

Statistical Methods

The effects of soil type, biochar type, and application rate of the biochar on the total CO_2-C evolution and their interactions were tested using a multivariate analysis of variance (MANOVA). Significant effects were identified when $p <$ 0.05. Multiple mean comparisons were performed using Fisher's protected least significant difference (LSD) procedures at $p < 0.05$. All the statistical analyses were performed with IBM SPSS Statistics, Version 22 (Somers, NY, USA).

RESULTS

Characteristics of Soils, Compost, and Biochars

Table 1 shows the characteristics of the studied soils. Lo soil is acidic and enriched with silt. Sp soil exhibits a texture similar to that of Lo soil but is pH neutral. The soil textures are silty clay loam and silt loam for Lo and Sp soils, respectively, based on Soil Taxonomy (Soil Survey Staff, 2010) [34]. By contrast, Ct soil exhibits high pH and EC levels, and it has a considerably higher proportion of sand than the other two soils. The Ct soil was classified as a sandy soil [34]. Low total C content was determined in all studied soils; it was below 0.2% in the Ct soil. The CECs of the Lo and Sp soils were approximately 15 cmol (+) kg^{-1}, which was considerably higher than that of the Ct soil. The CECs were consistent with the contents of clay and organic carbon in these soils. Regarding the total nitrogen content (TN) and inorganic nitrogen content (IN; NH_4^+-N + NO_3-N), the Ct soil exhibited a lower TN and IN than the other two soils did. Table 2 lists the properties of the bagasse compost and biochars produced at various temperatures. The pH value of the compost was 5.5, and the pH values of the two biochars were approximately 8.0. The total carbon (TC) of the compost and biochars was in the range of 30%–33%, and the compost contented more TN than the biochars did. The TN content in RHB-400 and RHB-700 were <0.5%. The C/N ratios of the two biochars were >70, which was considerably higher than that of the compost. The exchangeable K content of the compost and biochars were considerably higher than other exchangeable cations. In addition, the compost exhibited more exchangeable Ca and Mg than the biochars did.

Table 2 also reveals that the ROC contents were 3.2 g kg^{-1} and 2.1 g kg^{-1} in the RHB-400 and RHB-700, respectively. The biochars could contribute 64 mg and 42 mg ROC kg^{-1} soil at the application rate of 2% and 128 mg and 84 mg kg^{-1} ROC at the rate of 4%, respectively.

Table 1. Selected properties of the studied soils.

Properties	Soils		
	Laopi (Lo)	Shashuipu (Sp)	Choutseulun (Ct)
pH	4.43	6.90	8.25
EC $^+$ (dS m^{-1})	0.08	0.19	1.96
Sand (%)	14	17	95
Silt (%)	57	66	4.0
Clay (%)	29	17	1.0
Texture $^\#$	SiCL	SiL	S
OC (%)	1.78	1.38	0.12
$CaCO_3$ (g kg^{-1})	0.12	0.27	5.08
CEC (cmol(+) kg^{-1}) ¶	15.0	14.6	4.00
TN § (%)	0.16	0.19	0.01
NH_4^+-N (mg kg^{-1})	2.84	11.2	ND ^{e}
NO_3-N (mg kg^{-1})	14.2	47.8	8.33
ROC (g kg^{-1})	0.42	0.75	ND

$^+$: Electric conductivity; $^\#$: SiCL: silty clay loam; SiL: silt loam; S: sand; ¶: Cation exchange capacity: CEC (cmol(+) kg^{-1}); §:Total nitrogen content; : Readily oxidizable carbon; ^{e}: data not detected.

Table 2. Characteristics of the bagasse compost and the rice hull biochars in this study.

	Compost	RHB-400	RHB-700
pH (1:10 w/v)	5.50	7.99	8.03
TC (%)	30.2	31.0	32.9
TN (%)	1.08	0.41	0.35
C/N ratio	28	76	94
CEC (cmol(+) kg^{-1})	82.5	26.1	35.6
Exchangeable K (g kg^{-1})	6.94	7.01	7.02
Exchangeable Na (g kg^{-1})	0.44	0.28	0.24
Exchangeable Ca (g kg^{-1})	4.61	0.47	0.54
Exchangeable Mg (g kg^{-1})	2.12	0.22	0.23
Carbonate (g kg^{-1})	ND	1.52	1.88
ROC (g kg^{-1})	36.0	3.22	2.10

RHB-400 and RHB-700 are the rice hull biochars produced at 400 °C and 700 °C, respectively. Explanation of the abbreviations are the same of those in Table 1.

Carbon Dioxide Emissions from Soils

The CO_2 evolution rates and cumulative CO_2 emission of the treated soils during 70 days of incubation are shown inFigure 1. The CO_2 evolution rate was slightly lower in RHB-700 than in RHB-400 and exhibited a similar trend between the two biochars for a given treatment. As a representative, the results of RHB-400 treatments were shown in Figure 1.The CO_2evolution rates were obviously higher in the first two weeks for all tested soils than in the following period. The control (O) maintained a low CO_2 evolution rate and consequent cumulative CO_2 emission throughout the incubation period compared with the other treatments (Figure 1), which exhibited final accumulative CO_2 emission of 470, 594, and 213 mg CO_2-C kg soil^{-1} for the Lo, Sp, and Ct soils, respectively. Compared with control (O), application of the compost (O + C) considerably increased the cumulative emission of CO_2 in all studied soils, while co-application of biochars with the compost resulted in even higher values. Maximum amounts of CO_2 emission was observed in the treatment involving co-application of compost and 4% biochar, particularly for the RHB-

400and for the Sp soil. The amounts of the cumulative CO_2 emission of each treatment followed the order of Sp soil > Lo soil > Ct soil ($p < 0.05$) (Table 3).

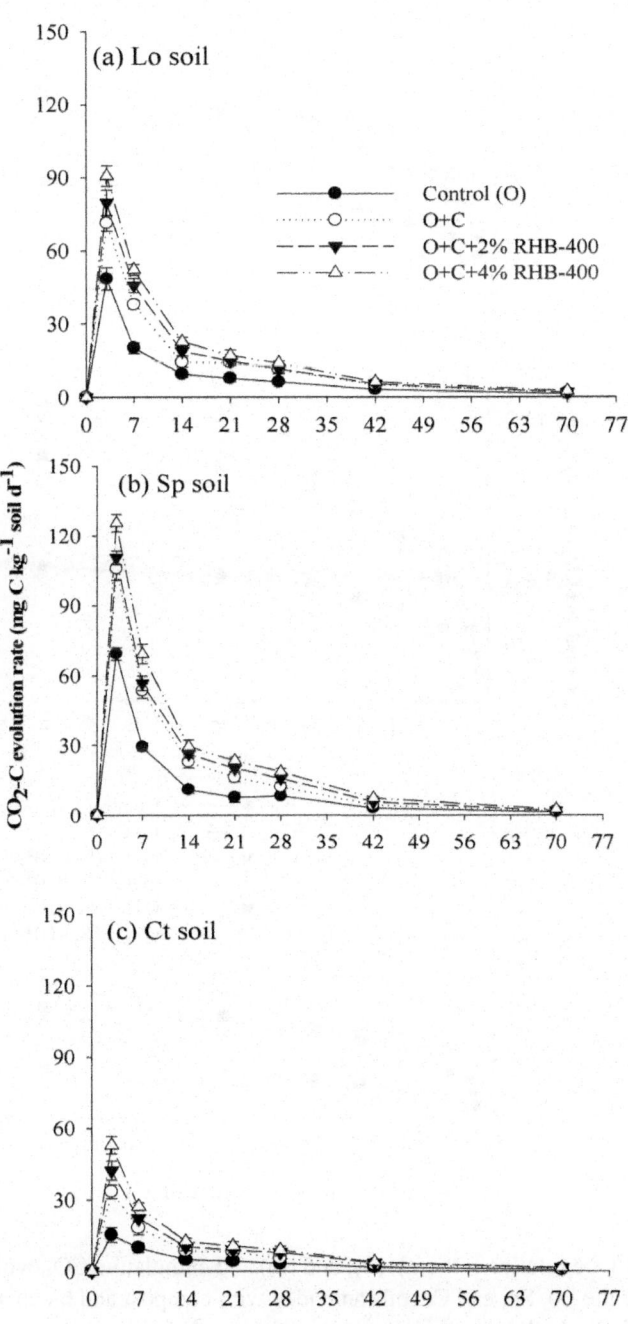

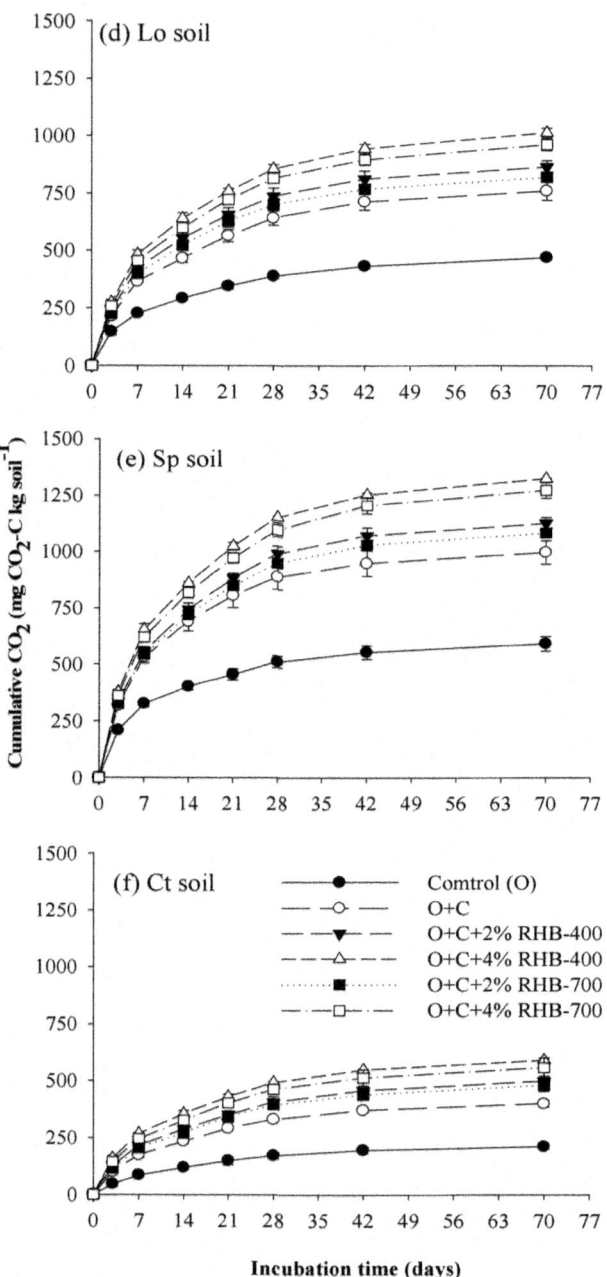

Figure 1. CO_2 evolution rate ((**a**), (**b**), and (**c**)) and cumulative CO_2 evolution ((**d**), (**e**), and (**f**)) for the Lo, Sp, and Ct soils amended with compost and biochars. The vertical error bars indicate the standard deviation. O (control): without compost and biochars; O + C: only compost (1%).

Table 3. Characteristics of the bagasse compost and the rice hull biochars in this study.

	Control	O + C	O + C + 2% RHB-400	O + C + 4% RHB-400	O + C + 2% RHB-700	O + C + 4% RHB-700
Lo	470 ± 4.00 [a]	761 ± 40.6 [a]	865 ± 27.5 [a]	1014 ± 20.2 [a]	821 ± 19.9 [a]	963 ± 25.7 [a]
Sp	594 ± 33.1 [b]	999 ± 51.7 [b]	1127 ± 26.4 [b]	1327 ± 2.12 [b]	1085 ± 11.8 [b]	1274 ± 34.9 [b]
Ct	212 ± 11.3 [c]	400 ± 15.0 [c]	499 ± 21.2 [c]	591 ± 8.09 [c]	482 ± 27.9 [c]	558 ± 43.9 [c]

Different letters along the column (different soil types) mean significant difference ($p < 0.05$) between each soil.

The three soils revealed the similar trends of differences in the ACM (%) among treatments (Figure 2). Application of the compost (O + C treatment) exhibited the highest ACM (%), namely 10.2%, 14.2%, and 6.6% for the Lo, Sp, and Ct soils, respectively, while the values were clearly lower in the treatments of co-application of the compost and biochars. Compared with the O+C treatment, the ACM (%) for the soils amended with compost and biochars significantly decreased by 66%–76%, 65%–76%, and 61%–72% for the Lo, Sp, and Ct soils, respectively. However, for a given biochar application rate, the pyrolization temperature of the biochar did not result in significant differences in the proportion of the ACM. For all treatments, the ACM was apparently higher in the Sp soil than in the other soils, and the Ct soil exhibited the lowest ACM. According to the MANOVA results (Table 4), the soil type and application rate significantly affected the cumulative CO_2 emission ($p < 0.001$), while biochar type had no significant effect. Moreover, a significant interaction between the soil type and the application rate was found ($p < 0.001$).

Table 4. Multivariate Analysis of Variance (MANOVA) of the total amounts of CO_2-C evolved after 70 day incubation for each treatment.

Parameter	mg C kg^{-1} soil		
	Freedom degree	F-value	Significance
Soil type (S)	2	2626.6	< 0.001
Biochar type (B)	1	0.26	0.6094
Application rate (R)	2	243.2	< 0.001
S × R	4	9.02	< 0.001
S × B	2	0.75	0.4802
B × R	2	0.20	0.8166
S × B × R	4	0.04	0.9960

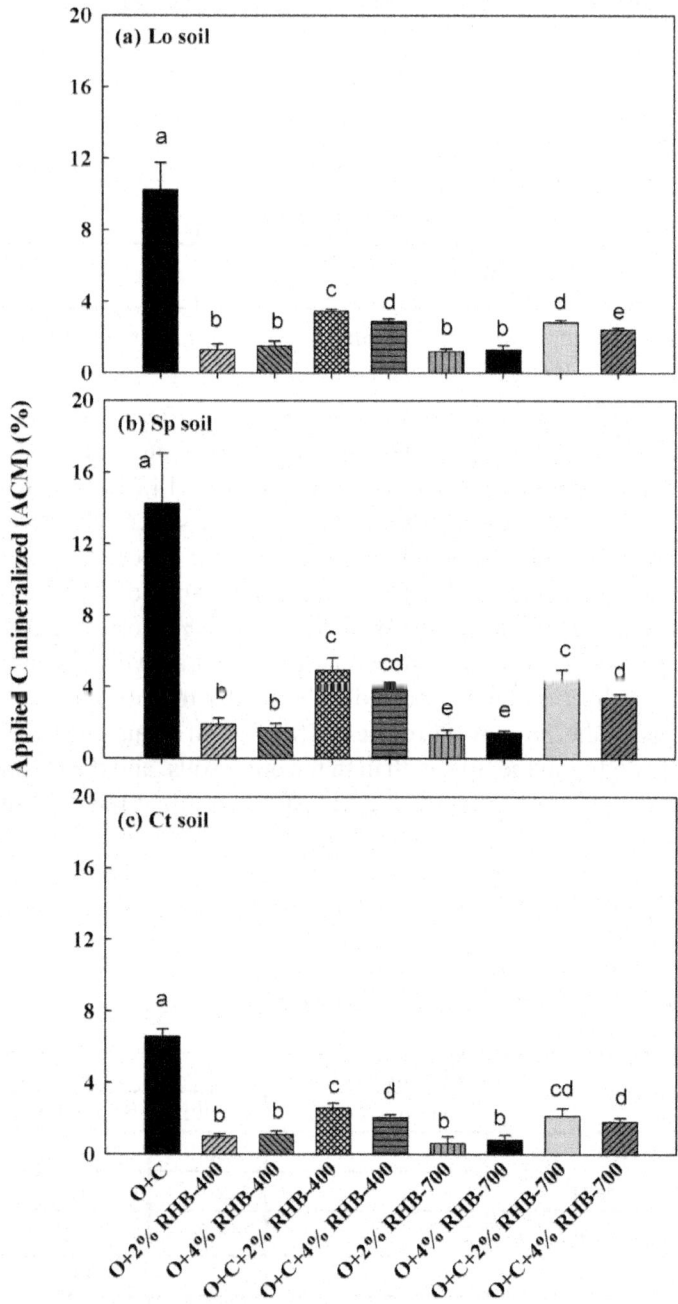

Figure 2. Percentage of the applied carbon mineralized (ACM %) of the Lo (**a**), Sp (**b**), and Ct (**c**) soils amended with compost (1%) and biochars at the end of the incubation. The vertical error bars indicate the standard deviation.

Kinetics of Carbon Mineralization

To clarify the interaction between compost and biochars, a comparison was conducted between the expected values and measured values of cumulative CO_2 emission in the treatment of co-applications (O + C + RHB) (Figure 3a,b). The expected values were calculated with the values from the individual applications of the compost and biochars as follows: (O + C) + (O + RHB) – O. Therefore, the differences between expected and measured values could be attributed to the effect of co-applications. In this study, the three soils exhibited a similar trend in the comparison. The trend found in the Sp soil was illustrated in Figure 3 as a representative of the three soils. The cumulative CO_2 emission curves of the measured values approximately reached a plateau while those of the expected values kept increasing, which indicates that the unstable C pool might decline by co-application.

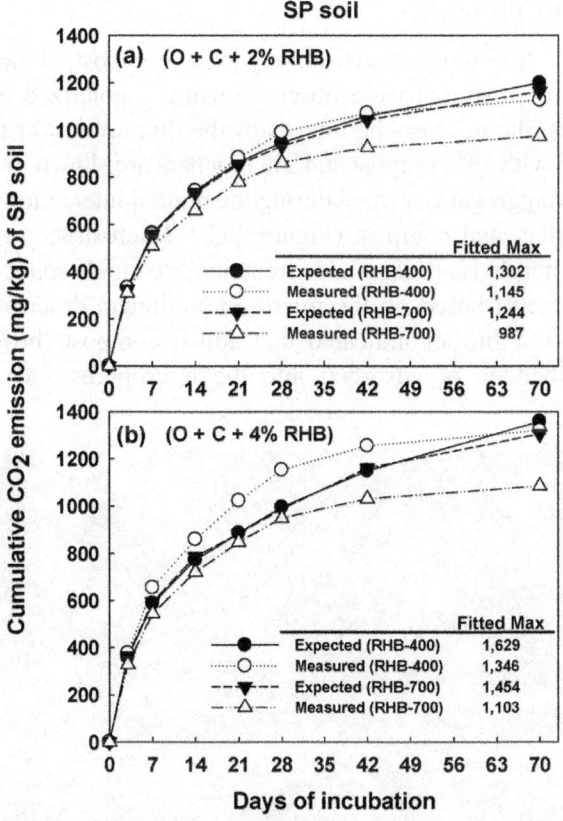

Figure 3. The expected and measured cumulative CO_2 emission in the Sp soil amended with the compost at the rate of 1% and biochars at the rates of 2% (**a**) and 4% (**b**).

Expected values were calculated from the treatments of control (O), compost only (O + C) and biochar only (O + RHB). Fitted Max: the maximum value estimated via two-pool kinetic model fitting.

Double-exponential model (two-pool kinetic model) was used to describe C mineralization of soils amended with biochar and compost. Molina *et al.* (1980) [35] has proposed a two-pool kinetic model of nitrogen mineralization, which was also successfully used to predict carbon mineralization [36]. We fitted the results with the model and estimated the maximum of unstable carbon pool as given in Figure 3. The results of Sp soil were revealed only because the similar trends were found among three studied soils. Co-application of biochar with compost obviously decreased the maximum of unstable carbon pool by 12.1%–17.4% and 20.7%–24.1% for RHB-400 and RHB-700 (Figure 3a,b), respectively.

Soil Micromorphology

To determine the interactions among the compost, biochar, and soil particles, micro-structures were observed using a polarized microscope. As representatives, the microscope images of the thin section of the Lo soil and Sp soil treated with 1% compost and 2% biochars are shown in Figure 4. After 70 days, macroaggregates formed during the mutual interaction among the soil particles, biochar, and compost (Figure 4a,b). Microstructure changed from single spaced porphyric (unamended treatment) to single spaced equal enaulic (biochar treatment) based on the micromorphological description guidelines [33]. Figure 4c–f further indicated that added compost (brown color) was obviously embedded or adsorbed into the micropores and surface of the biochar.

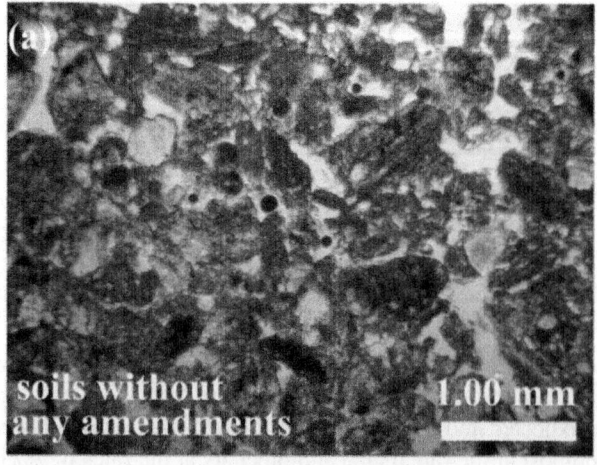

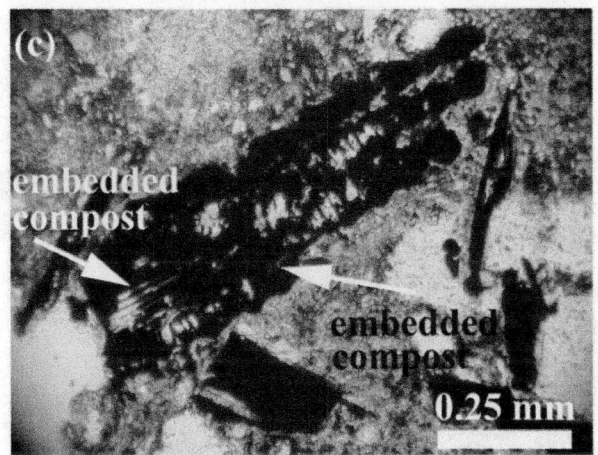

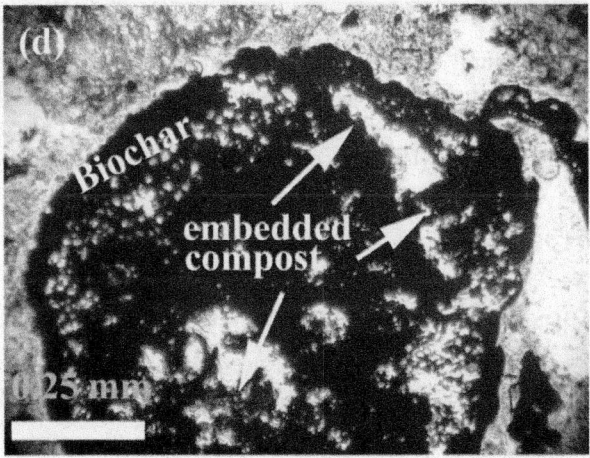

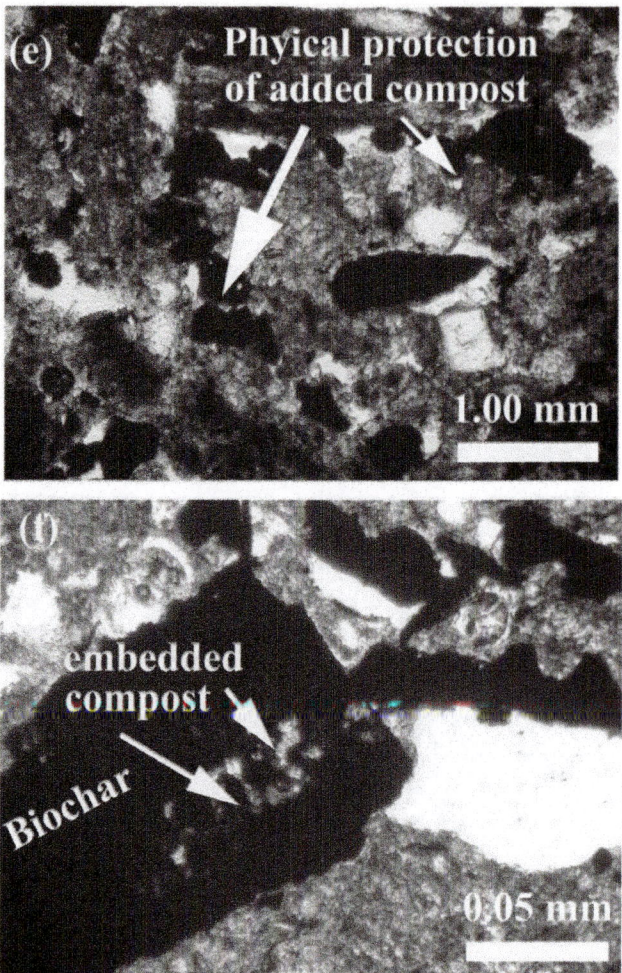

Figure 4. Microstructural observations in the biochar- and compost-amended soils (Lo soil and Sp soil) by using a polarized microscope: (**a**) un-amended Lo soil; (**b**) (**c**) the treatment of 2% RHB-400 + 1% compost in Lo soil, plain polarized light (PPL); (**d**)–(**f**) the treatment of 2% RHB-700 + 1% compost in Sp soil with PPL.

DISCUSSIONS

CO$_2$ Emissions from Soils Amended with the Compost and Biochars

Some researchers have suggested that biochar application in soils might facilitate the decomposition rate of organic matter to maintain nutrients for crops in temperate regions [9,10,11,18,35]. On contrary, in subtropical or tropical regions, it is better to stabilize OM from rapidly decomposing inducing financial lose while reducing compost application for land sustainability. Therefore, a new management strategy to slow down decomposition of organic amendments is necessary, particularly in subtropical and tropical regions.

The current results show that the treatments involving biochar exhibited higher cumulative CO_2 emission than biochar-free treatments (Figure 1). This could be due largely to the significantly higher emission of CO_2 in the first two weeks. Similar results have been reported by some studies [3,9,18], which indicated that labile C in biochars could effectively lead to an increase in CO_2 emission because of priming effects. Deenik *et al.* (2011) [37] demonstrated that biochars with high volatile matter (VM) contents also provide a source of bioavailable C, which stimulates microbial growth and increases C mineralization in soils. Therefore, we deduced that the increased CO_2 emission occurred after biochar addition because of (1) the mineralization of labile C (including VM) in the biochars, (2) interactive priming effects among the biochars, compost, and native SOM, and (3) the facilitation of soil aeration by biochar addition, which could be demonstrated by microstructure observation (Figure 4).

With incubation time, this study verified that added OM could be gradually stabilized through biochar addition. Figure 3 indicates that the cumulative CO_2 emission curves of the measured values approximately reached a plateau while those of the expected values kept increasing, which expressed that the unstable C pool might decline by co-application.

Furthermore, in this study, the biochar produced at a lower pyrolization temperature seemed to induce more cumulative CO_2 emissions in the biochar-amended soils than the biochar produced at a higher temperature did (Figure 1). This may be attributable to a greater proportion of recalcitrant C [12] and lower ROC content in RHB-700 than in RHB-400 (Table 2) [17,18]. We supposed that co-application of biochars with composts may be a better way to stabilized SOM and sequestrate carbon in the soils than individual application, especially for a biochar produced with higher temperatures.

To clarify the interaction among biochar, compost, and soil component, a micro-scale observation was carried out by polarized microscope. From our microstructure observation (Figure 4), a mechanism of SOM stabilization by biochar addition could be deduced as follows: soil structure was changed and some macro-aggregate were formed after biochar incorporation (Figure 4a;4b), which was also provided by our previous studies [38,39]. The formation of the new aggregates wrapped the biochar and compost in the aggregates, and therefore might prevent from rapidly decomposing by microbes (Figure 4c–f).

Accordingly, the decreases in unstable carbon pool (Figure 3) may also result from the sorption of compost-derived carbon onto the biochar, either within the biochar pores (Figure 4c,d,f) or onto the external biochar surfaces (Figure 4e). Cornelissen et al. (2005) [40] and Sobek et al. (2009) [41] reported that biochars exhibit extremely high adsorption affinity for organic matter and might suppress organic C mineralization. In addition, Kasozi et al. (2010) [42] reported that the organic matter sorption onto biochar surfaces is kinetically limited by slow diffusion into the subnanometer-sized pores dominating biochar surfaces. The various organomineral interactions lead to aggregations of clay particles and organic materials, which stabilizes both soil structure and the carbon compounds within the aggregates.

Effect of Soil Type on Carbon Mineralization with Compost and Biochar Amendment

According to the results shown in Figure 1, the differences in cumulative CO_2 emission between the treatments with and without biochar were approximately 59.9–252, 85.8–327, and 81.0–190 mg/kg for the Lo, Sp, and Ct soils, respectively. Siguaet al. (2014) [43] has incorporated several biochars into loamy and sandy soils, and the loamy soil exhibited a cumulative CO_2-C emission that was two to three-fold higher than that of the sandy soil, which was explained by the higher content of labile SOC in the loamy soil. In this study, the highest cumulative CO_2 emission was observed in the Sp soil for all treatments (Figure 1), which may be attributable to the higher ROC content of the Sp soil compared with other soils (Table 1).

Except for the effects of the initial labile C pool on the cumulative CO_2 emission, soil texture, and soil pH are also suggested to be critical control factors in carbon decomposition or CO_2 emission. Sissoko and Kpomblekou-A (2010) [44] indicated that the stabilizing effect of organic matter in soils contributed to the encapsulation between clay particles and entrapment of organic matter in small pores of aggregates, which are inaccessible to microbes. Qauuym et al. (2012) [11] revealed that charcoal is considerably more stable in Oxisol than in Alfisol, because Oxisol generally contains higher clay and Fe/Al oxide content. Furthermore, fine soil texture can physically protect SOM against decomposition by microbes through the adsorption of organic matter onto the inorganic clay surface and the entrapment of organic matter in small aggregates [15,45].However, the lowest rate and cumulative amount of CO_2 emission for all treatments were found in the Ct soil, which had much lower clay than the other two soils (Figure 1). Moreover, the ACM (%) was higher in the Sp soil for all treatments than the Lo and Ct soils (Figure 2), which indicates that carbon derived either from the compost or biochars decomposed more rapidly in the Sp soil despite its clay content not being the lowest.

The highest pH value (pH 8) of the Ct soil might alter the microbial population and therefore lead to a lower cumulative CO_2 emission than that of the other two soils. Soil pH value might influence the microbial activity and consequently limit the decomposition of applied organic amendments [46,47]. Therefore, the neutral Sp soil may be more suitable for microbial activity and consequently exhibited higher CO_2 emission than the other two soils. Our results implied that the efficiency of carbon stabilization caused by biochar application may be more sensitive to the soil pH than the clay content.

CONCLUSIONS

One of the best management practices (BMPs) for land sustainability in subtropics and tropics is long-term stabilization of SOM. Based on our results, the potential benefits of biochar application could reduce the C mineralization of the added compost through mutual interaction of biochars and compost and, thus, extend the efficiency of the compost application. Co-application of compost with biochar may be more efficient to sequester C than individual application into soils, especially for the biochar produced at high pyrolization temperature. In this study, applying 4% of both husk biochars produced at 400 and 700 °C to soils with 1% compost provided the highest efficiency in reducing the C loss in soils.

ACKNOWLEDGMENTS

The authors thank the Ministry of Science and Technology, Republic of China, for financially supporting this research under contract number NSC-101-2313-B-020-013-MY2. The authors are also grateful to Chuan-Chi Chien from the Industrial Technology Research Institute, Tainan, Taiwan, for providing the rice hull biochar.

AUTHOR CONTRIBUTIONS

Dr. Shih-Hao Jien designed all research, made all tables and figures and finished this paper writing; Chung-Chi Wang, who was the master student graduated from Dr. Jien's lab performed this research and analyzed the data; Dr. Chia-Hsing Lee and Tsung-Yu Lee provided their valuable opinions during the manuscript writing. All authors read and approved the final manuscript.

REFERENCES

1. Mekuria, W.; Noble, A.; Sengtaheuanghoung, O.; Hoanh, C.T.; Bossio, D.; Sipaseuth, N.; McCartney, M.; Langan, S. Organic and Clay-Based Soil Amendments Increase Maize Yield, Total Nutrient Uptake, and Soil Properties in Lao PDR. *Agroecol. Sustain. Food Syt.* 2014, *38*, 936–961.

2. Bolan, N.S.; Kunhikrishnan, A.; Choppala, G.K.; Thangarajan, R.; Chung, J.W. Stabilization of carbon in composts and biochars in relation to carbon sequestration and soil fertility. *Sci. Total Environ.* 2012, *424*, 264–270.

3. Troy, S.M.; Lawlo, P.G.; O'Flynn, C.J.; Healy, M.G. Impact of biochar addition to soil on greenhouse gas emissions following pigmanure application. *Soil Biol. Biochem.* 2013, *60*, 173–181.

4. Lehmann, J.; Czimczik, C.; Laird, D.; Sohi, S. Stability of biochar in the soil. In *Biochar for Environmental Management: Science and Technology*; Lehmann, J., Joseph, S., Eds.; Earthscan: London, UK, 2009; pp. 183–205.

5. Yuan, J.H.; Xu, R.K.; Zhang, H. The forms of alkalis in the biochar produced from crop residues at different temperatures. *Bioresour. Technol.* 2011, *102*, 3488–3497.

6. Zhao, X.; Wang, J.W.; Xu, H.J.; Zhou, C.J.; Wang, S.Q.; Xin, G.X. Effects of crop-straw biochar on crop growth and soil fertility over a wheat-millet rotation in soils of China. *Soil Use Manag.* 2014, *30*, 311–319.

7. Gaunt, J.; Lehmann, J. Energy balance and emissions associated with biochar sequestration and pyrolysis bioenergy production. *Environ. Sci. Technol.* 2008, *42*, 4152–4158.

8. Laird, D.A. The charcoal vision: A win-win-win scenario for simultaneously producing bioenergy, permanently sequestering carbon, while improving soil and water quality. *Agron. J.* 2008, *100*, 178–181.

9. Rogovska, N.; Laird, D.; Cruse, R.; Fleming, P.; Parkin, T.; Meek, D. Impact of Biochar on Manure Carbon Stabilization and Greenhouse Gas Emissions. *Soil Sci. Soc. Am. J.* 2011, *75*, 871–879.

10. Awad, Y.M.; Blagodatskaya, E.; Ok, Y.S.; Kuzyakov, Y. Effects of polyacrylamide, biopolymer, and biochar on decomposition of soilorganic matter and plant residues as determined by ^{14}C and enzyme activities. *Eur. J. Soil Biol.*2012, *48*, 1–10.

11. Awad, Y.M.; Blagodatskaya, E.; Ok, Y.S.; Kuzyakov, Y. Effects of polyacrylamide, biopolymer and biochar on the decomposition of ^{14}C-labelled maize residues and on their stabilization in soil aggregates.

Eur. J. Soil Sci. 2013, *64*, 488–499.

12. Qayyum, M.F.; Steffens, D.; Reisenauer, H.P.; Schubert, S. Biochars influence differential distribution and chemical composition of soil organic matter. *Plant Soil Environ.* 2014, *60*, 337–343.

13. Fernández, J.M.; Nieto, M.A.; López-de-sá, E.G.; Gascó, G.; Méndez, A.; Plaza, C. Carbon dioxide emmisions from semi-arid soils amended with biochar alone or combined with mineral and organic fertilizers. *Sci. Total Environ.* 2014,*482–483*, 1–7.

14. Kuzyakov, Y.; Subbotina, I.; Chen, H.; Bogomolova, I.; Xu, X. Black carbon decomposition and incorporation into soil microbeal biomass estimated by ^{14}C labeling. *Soil Biol. Biochem.* 2009, *41*, 210–219.

15. Van Veen, J.A.; Kuikman, P.J. Soil structure aspects of decomposition of organic matter by micro-organisms.*Biogeochemistry* 1990, *11*, 213–233.

16. Hamer, U.; Marschner, B.; Brodowski, S.; Amelung, W. Interactive priming of black carbon and glucose mineralisation. *Org. Geochem.* 2004, *35*, 823–830.

17. Zimmerman, A.R.; Gao, B.; Ahn, M.Y. Positive and negative carbon mineralization priming effects among a variety of biochar-amended soils. *Soil Biol. Biochem.* 2011, *43*, 1169–1179.

18. Keith, A.; Singh, B.; Singh, B.P. Interactive Priming of Biochar and Labile Organic Matter Mineralization in a Smectite-Rich Soil. *Environ. Sci. Technol.* 2011, *45*, 9611–9618.

19. Streubel, J.D.; Collins, H.P.; Garcia-Perez, M.; Tarara, J.; Granatstein, D.; Kruger, C.E. Influence of contrasting biochar types on five soils at increasing rates of application. *Soil Sci. Soc. Am. J.* 2011, *75*, 1402–1413.

20. Thomas, G.W. Soil pH and soil acidity. In *Methods of Soil Analysis: Soil Science Society of America Book Series 5 Part 3—Chemical Methods*; Sparks, D.L., Ed.; ASA and SSSA: Madison, WI, USA, 1996; pp. 487–488.

21. Rhoades, J.D. Soluble salts. In *Methods of Soil Analysis Part 2—Chemical and Microbiological Properties*; Page, A.L., Ed.; ASA and SSSA: Madison, WI, USA, 1982; pp. 167–179.

22. Gee, G.W.; Bauder, J.W. Particle-size analysis. In *Methods of Soil Analysis Part 1—Physical and Mineralogical Methods*; Klute, A., Ed.; ASA and SSSA: Madison, WI, USA, 1986; pp. 383–411.

23. Sumner, M.E.; Miller, W.P. Cation exchange capacity and exchange coefficients. In *Methods of Soil Analysis: Soil Science Society of America Book Series 5 Part 3—Chemical Methods*; Sparks, D.L., Ed.; ASA and

SSSA: Madison, WI, USA, 1996; pp. 1218–1220.

24. Nelson, D.W.; Sommers, L.E. Total carbon, organic carbon, and organic matter. In *Methods of Soil Analysis: Soil Science Society of America Book Series 5 Part 3—Chemical Methods*; Sparks, D.L., Ed.; ASA and SSSA: Madison, WI, USA, 1996; pp. 961–1010.

25. Bremner, J.M.; Mulvaney, C.S. Nitrogen-total. In *Methods of Soil Analysis, Part 2—Chemical and Microbiological Properties*; Page, A.L., Ed.; ASA and SSSA: Madison, WI, USA, 1982; pp. 595–624.

26. Mulvaney, R.L. Nitrogen-Inorganic forms. In *Methods of Soil Analysis: Soil Science Society of America Book Series 5 Part 3—Chemical Methods*; Sparks, D.L., Ed.; ASA and SSSA: Madison, WI, USA, 1996; pp. 1123–1184.

27. Loeppert, R.H.; Suarez, D.L. Carbonate and gypsum. In *Methods of Soil Analysis Part 2—Chemical and Microbiological Methods*, 2nd ed.; Agronomy Monograph 9; America Society of Agronomy and Soil Science Society of America: Madison, WI, USA, 1982; pp. 437–451.

28. Blair, G.J.; Lefroy, R.D.B.; Lisle, L. Soil carbon fractions based on their degree of oxidation, and the development of a carbon management index for agricultural systems. *Aust. J. Agric. Res.* 1995, *46*, 1459–1466.

29. Conteh, A.; Lefroy, R.D.B.; Blair, G.J. Dynamics of organic matter in soils as determined by variations in $^{13}C/^{12}C$ isotopic ratios and fractionation by ease of oxidation. *Aust. J. Soil Res.* 1997, *35*, 881–890.

30. Zibilske, L.M. Carbon mineralization. In *Methods of Soil Analysis: Soil Science Society of America Book Series 5 Part 2—Microbial and Biochemical Properties*; Weaver, R.W., Ed.; ASA and SSSA: Madison, WI, USA, 1994; pp. 835–863.

31. Novak, J.M.; Busscher, D.W.; Watts, D.W.; Laird, D.A.; Ahmedna, M.A.; Niandou, M.A.S. Short-term CO_2 mineralization after additions of biochar and switchgrass to a TypicKandiudult. *Geoderma* 2010, *154*, 281–288.

32. Riberio, H.M.; Fanqueiro, D.; Alves, F.; Vasconcelos, E.; Coutinho, J.; Bol, R.; Cabral, F. Carbon-mineralization kinetics in an organically managed Cambic Arenosol amended with organic fertilizers. *J. Plant Nutr. Soil Sci.* 2010, *173*, 39–45.

33. Stoops, G. *Guidelines for Analysis and Description of Soil and Regolith Thin Sections*; Soil Science Society of Amenrica, Inc.: Madison, WI, USA, 2003.

34. Soil Survey Staff. *Keys to Soil Taxonomy, 11th edn USDA-NRCS, Agricultural Handbook No. 436*; US Government Printing Office:

Washington, DC, USA, 2010.

35. Molina, J.A.E.; Clapp, C.E.; Larson, W.E. Potentially mineralizable nitrogen in soil: the simple exponential model does not apply to the first 12 weeks of incubation. *Soil Sci. Soc. Am. J.* 1980, *44*, 442–443.

36. Liang, B.Q.; Lehmann, J.; Solomon, D.; Kinyangi, J.; Grossman, J.; O'Neill, B.; Skjemstad, J.O.; Thies, J.; Luizão, F.J.; Petersen, J.; *et al.* Black carbon increases cation exchange capacity in soils. *Soil Sci. Soc. Am. J.* 2006, *70*, 1719–1730.

37. Deenik, J.L.; Diarra, A.; Uehara, G.; Campell, S.; Sumiyoshi, Y.; Antal, M.J., Jr. Charcoal ash and volatile matter effects on soil properties and plant growth in an acid Ultisol. *Soil Sci.* 2011, *176*, 336–345.

38. Jien, S.H.; Wang, C.S. Effects of biochar on soil properties and erosion potential in a highly weathered soil. *Catena*2013, *110*, 225–233.

39. Hseu, Z.Y.; Jien, S.H.; Chien, W.S.; Liou, R.C. Impacts of biochar on physical properties and erosion potential of a mudstone slopeland soil. *Sci. World J.* 2014.

40. Cornelissen, G.; Gustafsson, O.; Bucheli, T.D.; Jonker, M.T.O.; Koelmans, A.A.; VanNoort, P.C.M. Extensive sorption of organic compounds to black carbon, coal, and kerogen in sediments and soils: Mechanisms and consequences for distribution, bioaccumulation, and biodegradation. *Environ. Sci. Technol.* 2005, *39*, 6881–6895.

41. Sobek, A.; Stamm, N.; Bucheli, T.D. Sorption of phenyl urea herbicides to black carbon. *Environ. Sci. Technol.* 2009, *43*, 8147–8152.

42. Kasozi, G.N.; Zimmerman, A.R.; Nkedi-Kizza, P.; Gao, B. Catechol and humic acid sorption onto a range of laboratory-produced black carbons (biochars). *Environ. Sci. Technol.* 2010, *44*, 6189–6195.

43. Sigua, G.C.; Novak, J.M.; Watts, D.W.; Cantrell, K.B.; Shumaker, P.D.; Szogi, A.A.; Johnson, M.G. Carbon mineralization in two ultisols amended with different sources and particle sizes of pyrolyzed biochar. *Chemosphere*2014, *103*, 313–321.

44. Sissoko, A.; Kpomblekou-A, K. Carbon decomposition in broiler litter-amended soils. *Soil Biol. Biochem.* 2010, *42*, 543–550.

45. Sørensen, L.H. Size and persistence of the microbial biomass formed during the humification of glucose, hemicellulose, cellulose, and straw in soils containing different amounts of clay. *Plant Soil* 1983, *75*, 121–130.

46. Motavalli, P.P.; Palm, C.A.; Parton, W.J.; Elliott, E.T.; Frey, S.D. Soil pH and organic C dynamics in tropical forest soils: evidence from laboratory and simulation studies. *Soil Biol. Biochem.* 1995, *27*, 1589–1599.

47. Huang, C.C.; Chen, Z.S. Carbon and nitrogen mineralization of sewage sludge compost in soils with different initial pH. *Soil Sci. Plant Nutr.* 2009, *55*, 715–724.

Chapter 3

LONG-TERM GRAZING EXCLUSION IMPROVES THE COMPOSITION AND STABILITY OF SOIL ORGANIC MATTER IN INNER MONGOLIAN GRASSLANDS

Chunyan Wang[1,2], Nianpeng He[1], Jinjing Zhang[3], Yuliang Lv[2], Li Wang[3]

[1]Key Laboratory of Ecosystem Network Observation and Modeling, Institute of Geographic Sciences and Natural Resources Research, Chinese Academy of Sciences, Beijing, 100101, China

[2]College of Geographical Science, Southwest University, Chongqing, 400715, China

[3]College of Resource and Environmental Science, Jilin Agricultural University, Changchun, 130118, China

ABSTRACT

Alteration of the composition of soil organic matter (SOM) in Inner Mongolian grassland soils associated with the duration of grazing exclusion (GE) has been considered an important index for evaluating the restoring effects of GE practice. By using five plots from a grassland succession series from free grazing to 31-year GE, we measured the content of soil organic carbon (SOC), humic acid carbon (HAC), fulvic acid carbon (FAC), humin carbon (HUC), and humic acid structure to evaluate the changes in SOM composition. The results showed that SOC, HUC, and the ratios of HAC/FAC and HAC/extractable humus carbon (C) increased significantly with prolonged GE duration, and their relationships can be well fitted by positive exponential equations, except for FAC. In contrast, the HAC content increased logarithmically with prolonged GE duration. Long-term GE enhanced the content of SOC and soil humification, which was obvious after more than 10 years of GE. Solid-state ^{13}C nuclear magnetic resonance spectroscopy showed that the ratios of alkyl C/O-alkyl C first decreased, and then remained stable with prolonged GE. Alternately, the ratios of aromaticity and hydrophobicity first increased, and then were maintained at relatively stable levels. Thus, a decade of GE improved the composition and structure of SOM in semiarid grassland soil and made it more stable. These findings provide new evidence to support the positive effects of long-term GE on soil SOC sequestration in the Inner Mongolian grasslands, in view of the improvement of SOM structure and stability.

INTRODUCTION

Soil organic matter (SOM) plays important roles in retaining and supplying plant nutrients, and in improving soil aggregation and erodibility [1,2]. SOM, as the largest carbon (C) pool in terrestrial ecosystems, has been commonly divided into active, slow, and passive C fractions according to the turnover time [3]. Six et al. [4] divided SOM into protected or unprotected fractions to explore the underlying mechanisms of decomposition. These fractions have some overlap in stabilization mechanisms, such that the unprotected pool represents the active fractions and part of the slow pool, and the biochemically protected pool is comparable to the passive pool to some extent.

Some studies have investigated changes in SOM composition and stability in agriculture ecosystems by mainly evaluating humic substances and other organic macromolecules [5]. SOM components related to soil quality are closely associated with soil humified fractions [6], which can improve soil buffering capacity, moisture retention, and micronutrient supply [7]. Changes in soil humus are supposed to be the most effective component and represent the stability of soil structure and resistance to erosion [8]. According to its classical classification, soil humus can be divided into humic acid (HA), fulvic acid (FA), and humin (HU). Different components of soil humus have specific contributions towards soil fertility according to their humus composition and chemical structure [9].

Few studies have investigated changes in SOM composition and structure, although soil C sequestration resulting from land-use change or management of forest and grassland has been evaluated [10,11]. In Inner Mongolian grasslands (78.8×10^6 ha), the practice of grazing exclusion (GE) has been deemed as an effective approach to restore these degraded grasslands. At the same time, some studies have demonstrated that long-term GE has tremendous potential for increasing soil C and nitrogen storage in temperate grasslands in northern China [12–16]. However, it is still unclear how SOM composition and structure change dynamically with the duration of GE.

In this study, we used a grassland restoration chronosequence with five GE durations (0–31 year) in Inner Mongolia to investigate the dynamics of SOM composition after GE. Furthermore, we used solid-state ^{13}C cross-polarization magic spinning nuclear magnetic resonance spectroscopy (CPMAS NMR) to explore changes in HA structure. The main objectives of the present study were to: 1) investigate the influences of long-term GE on SOM composition in semiarid grassland soils, and 2) explore changes in SOM stability with long-term GE.

MATERIAL AND METHODS

Study Sites

The experimental plots belong to typical temperate grassland at the Inner Mongolia Grassland Ecosystem Research Station (IMGERS) of the Chinese Academy of Sciences (43°33′N,116°40′E), which has a typical semi-arid continental climate. The mean annual temperature is 1.1°C. The annual precipitation is approximately 345 mm, 70% rainfall occurring in June, July, and August. The soil is chestnut, which is equivalent to Calcic Orthic Aridisol in the US soil taxonomy classification system, and it developed from Aeolian sediments. The soils are characterized by rich sand content with the range of sand from 60% to 75% [17]. The vegetation consists predominantly of grassland plants, such as *Leymus chinensis* (44.5%, relative biomass), *Stipa grandis* (34.0%), and *Cleistogenes squarrosa* (8.7%) [15].

   Five experimental plots were selected based on the preexisting experimental plots of IMGERS. The plots were designated as GE0, GE4, GE7, GE11, and GE31. Plot GE0 had been exposed to long-term grazing by sheep and was in a slightly degraded condition in terms of plant community and diversity. Plots GE4, GE7, GE11, and GE31 were established in 2008, 2004, 1999, and 1979, respectively, by fencing off a section of previous grazing grasslands. These GE plots ranged from 0.8 ha to 24 ha in area, and had similar vegetation and topography across a 2-km area. Changes in soil properties in these plots (as presented in Table 1) therefore mainly resulted from the influence of grazing intensity and GE duration on new organic matter input by plants and SOM turnover.

Table 1. Changes in the selected soil properties in the grazing-exclusion grassland chronosequence.

Grassland type	Aboveground biomass (gm⁻²)	Litter (g m⁻²)	SOC‡ (g kg⁻¹)	TN (g kg⁻¹)	TP (g kg⁻¹)	PH
GE0†	60.28 ± 20.60 ^{b5}	30.53 ± 13.83^c	14.36 ± 1.26^c	1.41 ± 0.01^b	0.22 ± 0.02^c	8.16 ± 0.29^a
GE4	162.25 ± 14.97^a	62.85 ± 7.51^b	14.31 ± 0.61^c	1.60 ± 0.01^a	0.27 ± 0.01^b	8.07 ± 0.11^a
GE7	166.18 ± 13.27^a	75.17 ± 12.37^b	15.03 ± 0.96^c	1.64 ± 0.02^a	0.30 ± 0.01^a	7.92 ± 0.16^a
GE11	171.64 ± 9.64^a	82.84 ± 18.27^b	17.23 ± 1.27^b	1.72 ± 0.01^a	0.29 ± 0.01^a	7.66 ± 0.19^a
GE31	148.93 ± 41.27^a	121.12 ± 32.69^a	19.95 ± 0.27^a	1.42 ± 0.07^c	0.28 ± 0.01^b	7.19 ± 0.29^b
F	20.508	14.947	18.731	210.606	50.698	4.84
P	<0.001	<0.001	<0.001	<0.001	<0.001	0.007

† GE0, free grazing; GE4, 4-year grazing exclusion; GE7, 7-year grazing exclusion; GE11, 11-year grazing exclusion; GE31, 31-year grazing exclusion.
‡ SOC, soil organic carbon; TN, Soil total nitrogen; TP, Soil total phosphorus.
§ Data were represented as mean ± SD (n = 4). The same superscript letters within each column indicated no significant difference at $P < 0.05$.

doi:10.1371/journal.pone.0128837.t001

Field Sampling

In each experimental plot, an east-west transect was established with four equal-sized replicate blocks (20 × 20 m each). Field sampling was conducted in July 2011. In each block, one sampling quadrat (each 1 m × 1 m) was first established to investigate aboveground biomass with all the plant species combined. Litter was subsequently collected. In each block, approximately 10 soil cores were taken randomly to a depth of 20 cm using a soil auger (8 cm in diameter), and mixed as a sample. Each sample was air-dried in a ventilation room, sieved using 2-mm sieves, and cleared of visible roots and organic debris by hand for further analysis.

Laboratory Analysis

The content of organic C in all samples was measured by using the modified Mebius method [18]. Total soil nitrogen (TN) was measured with a modified Kjeldahl wet digestion procedure [19], using a 2300 Kjeltec Analyzer Unit (FOSS Tecator, Hoganas, Sweden). Total phosphorus (TP) was determined by the ammonium molybdate method after persulfate oxidation [20]. Soil pH was determined using a pH meter and a slurry of soil mixed with distilled water (1:2.5). In this study, the measurements for soil properties were conducted in four replicates.

Humus Composition Analysis

Soil humus composition was analyzed as proposed by Kumada [21] with minor modifications [22]. Briefly, a 5-g soil sample was passed through a 60-mesh sieve and placed in 100 mL centrifuge tubes. Distilled water (80 mL) was then added to each tube and the tubes were shaken for 1 h at 70°C in a thermostatic water bath oscillator. The mixture was centrifuged at 3500 r min^{-1} for 15 min, and the supernatant was discarded. The residue, which was the precipitate in the centrifuge tube, was washed twice with distilled water. Subsequently, a 30 mL mixture of 0.1 mol L^{-1} NaOH and 0.1 mol L^{-1} sodium pyrophosphate was added to the soil residue (pH 13), shaken for 1 h at 70°C and then centrifuged at 3500 r min^{-1} for 15 min. The supernatant was filtered into a 50 ml volumetric flask. The residue was washed twice with 20 mL of the above mixture (10 mL every time). The supernatant from the second centrifugation step was also filtered into the same 50 mL volumetric flask to a final volume of 50 mL. The solution contained extractable humic substances. The residue in the centrifuge tube was incubated with distilled water at 55°C and passed through a 60-mesh sieve to provide HU. To 30 mL of the humic substance solution, 0.5 mol L^{-1} H$_2$SO$_4$ was added and the pH was adjusted to 1.0–1.5. The mixture was subjected to 60–70°C for 1.5 h, and then left overnight. The following day, the solution was filtered into a 50 mL volumetric flask to obtain FA after the volume was determined. The precipitate on the filter paper was washed three times with 0.25 mol L^{-1} H$_2$SO$_4$ and dissolved in a 50 mL volumetric flask using 0.05 mol L^{-1} NaOH to obtain HA, after adding distilled water to volume. The C contents of extractable humic substances, HU (HUC), and HA (HAC) were determined by the K$_2$Cr$_2$O$_7$ method [18], whereas the C content of FA (FAC) was calculated by subtracting HAC from the extractable humus substance content [23].

Humic Acid Measurement

Isolation and purification of HA were conducted following previous described methods [24,25] with minor modifications [26]. Briefly, 100 g of the soil sample was first suspended in distilled water and 0.05 mol L^{-1} HCl to remove poorly decomposed light fractions and carbonates. The soil samples were then extracted using a solution of 0.1 mol L^{-1} NaOH and 0.1 mol L^{-1} Na$_4$P$_2$O$_7$ with 5% (w/v) Na$_2$SO$_4$·10H$_2$O at 25°C for 48 h. The extraction procedure was repeated three times on the residues until the supernatant was colorless. The combined alkaline supernatants were acidified to pH 1.0 with 6 mol L^{-1} HCl to separate HA. After three cycles of dissolution in 0.1 mol L^{-1} NaOH and re-precipitation with 6 mol L^{-1}HCL, HA was shaken five times in a 0.5% (v/v) HCl-HF solution, dialyzed against distilled water until it was Cl-free, and finally freeze-dried. The solid-state ^{13}C CPMAS NMR spectra were used to measure SOM composition on a Bruker (Switzerland) spectrometer operating at 100.61 MHz, equipped with a 4 mm probe head. The conditions were as follows: spinning rate 5 kHz, contact time 4 ms, recycle delay time 0.5 s, line broadening 100 Hz, and zero-filling 3072 data points. The spinning side band was corrected according to Conte et al. [27]. According to the main chemical shift regions, spectra were divided into four regions [28]: alkyl C (0–50 ppm), O-alkyl C (50–110 ppm), aromatic C (110–160 ppm), and carbonyl C (160–200 ppm). As the methods described by Dai et al. [29] and Zhang et al. [26], aromaticity and hydrophobicity were calculated as follows:

$$\text{Aromaticity} \ (\%) = \left[\frac{\text{Aromatic} \quad C}{\text{Aromatic} \quad C \ + \ \text{Alkyl} \quad C \ + \ O - \text{alkyl} \quad C} \right] \times 100 \tag{1}$$

$$\text{Hydrophobicity} \ = \ \frac{\text{Alkyl} \quad C \ + \ \text{Aromatic} \quad C}{O - \text{alkly} \quad C \ + \ \text{Carbonyl} \quad C} \tag{2}$$

Statistical Analyses

One-way analysis of variance (one-way ANOVA) with Duncan tests was used to evaluate the differences in soil properties and SOM composition among different grasslands. Pearson correlations were evaluated between different SOM compositions. Regression analyses were conducted to test the relationships between SOM composition and GE duration. Statistical significance was defined as $P = 0.05$. All statistical analyses were performed using SPSS (version 13.0).

RESULTS

Changes in Soil Properties

There were significant increases in the aboveground biomass, litter, SOC, TN, and TP after GE, and these parameter were significantly different between grazing grassland (GE0) and long-term GE grasslands (all $Ps < 0.001$; Table 1). Moreover, the content of SOC increased exponentially with the duration of GE ($R^2 = 0.79$, $P < 0.001$)(Fig 1A). The contents of TN and TP in soils first increased and then decreased to some extent with prolonged GE. Soil pH decreased from 8.10 in GE0 to 7.19 in GE31, but it was not significantly different among the 4 GE grasslands.

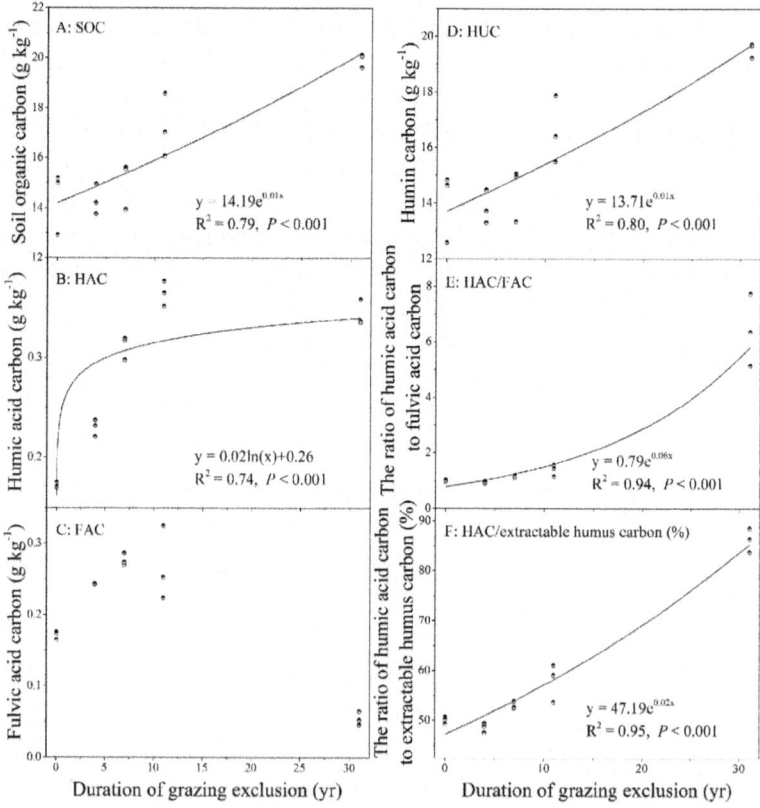

Figure 1. Relationships between soil organic carbon (SOC), humic acid carbon (HAC), fulvic acid carbon (FAC), humic acid carbon (HUC), HAC/FAC, HAC/extractable humus carbon with the duration of grazing exclusion.

Changes in SOM Composition

The content of different SOM components varied significantly among the five plots (all $Ps < 0.001$; Table 2). In detail, the content of HAC increased from 0.17 g kg^{-1} in GE0 to 0.36 g kg^{-1}, and the relationship between HAC and GE duration was be well fitted by a logarithmic equation ($R^2 = 0.74$, $P < 0.001$) (Fig 1B). Additionally, HUC, HAC/extractable humus C, and HAC/FAC all increased exponentially with the duration of GE ($R^2 = 0.80$, $P < 0.001$ for HUC; $R^2 = 0.95$, $P < 0.001$ for HAC/extractable humus C; $R^2 = 0.94$, $P < 0.001$ for HAC/FAC). In contrast, FAC did not have a similar pattern, as it first increased and then decreased to some extent with the prolonged GE (Table 2).

Table 2. Changes in the SOM composition along the grazing-exclusion grassland chronosequence.

Grassland type	HAC[‡] (g kg^{-1})	FAC (g kg^{-1})	HUC (g kg^{-1})	Extractable humus C (g kg^{-1})	HAC/FAC	HAC/ extractable humus C (%)
GE0[†]	0.17 ± 0.01[e§]	0.17 ± 0.01[b]	14.02 ± 1.25[c]	0.34 ± 0.01[c]	1.00 ± 0.03[b]	50.04 ± 0.70[cd]
GE4	0.23 ± 0.01[d]	0.24 ± 0.01[a]	13.83 ± 0.60[c]	0.47 ± 0.01[b]	0.95 ± 0.04[b]	48.58 ± 0.97[d]
GE7	0.31 ± 0.01[c]	0.28 ± 0.01[a]	14.45 ± 0.96[c]	0.59 ± 0.02[a]	1.13 ± 0.03[b]	52.96 ± 0.77[c]
GE11	0.36 ± 0.01[a]	0.27 ± 0.05[a]	16.60 ± 1.20[b]	0.63 ± 0.06[a]	1.39 ± 0.21[b]	57.96 ± 3.84[b]
GE31	0.34 ± 0.01[b]	0.05 ± 0.01[c]	19.55 ± 0.26[a]	0.40 ± 0.01[c]	6.42 ± 1.30[a]	86.25 ± 2.41[a]
F	180.731	43.755	20.219	48.249	49.193	160.792
P	<0.001	<0.001	<0.001	<0.001	<0.001	<0.001

[†] GE0, free grazing; GE4, 4-year grazing exclusion; GE7, 7-year grazing exclusion; GE11, 11-year grazing exclusion; GE31, 31-year grazing exclusion.
[‡] HAC, Humic acid carbon; FAC, Fulvic acid carbon; HUC, Humin carbon; Extractable humus C, Extractable humus carbon;HAC/FAC, The ratio of humic acid carbon to fulvic acid carbon; HAC/extractable humus C, The ratio of humic acid carbon to extractable humus carbon.
[§] Data were represented as mean ± SD (n = 3). The same superscript letters within each column indicated no significant difference at $P < 0.05$.

doi:10.1371/journal.pone.0128837.t002

HUC was not significantly different among GE0, GE4, and GE7, but it increased exponentially with the duration of GE ($R^2 = 0.80$, $P < 0.001$; Fig 1 and Table 2). The ratio of HAC/extractable humus C was lowest in GE4 (48.58%) and highest in GE31 (86.25%), and it increased exponentially with the duration of GE ($R^2 = 0.95$, $P < 0.001$; Fig 1F). Furthermore, the ratios of HAC/FAC also increased exponentially with the duration of GE ($R^2 = 0.94$, $P < 0.001$; Fig 1E).

Relationships among C Content in Different Components

SOC, HAC, and HUC were positively correlated with each other (Table 3). Moreover, HAC/extractable humus C and HAC/FAC had significantly positive correlations with SOC. HUC, HAC/extractable humus C, and HAC/FAC had significantly positive correlations with each other, whereas FAC showed negative correlations with other components (Table 3).

Table 3. Pearson correlation of organic carbon among different SOM components.

	SOC	HAC	FAC	HUC	HAC/FAC	HAC/extractable humus C
SOC[†]	1					
HAC	0.708**	1				
FAC	-0.557*	0.001	1			
HUC	0.999**	0.676**	-0.595*	1		
HAC/FAC	0.823**	0.464	-0.848**	0.840**	1	
HAC/ Extractable humus C	0.870**	0.575*	-0.813**	0.883**	0.973**	1

† SOC, soil organic carbon; HAC, humic acid carbon; FAC, Fulvic acid carbon; HUC, Humin carbon; Extractable humus C, Extractable humus carbon;
HAC/FAC, The ratio of humic acid carbon to fulvic acid carbon; HAC/extractable humus C, The ratio of humic acid carbon to extractable humus carbon.
* $P < 0.05$ and
**$P < 0.01$.

doi:10.1371/journal.pone.0128837.t003

Changes in the Structure of Humic Acid

The structures of HA, as shown in Fig 2, were similar among the different plots. In detail, the contents of alkyl C and O-alkyl C first increased and then decreased with prolonged GE duration (Table 4).

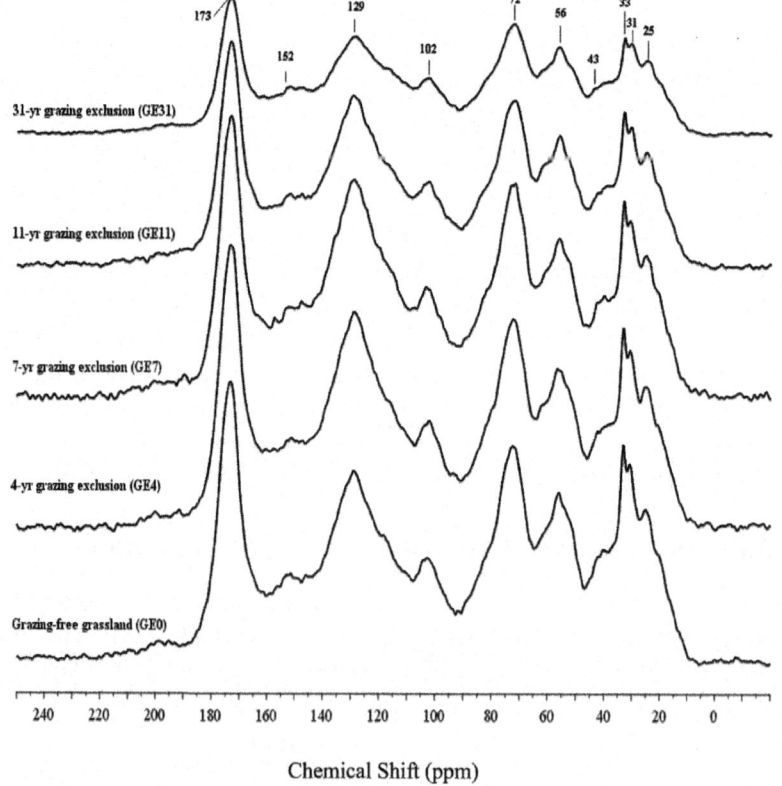

Figure 2. Solid-state[13]C CPMAS NMR spectra for humic acid (HA) under grazing-exclusion grassland chronosequence.

Aromatic C was lowest in GE0 (28.65%) and highest in GE4 (31.56%). The content of carbonyl C was significantly lower in GE31 than in other GE grasslands, but it was not significantly different among the plots of GE0, GE4, GE7, and GE11 (Table 4). The ratio of alkyl to O-alkyl decreased with GE, it was 0.03, 0.06, 0.04, and 0.04 in GE4, GE7, GE11, and GE31, respectively. The ratio of hydrophobic C to hydrophilic C increased with increasing GE duration, and reached relative equilibrium at decade of GE application.

Table 4. Relative distribution (%) of organic carbon in HA by ^{13}C CPMAS NMR.

	Alkyl C (0–50ppm) (%)	O-alkyl C (50–110 ppm) (%)	Aromatic C (110–160 ppm) (%)	Carbonyl C (160–210 ppm) (%)	Aromaticity[‡] (%)	Alkyl C/O-alkyl C	Hydrophobicity[§]
GE0[†]	19.33	34.45	28.65	17.57	0.35	0.56	0.92
GE4	17.84	33.45	31.56	17.15	0.38	0.53	0.98
GE7	17.39	34.45	31.44	16.72	0.38	0.50	0.95
GE11	17.64	34.04	30.69	17.64	0.37	0.52	0.94
GE31	18.66	35.57	29.82	15.95	0.35	0.52	0.94

[†] GE0, free grazing; GE4, 4-year grazing exclusion; GE7, 7-year grazing exclusion; GE11, 11-year grazing exclusion; GE31, 31-year grazing exclusion.
[‡] Aromaticity = Aromatic C/(Alkyl C+O-alkyl C+Aromatic C)×100%.
[§]Hydrophobicity = (Alkyl C + Aromatic C)/(O-alkyl C + Carbonyl C).

doi:10.1371/journal.pone.0128837.t004

DISCUSSION

Long-term Grazing Exclusion Enhances Soil C Storage in Semiarid Grasslands

The content of SOC in the surface soil increased with GE, and the exponential equations well fitted the changes associated with the duration of GE. Our findings showed that SOC content in grasslands increased slowly in the first phase of GE, and faster after a decade of GE. The results were consistent with our previous study [30], suggesting that long-term GE can be conducive to enhancing SOC content. The change in SOC depended on the balance between SOM decomposition and new SOM input. The practice of GE promoted the restoration of grassland vegetation and directly resulted in increased SOM input from litter and roots [31,32]. In this study, the litter and aboveground biomass in these GE grasslands were significantly higher than those of grazing grasslands (Table 1). Additionally, the practice of GE decreased SOM decomposition by maintaining a better soil aggregate structure through exclusion of livestock stamping [33,34], and the higher height and density of the aboveground vegetation improved soil surface roughness, thereby reducing soil erosion by wind and water in these GE grasslands [35,36]. Furthermore, He et al. [37] reported that higher litter accumulation in the soil surface resulted in a lower soil temperature (2–3°C lower) in the long-term GE grasslands. Lower soil temperature may reduce the decomposition of SOM and benefit the accumulation of SOC to some extent [38].

Long-term Grazing Exclusion Improves Soil Humification and SOM stability

The composition of SOM varied among different plots, and long-term GE improved soil humification and SOM stability to some extent. Changes in SOM input characteristics (e.g., input, C/N ratio, and the content of protein and polysaccharides) and soil temperature and moisture may influence SOM breakdown and formation [34,39]. HAC increased logarithmically with the duration of GE (Fig 1B), that is, HAC increased initially and then, attained stability after a decade of GE. Dou [40] proposed that hypothermia decreased the formation of HA. Additionally, higher soil moisture could reduce SOM decomposition, and the reduced microbial activity could reduce the decomposition of HA. Moreover, excessive moisture will prevent further condensation of HA [41]. Therefore, lower soil temperature and higher moisture in the long-term GE grasslands [37] should be the main reasons for the alteration of HA. Sheng and Zhao [42] demonstrated that plant biomass and the content of HA were positively correlated in semi-arid habitat conditions because lower plant biomass and coverage in favor of higher O_2 exchange between soil and atmosphere resulted in oxidative degradation of HA.

HUC increased exponentially with the duration of GE in Inner Mongolian grasslands. Yang et al. [43] found that an increase in the proportion of HA and HU in the presence of grass cover resulted in higher soil C sequestration potential. Moreover, Seddaiu et al. [11] demonstrated that the content of HUC can indicate the stability of SOM. Based on the findings that HUC and SOC have positive correlations in long-term GE grasslands, we assumed that the stability of SOM might be enhanced by long-term GE to some extent. In this study, the content of SOC and HUC increased exponentially with the duration of GE. The finding that HU and HUC did not arrive at the equilibrium after the 3-decade GE indicated that the recovery of recalcitrant fractions in addition to the total SOM pools requires a longer duration [44]. Thus, long-term GE not only increased SOM content but also made it more stable [4].Compared with FA, HA has higher molecular weight and the degree of polymerization, and the latter is associated with the humification rate. A higher HA/FA ratio indicates higher humification degree [45], and hence HA/FA is used as an index to determine soil humification degree and molecular complexity [46]. The ratios of HAC/extractable humus C have been used as an indicator for the degree of humification, where a higher ratio implies larger molecular weight, more complex molecular structure, and higher quality of HA [47]. The higher correlation of HAC/FAC and HAC/extractable humus C ratio reported here confirm that both measure represent the humification degree well. Furthermore, our findings that the ratios of HAC/FAC and HAC/extractable humus C increased exponentially with the duration of GE (Fig 1) imply higher degree of humification for the SOM.

Grazing Exclusion Alters the Composition of Humic Acid

HA is the most active component of humus, and its high cation exchange capacity enables soil fertilizer retention. It is also an organic binder that regulate the formation of soil structure [48]. A similar HA skeleton was observed in these GE grasslands (Fig 2), although there were some small alterations in the different components (Fig 2 and Table 4). Short-term GE decreased alkyl C and O-alkyl C and increased aromatic C. However, long-term GE increased the content of aliphatic C but decreased the content of aromatic C. Inconsistent changes in different HA components with the practice of GE led to the observed increases in aromaticity and the ratio of hydrophobic C/hydrophilic C, and the observed decrease in the ratio of alkyl C/O-alkyl C.

It was generally considered that alkyl C was derived from original plant biopolymers (such as cutin, suberin, and waxes) or from metabolic products of soil microorganisms, which comprise the most persistent fraction of SOM [49,50], whereas O-alkyl C (e.g., carbohydrates and polysaccharides) was easily decomposed; therefore, alkyl C/O-alkyl C is commonly used as an index of decomposability of SOM. The higher hydrophobicity of humic substances is indicates higher stability of SOM [51,52]. The ratio of aromaticity has been used to indicate the degree of aromaticity and aliphatic properties [53], with larger ratios indicating a more aromatic and less aliphatic humic substance. We therefore assumed that the soil structure in these long-term GE grasslands became more stable with stronger aliphatic properties and weaker aromaticity.

CONCLUSION

Long-term GE significantly influences SOM composition. The contents of SOC, HUC, and the ratios of HAC/extractable humus C and HAC/FAC increase exponentially with the duration of GE, and HAC shows a significant logarithmic increase with prolonged GE. Based on the ratios of HAC/extractable humus C and HAC/FAC, we concluded that the humification degree increased in the 3-decade GE grasslands. Aromaticity, alkyl C/O-alkyl C ratio, and hydrophobicity decreased and HUC content increased in the long-term GE grasslands, which indicated the SOM was more stable. These findings provide new insights into the stability of increasing SOC storage in long-term GE grasslands in view of SOM composition and stability.

ACKNOWLEDGMENTS

We thank the Inner Mongolia Grassland Ecosystem Research Station for providing some data and for access to the experimental sites.

AUTHOR CONTRIBUTIONS

Conceived and designed the experiments: CW NH. Performed the experiments: CW LW. Analyzed the data: CW NH JZ. Contributed reagents/materials/ analysis tools: JZ. Wrote the paper: CW NH YL.

REFERENCES

1.   Tisdall JM, Oades JM (1982) Organic matter and water-stable aggregates in soils. Journal of Soil Science 33: 141–163. doi: 10.1111/j.1365-2389.1982.tb01755.x

2.   Breshears DD, Whicker JJ, Johansen MP, Pinder JE (2003) Wind and water erosion and transport in semi-arid shrubland, grassland and forest ecosystems: Quantifying dominance of horizontal wind-driven transport. Earth Surface Processes and Landforms 28: 1189–1209. doi: 10.1002/esp.1034

3.   Parton WJ, Schimel DS, Cole CV, Ojima DS (1987) Analysis of factors controlling soil organic matter levels in great-plains grasslands. Soil Science Society of America Journal 51: 1173–1179. doi: 10.2136/sssaj1987.03615995005100050015x

4.   Six J, Conant RT, Paul EA, Paustian K (2002) Stabilization mechanisms of soil organic matter: Implications for C-saturation of soils. Plant and Soil 241: 155–176. doi: 10.1023/a:1016125726789

5.   Stevenson FJ (1994) Humus chemistry: genesis, composition, reactions: John Wiley & Sons.

6.   Papini R, Valboa G, Favilli F, L'Abate G (2011) Influence of land use on organic carbon pool and chemical properties of Vertic Cambisols in central and southern Italy. Agriculture, ecosystems & environment 140: 68–79. doi: 10.1016/j.agee.2010.11.013

7.   Guimaraes DV, Gonzaga MIS, da Silva TO, da Silva TL, Dias ND, Matias MIS. (2013) Soil organic matter pools and carbon fractions in soil under different land uses. Soil & Tillage Research 126: 177–182. doi: 10.1016/j.still.2012.07.010

8.   Piccolo A, Conte P, Spaccini R, Mbagwu J (2005) Influence of land use on the characteristics of humic substances in some tropical soils of Nigeria.

European journal of soil science 56: 343–352. doi: 10.1111/j.1365-2389.2004.00671.x

9. Watanabe A, Sarno , Rumbanraja J, Tsutsuki K, Kimura M (2001) Humus composition of soils under forest, coffee and arable cultivation in hilly areas of south Sumatra, Indonesia. European Journal of Soil Science 52: 599–606. doi: 10.1046/j.1365-2389.2001.00410.x

10. Yang ZH, Singh BR, Sitaula BK (2004) Soil organic carbon fractions under different land uses in mardi watershed of Nepal. Communications in Soil Science and Plant Analysis 35: 615–629. doi: 10.1081/css-120030347

11. Seddaiu G, Porcu G, Ledda L, Roggero PP, Agnelli A, Corti G. (2013) Soil organic matter content and composition as influenced by soil management in a semi-arid Mediterranean agro-silvo-pastoral system. Agriculture Ecosystems & Environment 167: 1–11. doi: 10.1016/j.agee.2013.01.002

12. He N, Zhang Y, Dai J, Han X, Baoyin T, Yu G. (2012) Land-use impact on soil carbon and nitrogen sequestration in typical steppe ecosystems, Inner Mongolia. Journal of geographical sciences 22: 859–873. doi: 10.1007/s11442-012-0968-4

13. Zhou Z, Sun OJ, Huang J, Li L, Liu P, Han X. (2007) Soil carbon and nitrogen stores and storage potential as affected by land-use in an agro-pastoral ecotone of northern China. Biogeochemistry 82: 127–138. doi: 10.1007/s10533-006-9058-y

14. Wiesmeier M, Steffens M, Mueller C, Kölbl A, Reszkowska A, Peth S, et al. (2012) Aggregate stability and physical protection of soil organic carbon in semi-arid steppe soils. European journal of soil science 63: 22–31. doi: 10.1111/j.1365-2389.2011.01418.x

15. He NP, Yu Q, Wu L, Wang YS, Han XG (2008) Carbon and nitrogen store and storage potential as affected by land-use in a *Leymus chinensis* grassland of northern China. Soil Biology & Biochemistry 40: 2952–2959. doi: 10.1016/j.soilbio.2008.08.018

16. He NP, Han XG, Yu GR, Chen QS (2011) Divergent changes in plant community composition under 3-decade grazing exclusion in continental Steppe. PlosOne 6(11): e26506. doi: 10.1371/journal. pone.0026506. pmid:22073169

17. He NP, Wu L, Wang YS, Han XG (2009) Changes in carbon and nitrogen in soil particle-size fractions along a grassland restoration chronosequence in northern China. Geoderma 150: 302–308. doi: 10.1016/j.geoderma.2009.02.004

18. Nelson D, Sommers L, Page A, Miller R, Keeney D (1982) Total carbon, organic carbon, and organic matter. In: Page A.L., Miller R.H., and Keeney D.R., editors, Methods of soil analysis. ASA and SSSA, Madison, WI.

19. Gallaher R, Weldon C, Boswell F (1976) A semiautomated procedure for total nitrogen in plant and soil samples. Soil Science Society of America Journal 40: 887–889. doi: 10.2136/sssaj1976.03615995004000060026x

20. Kuo S (1996) Phosphorus. In:DL et al (eds) Methods of soil analysis. Part 3. Chemical methods. Soil Science Society of America and American Society of Agronomy, Madison.

21. Kumada K (1988) Chemistry of soil organic matter: Elsevier.

22. Zhang JJ, Hu F, Li HX, Gao Q, Song XY, Ke X, et al. (2011) Effects of earthworm activity on humus composition and humic acid characteristics of soil in a maize residue amended rice-wheat rotation agroecosystem. Applied Soil Ecology 51: 1–8. doi: 10.1016/j.apsoil.2011.08.004

23. Lao J (1988) Handbook of Soil Agro-Chemistry Analysis. Beijing: Agriculture Publishing House.

24. Piccolo A, Zaccheo P, Genevini P (1992) Chemical characterization of humic substances extracted from organic-waste-amended soils. Bioresource technology 40: 275–282. doi: 10.1016/0960-8524(92)90154-p

25. Dou S, Tan S, Xu X, Chen E (1991) Effect of pig manure application on the structural characteristics of humic acid in brown soil. Pedosphere 1: 345–354. pmid:1647301

26. Zhang JJ, Dou S, Song XY (2009) Effect of long-term combined nitrogen and phosphorus fertilizer application on 13C CPMAS NMR spectra of humin in a Typic Hapludoll of northeast China. European Journal of Soil Science 60: 966–973. doi: 10.1111/j.1365-2389.2009.01191.x

27. Conte P, Piccolo A, van Lagen B, Buurman P, de Jager PA (1997) Quantitative differences in evaluating soil humic substances by liquid- and solid-state C-13-NMR spectroscopy. Geoderma 80: 339–352. doi: 10.1016/s0016-7061(97)00059-1

28. Huan ZQ, Xu ZH, Chen CR, Boyd S (2008) Changes in soil carbon during the establishment of a hardwood plantation in subtropical Australia. Forest Ecology and Management 254: 46–55. doi: 10.1016/j.foreco.2007.07.021

29. Dai KH, Johnson CE, Driscoll CT (2001) Organic matter chemistry and dynamics in clear-cut and unmanaged hardwood forest ecosystems. Biogeochemistry 54: 51–83.

30. Wu L, He N, Wang Y, Han X (2008) Storage and dynamics of carbon and nitrogen in soil after grazing exclusion in Leymus chinensis grasslands of northern China. Journal of Environmental Quality 37: 663–668. pmid:18396553 doi: 10.2134/jeq2007.0196

31. Gao YZ, Giese M, Lin S, Sattelmacher B, Zhao Y, Brueck H. (2008) Belowground net primary productivity and biomass allocation of a grassland in Inner Mongolia is affected by grazing intensity. Plant and Soil 307: 41–50. doi: 10.1007/s11104-008-9579-3

32. Gao Y, Giese M, Han X, Wang D, Zhou Z, Brueck H, et al. (2009) Land use and drought interactively affect interspecific competition and species diversity at the local scale in a semiarid steppe ecosystem. Ecological research 24: 627–635. doi: 10.1007/s11284-008-0532-y

33. Steffens M, Kolbl A, Kogel-Knabner I (2009) Alteration of soil organic matter pools and aggregation in semi-arid steppe topsoils as driven by organic matter input. European Journal of Soil Science 60: 198–212. doi: 10.1111/j.1365-2389.2008.01104.x

34. Wiesmeier M, Steffens M, Mueller CW, Kolbl A, Reszkowska A, Peth S, et al. (2012) Aggregate stability and physical protection of soil organic carbon in semi-arid steppe soils. European Journal of Soil Science 63: 22–31. doi: 10.1111/j.1365-2389.2011.01418.x

35. Hoffmann C, Funk R, Li Y, Sommer M (2008) Effect of grazing on wind driven carbon and nitrogen ratios in the grasslands of Inner Mongolia. Catena 75: 182–190. doi: 10.1016/j.catena.2008.06.003

36. Hoffmann C, Funk R, Wieland R, Li Y, Sommer M (2008) Effects of grazing and topography on dust flux and deposition in the Xilingele grassland, Inner Mongolia. Journal of Arid Environments 72: 792–807. doi: 10.1016/j.jaridenv.2007.09.004

37. He N, Han X, Yu G, Chen Q (2011) Divergent changes in plant community composition under 3-decade grazing exclusion in continental steppe. PloS one 6: e26506. doi: 10.1371/journal.pone.0026506. pmid:22073169

38. He NP, Wang RM, Dai JZ, Gao Y, Wen XF, Yu GR. (2013) Changes in the temperature sensitivity of SOM decomposition with grassland succession: Implications for soil C sequestration. Ecology and Evolution 3: doi: 10.1002/ece1003.1881.

39. Doane TA, Devêvre OC, Horwáth WR (2003) Short-term soil carbon dynamics of humic fractions in low-input and organic cropping systems. Geoderma 114: 319–331. doi: 10.1016/s0016-7061(03)00047-8

40. Dou S (2010) Soil organic matter: Beijing: Sciences Press.

41. Peng F, Wu J (1965) Composition of humus in paddy soils. Acta Pedologica Sinica 13: 208–215.

42. Sheng X, Zhao Y (1997) Impact of grassland biomass on soil organic matter. Chinese Journal of Soil Science 28: 244–245.

43. Yang Z, Singh BR, Sitaula BK (2004) Fractions of organic carbon in soils under different crop rotations, cover crops and fertilization practices. Nutrient Cycling in Agroecosystems 70: 161–166. doi: 10.1023/b:fr es.0000048479.30593.ea

44. Burke IC, Lauenroth WK, Coffin DP (1995) Soil organic-matter recovery in semiarid grasslands: Implications for the conservation reserve program. Ecological Applications 5: 793–801. doi: 10.2307/1941987

45. Yang Z, Singh B, Sitaula B (2004) Soil organic carbon fractions under different land uses in Mardi watershed of Nepal. Communications in soil science and plant analysis 35: 615–629. doi: 10.1081/css-120030347

46. Doran JW (1980) Soil microbial and biochemical changes associated with reduced tillage. Soil Science Society of America Journal 44: 765–771. doi: 10.2136/sssaj1980.03615995004400040022x

47. Dou S, Jiang Y (1988) Effect of application of organic materials on the properties of humic substances in organo-mineral complexes of soil—II. effect of organic materials on the humus composition and opticalcharacteristics of humic acids in organo-mineralcomplexes. Acta Pedologica Sinica 25: 252–261.

48. Shang S (2012) Soil organic carbon fractions and their structures under different types of forest in subtropics: Zhenjiang A&F University.

49. Dou S, Zhang JJ, Li K (2008) Effect of organic matter applications on ^{13}C-NMR spectra of humic acids of soil. European Journal of Soil Science 59: 532–539. doi: 10.1111/j.1365-2389.2007.01012.x

50. Ussiri DAN, Johnson CE (2003) Characterization of organic matter in a northern hardwood forest soil by C^{-13} NMR spectroscopy and chemical methods. Geoderma 111: 123–149. doi: 10.1016/s0016-7061(02)00257-4

51. Piccolo A, Mbagwu JSC (1999) Role of hydrophobic components of soil organic matter in soil aggregate stability. Soil Science Society of America Journal 63: 1801–1810. doi: 10.2136/sssaj1999.6361801x

52. Spaccini R, Mbagwu JSC, Conte P, Piccolo A (2006) Changes of humic substances characteristics from forested to cultivated soils in Ethiopia. Geoderma 132: 9–19. doi: 10.1016/j.geoderma.2005.04.015

53. Zhang T, Li YF, Chang SX, Jiang PK, Zhou GM, Liu J, et al. (2013)

Converting paddy fields to Lei bamboo (Phyllostachys praecox) stands affected soil nutrient concentrations, labile organic carbon pools, and organic carbon chemical compositions. Plant and Soil 367: 249–261. doi: 10.1007/s11104-012-1551-6

Chapter 4

MOISTURE AND NUTRIENT STORAGE CAPACITY OF CALCINED EXPANDED SHALE

John J. Sloan[1], Peter A.Y. Ampim[1], Raul I. Cabrera[1], Wayne A. Mackay[2], and Steve W. George[3]

[1]Texas AgriLife Research; Dallas, TX, USA

[2]Mid-Florida Research & Education Center, Apopka, FL, USA

[3]Texas AgriLife Extension Service; Dallas, TX, USA

INTRODUCTION

Expanded shale (EXSH) is an important and increasingly popular soil conditioner with several horticultural applications, including its use as a soil amendment for clay textured soils (Sloan et al., 2002), as an ingredient in plant growing media (Sloan et al., 2010) or green roof substrates (Ampim et al., 2010). It is a lightweight material produced by firing mined lumps of shale at high temperatures in a rotary kiln in a process similar to that of clay ceramics. The resulting product can be screened to create various size fractions depending on the intended use. For example, Texas Industries (TXI) of North Texas, USA produces five size fractions of expanded shale that includes the following ranges, from smallest to largest, 0.07 to 0.60 mm, 0.60 to 2.0 mm, 2.0 to 4.8 mm, 4.8 to 6.4 mm, 6.4 to 9.5 mm, 9.5 to 12.7mm, 12.7 to 15.9 mm Expanded shale aggregates are suitable as components of planting media and soil amendments because, unlike most minerals, they are porous, stable, and resistant to decomposition (Ferguson, 2005). Expanded shale is believed to beneficially modify growth media properties by enhancing overall aeration, improving water and nutrient holding and release capacities, and promoting optimum plant growth (Blunt, 1988; Dunnett and Kingsbury, 2008).

Sloan et al. (2002) found that expanded shale consistently improved overall plant performance better than quartz sand, sphagnum peatmoss and cottonseed hull when they were used as amendments for poorly-drained Austin silty clay soils suggesting its superiority as a soil conditioner for the production of horticultural crops on soils with poor tillage characteristics. In a similar way, Nash et al. (1990) found that a potting medium comprising a mixture of peat moss and expanded shale increased the growth and quality of petunia and impatiens. Smalley et al. (1993) also found that amending soils with products containing expanded shale did not hamper plant performance. Though growth index and plant dry weight of Salvia (Salvia splendens) and Vinca (Catharanthus roseus) increased with increasing fertilizer levels for all their treatments, the greatest performance of these plants were observed for treatments amended with the product containing expanded shale, granite sand and composted poultry litter. In another application, Forbes et al. (2004 and 2005) discovered that expanded shale is a potential sorbent for phosphorus in subsurface flow wetlands because of its high hydraulic conductivity, large surface area and P sorption capacities. Forbes and et al. (2004) found that expanded shale retained 164 ± 110 g P/m^2/yr in pilot-scale wetlands. Regardless of the generally good experimental reports for expanded shale and/or mixtures containing it, there is a need to understand the mechanisms controlling the basic water and nutrient retention properties of expanded shale and to ascertain its suitability for horticultural and soil amendment uses because expanded shale products can vary depending on geologic origin of the raw materials and differences in the production process (Ferguson, 2005).

In the United States of America, expanded shale is manufactured by multiple companies, most of which are members of the Expanded Shale, Clay, and Slate Institute (ESCSI). Expanded shale has a third to half the weight of regular rock or sand so it is easy to transport but has enough weight to avoid being carried away by water or wind. It is inert, inorganic and therefore not expected to degrade or react with agricultural or horticultural chemicals (TXI, 2009). Given the increasing interest in using EXSH as a soil amendment or as an ingredient in growing media, it is important to understand its interaction with water and nutrients. The first step towards realizing this goal is to thoroughly understand the basic properties of the material itself. The objective of this study therefore was to evaluate the dynamics of water and nutrient adsorption by EXSH.

MATERIALS AND METHODS

Basic Chemical Properties

pH and Electrical Conductivity

Untreated EXSH (10 g) was equilibrated with deionized water (10 mL) for 60 min on a reciprocal shaker and pH was measured in the supernatant with a hydrogen-specific combination electrode. Additionally, pH of the EXSH was measured using 0.01 M CaCl$_2$ as the equilibrating solution in order to minimize the possible effect of variable background salt concentrations due to soluble components in the EXSH (ASTM D 4972-89). Electrolytic conductivity (EC) of the EXSH was measured by equilibrating 20 g of EXSH with 20 mL of deionized water for 60 min on a reciprocal shaker. Then the supernatant was filtered through Whatman No. 2 filter paper and EC was measured with a dip-type electrode. The filtered supernatant was saved for determination of soluble constituents.

Soluble and Labile Elements

EXSH constituents that dissolve easily in water are immediately available to plants and microbes when applied to soil provided they are not immobilized in the soil. To determine the amounts of soluble elements in EXSH, the supernatants saved from the EC analysis were analyzed for Na, K, Ca, and Mg by flame atomic absorption spectrophotometry. In addition to soluble elements, labile elements in the EXSH were measured by extracting 20 g samples of EXSH with 20 mL of a buffered (pH 5) 1 mol L^{-1} acetic acid solution (Gibson and Farmer, 1986). Labile elements are not immediately soluble in water, but they are potentially available to plants and soil microorganisms when applied to soil.

Calcium Carbonate Equivalent

Calcium carbonate equivalent (CCE) is the measure of a material's ability to neutralize acidity relative to pure calcium carbonate. The comparison is useful for determining how a material will affect soil pH when mixed with soil. The CCE of EXSH was estimated by suspending 10 g of expanded shale in 50 mL deionized water and then titrating with 0.1 mole L^{-1} HCl to a pH 7 endpoint (modified from AOAC, 2005). The amount of pure $CaCO_3$ required to neutralize an equal amount of acidity was calculated and then CCE was determined by dividing the weight of calculated pure $CaCO_3$ by the actual weight of EXSH. Cation Exchange Capacity (CEC): Methods that consist of extracting and summing exchangeable cations are not suitable for expanded shale because the manufacturing process produces significant amounts of Ca, Mg, K, and Na oxides that readily dissolve in water. On the other hand, expanded shale is essentially devoid of elements that were present in low concentrations in the original material and were subsequently volatilized during the heating process. One such element is nitrogen. Therefore, the ammonium saturation method was used to measure the CEC of the expanded shale (Chapman, 1965). For this method, the exchange complex of prewashed EXSH was saturated with NH_4^+, followed by replacement and extraction of NH_4^+ with Na^+. The extracted NH_4^+ was then quantified using a standard method consisting of alkaline steam distillation, boric acid capture, and titration with standard acid.

Water Uptake

Water uptake experiments were designed to evaluate uptake and retention of water by individual EXSH aggregates rather than a bulk amount of the material (Fig.1). The volume of water measured in this manner includes only the internal pores of each EXSH aggregate. It should be noted that measuring the water holding capacity of a bulk volume of EXSH would give a higher value because it would also include water held not only in the internal pores, but also in the pore spaces created between adjacent aggregates.

Figure 1. Magnified view of a single expanded shale particle showing a highly porous exterior surface with range of pore sizes continuing into the interior of the particle.

Maximum Water Holding Content

To measure maximum water holding capacity, a quantity of EXSH was submerged in water for 72 h in order to completely saturate individual aggregates. At the end of that time, EXSH was removed from the water, weighed to determine saturated weight, and dried in an oven at 105°C for 48 hours or until a constant weight was obtained. The maximum water holding content was calculated as the difference between the saturated weight and the oven-dry weight divided by the oven-dry weight (w/w).

Water Adsorption Rate

A laboratory experiment was designed to quantify the rate at which water was absorbed by individual EXSH aggregates. Randomly selected individual EXSH aggregates were spread onto water-saturated filter paper. The filter paper was maintained in a saturated condition by the use of paper wicks connected to a water reservoir. This allowed unrestricted diffusion of water from the saturated filter paper into the EXSH aggregates through the point of direct contact. EXSH was left on the water-saturated filter paper for periods ranging from 2.5 minutes to 72 hours. After the specified time, EXSH was removed from the filter paper and oven-dried to determine water content.

Nutrient Release from Fertilizer-Treated Expanded Shale

A bulk quantity of EXSH was soaked overnight in deionized water to remove soluble nutrients inherent to the material and then oven-dried for 24 h at 60°C. After drying, a quantity of this pre-washed EXSH was nutrient-loaded by soaking for 48 h in a soluble fertilizer solution containing 0.56 g L^{-1} NH_4-N, 1.0 g L^{-1} NO_3-N, 2.44 g L^{-1} urea-N, 1.76 g L^{-1} P, and 3.32 g L^{-1} K. Upon removal from the nutrient solution, the EXSH was rinsed with water to remove surficial deposits of nutrient solution and then allowed to air dry. Untreated and nutrient-loaded EXSH were mixed in ratios of 100:0, 75:25, 50:50, 25:75, and 0:100 (v/v) to create growing media with increasing proportions of nutrient-loaded EXSH. In order to assess the bioavailability of nutrients in the fertilizer-treated EXSH growing media, each blend was sequentially extracted six times with water and the water extracts were analyzed for selected fertilizer-added nutrients. More specifically, a 25 g quantity of each EXSH treatment was placed in a 125 mL flask, immersed in 40 mL of deionized water, and agitated on a reciprocal shaker for 30 min. Then the supernatant was filtered through Whatman No. 2 filter paper and analyzed for NH_4-N, NO_3-N and orthophosphate-P. This process was repeated five more times for a total of six extracts.

Bioavailability of Adsorbed Nutrients

Nutrient bioavailability in a growing medium is best demonstrated by a plant's ability to extract and utilize essential elements. A greenhouse experiment was initiated to test the bioavailability of fertilizer-applied nutrients in the EXSH growing blends described above. Four replicates of each expanded shale blend that contained 0, 25, 50, 75, or 100% nutrientloaded EXSH were placed in 500 cm^3 greenhouse pots. The EXSH was covered with a thin layer of acid-washed sand so that Romaine lettuce seeds (Lactuca sativa) planted on the surface of the growing media would have an adequate germination bed. Approximately 20 seeds were sowed onto the sand layer followed by misting with deionized water twice a day to initiate seed germination. After emergence of lettuce seedlings, the number of plants was thinned to 5 plants per greenhouse pot. Plants were watered daily with deionized water, but no additional nutrients were supplied. After one month, above ground plant tissue was harvested and dried to determine yield. Plant tissue was analyzed for macro- and microelements.

RESULTS AND DISCUSSION

Basic Chemical Properties

Electrical Conductivity (EC)

A 1:1 water extract of EXSH produced an EC of 1.6 dS m^{-1} (Table 1). Electrical conductivity (EC) is an indirect measurement of the soluble salt concentration of a material. Excessive salt is detrimental to plant growth. Several researchers have proposed different EC ranges that are limiting for plant growth. Mengel and Kirkby (1982) reported that EC values of less than 2 dS/m in soil will have negligible salinity effects on plant growth while Wright (1986) recommended an EC value of 2 dS/m for healthy and vigorous plant growth. In contrast, Gajdos (1997) indicated that EC values exceeding 1-3 dS/m could be detrimental to plants. In containers, however, Lemaire et al. (1985) found that only EC values greater than 3.5 dS/m were too high for vigorous plant growth. Based on published research findings (Lemaire et al., 1985; Mengel and Kirkby, 1982; Wright, 1986), the EC value measured for EXSH in this study will likely not create salinity problems when added to soil or potting media.

pH

Table 1 shows pH values for EXSH measured in water and in a 0.01 mole L^{-1} dilute $CaCl_2$ solution. The $CaCl_2$ solution was used to eliminate the effects of background salt concentrations in the EXSH. The dilute calcium chloride method for pH measurement is widely trusted to produce the most reliable results close to the actual pH in the root zone (Handrek and Black, 1994). As a result, the close agreement between the two values (pH 8.3 for water versus pH 8.5 for $CaCl_2$) indicates minimal effects due to soluble salts and suggests that both values accurately represent the pH of EXSH. The measured pH values for EXSH reflect the presence of calcium, magnesium, potassium, and sodium oxides (CaO, MgO, K_2O, and Na_2O). These oxides were formed as a result of the high temperatures used to manufacture the expanded shale. When EXSH is exposed to water, these oxides hydrolyze to form hydroxides, which can raise pH to as high as 13 in the equilibrium solution. The fact that the equilibrium pH of the EXSH tested in this study only increased to 8.3 suggests that the material contains only relatively small quantities of reactive oxides. Expanded shale materials with higher quantities of reactive oxides would have higher equilibrium pH values.

Calcium Carbonate Equivalent (CCE)

A material's capacity to neutralize acidity indicates its potential effect on pH when mixed with soil or other growing media materials, such as sphagnum peat moss, compost, sand, vermiculite, etc. Calcium carbonate equivalence can be an important value for alkaline materials, such as EXSH, that will be blended with other growing media ingredients because it predicts the how the EXSH will affect the final pH of the growing media. The calcium carbonate equivalent (CCE) shows the material's effectiveness relative to pure calcium carbonate, which is the standard material for neutralizing soil acidity. The EXSH analyzed in this study had a CCE of 0.2 to 0.3 percent. In other words, EXSH is less than 1% effective at neutralizing soil acidity as calcium carbonate limestone. This indicated that although measured pH values for EXSH were slightly above the broad range of 5.5 to 8 known to support plant growth (Bunt, 1988), it is unlikely that its addition to soil, peat moss, compost, or other growing media ingredients would result in long term increases in the pH of the growing medium. However, Sloan et al. (2010) reported that the pH of various organic growing media, including sphagnum peat moss, pine bark, compost, and biosolids, were slightly elevated when the proportion of EXSH exceeded 30% (v/v), especially when the initial pH was less than 5.0.

Table 1. Basic chemical and physical properties of expanded shale (ExSh)

Parameter	Measured Value	
	Mean	SD†
Electrical conductivity, dS m⁻¹	1.6	(± 0.2)
pH		
(1:1 – water)	8.25	(± 0.62)
(1:1 – 0.01 mol L⁻¹ CaCl₂)	8.47	(± 0.21)
Calcium carbonate equivalent (CCE), %	0.2 – 0.3	
Water holding capacity, %	37.8	(± 1.35)
Cation exchange capacity (CEC), cmole kg⁻¹	2.75	(±0.15)
Extractable elements, mg/kg		
Water soluble		
Sodium (Na)	55.0	(± 6.5)
Potassium (K)	64.2	(± 6.0)
Magnesium (Mg)	29.0	(± 2.8)
Calcium (Ca)	88.3	(± 10.6)
Acid soluble§		
Sodium (Na)	87.3	(± 5.3)
Potassium (K)	126.8	(± 21.0)
Magnesium (Mg)	660	(± 40)
Calcium (Ca)	843	(± 59)

† Standard deviation of the mean. Shown in parentheses in table
§ 1 mol L⁻¹ sodium acetate buffered to pH 5 with acetic acid.

It is likely that EXSH will have a positive impact on strongly acidic growing media, such as peat moss and pine bark due to it alkaline pH, but that it will have only a small impact on growing media with neutral to alkaline pH values, such as compost and biosolids.

Soluble and Labile Elements

Expanded shale was extracted with water to measure the concentration of soluble elements and with acidified sodium acetate to measure potentially plant-available (labile) elements. Concentrations of water-soluble sodium (Na), potassium (K), magnesium (Mg), and calcium (Ca) in EXSH were small and similar in magnitude (Table 1). Measured concentrations for these elements are by far lower than guideline values established for cultivation which are < 857 mg/kg for Ca, provide small amounts of immediately plant-available K, Mg, and Ca plus larger amounts of more slowly available Mg and Ca. However, the inherent concentrations of these elements in EXSH are too low to significantly affect plant nutrition – either beneficially or detrimentally when it is added to soil.

Water Uptake

Two facets of water uptake by EXSH were investigated: 1) the maximum water-holding capacity and 2) the rate of water uptake. The maximum water holding capacity provides information on the total porosity in EXSH aggregates. After soaking in water for at least 72 hr, the EXSH tested in our study contained 37.8% water (Table 1). This value includes the amount of water in EXSH when 100% of the pores were saturated plus the small amount of water on the surfaces of its aggregates. The water-holding capacity of EXSH shows that it can be a significant source of water storage in soil. However, a typical silt loam soil will have a greater available water holding capacity than EXSH. Therefore, EXSH would be an appropriate soil amendment for soils that retain excess water and exhibit poor drainage characteristics. Sloan et al. (2002) reported that 3 to 6 mm diameter expanded shale, similar to that used in this study, effectively increased the root mass of pansy plants (Viola x wittrockiana) compared to un-amended clay soil during an excessively wet growing season.

The rate at which EXSH can absorb water is perhaps more important than its total water holding capacity. The water uptake rate provides information on the size of pores in the EXSH aggregates and can be related to the amount of water that will be available to plants. Figure 2 shows that EXSH absorbed 15% of its weight in water within 10 minutes and 20% within 2 hours. The water uptake rate slows considerably beyond 2 hours, but continued to increase to a maximum of 36% at 150 hours.

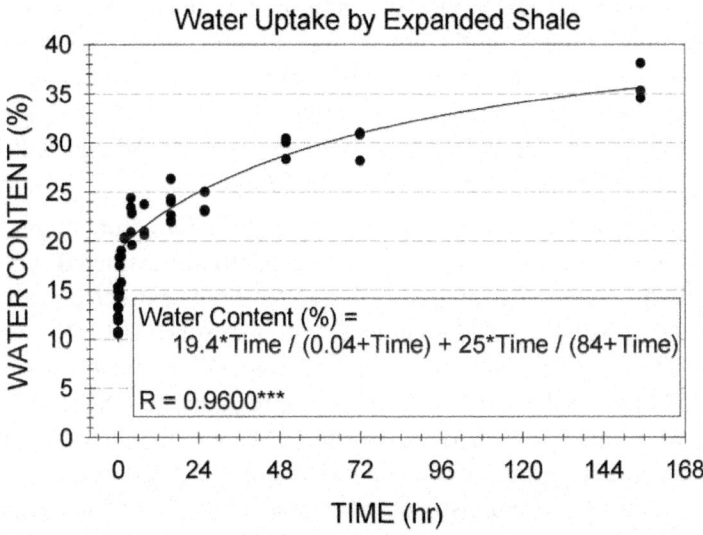

Figure 2. Rate of water diffusion into expanded shale (EXSH) from a saturated porous medium. Data was fitted to a double rectangular hyperbolic function to create a predictive uptake equation. *** Significant at the 0.001 level of probability.

The experiment was designed so that water could only enter the EXSH aggregates via capillary diffusion. When an EXSH aggregate was placed in contact with a saturated porous medium, water quickly diffused throughout the aggregate's surface layer until the exterior pores of the aggregate were saturated. A magnified view of an individual expanded shale aggregate shows the exterior pores to be relatively large and evenly distributed around the surface (Fig. 1). Water is held in these pores at relatively low tension and is therefore easily available for plant uptake, but it is also susceptible to rapid evaporation. This water accounts for 15 to 20% (w/w) of the total water holding capacity of EXSH. The remaining 16 to 21% of the total water is held in the much smaller interior pores of the EXSH aggregates. This water is held at relatively high tensions and is therefore mostly unavailable for plant use.

Nutrient Release from Fertilizer-treated EXSH

The porous nature of EXSH allows it to absorb water and therefore, any chemical constituents dissolved in water will also be absorbed by the EXSH aggregate. For this experiment, EXSH was loaded with nutrients by soaking it in a fertilizer solution for 48 hours, followed by rinsing in deionized water, and then air-drying. The reason for rinsing the fertilizer-impregnated EXSH was to ensure that nutrients released from the aggregates were from the internal pores and not residual surface deposits of fertilizer. Based on results from the water uptake experiment (Fig. 2), 5 to 10% of the pore volume was not saturated with fertilizer solution because the EXSH was soaked for only 48 hours whereas 150 hours was needed to attain complete saturation. Also, it is likely that 10 to 15% of the fertilizer was removed from the exterior EXSH during the rinsing process. These two factors probably reduced the total nutrient-supplying capacity of the fertilizer-treated EXSH by 15 to 25%.

A portion of each fertilizer-treated EXSH medium containing 0, 25, 50, 75, or 100% nutrientloaded EXSH was sequentially extracted with water to determine the kinetics of nutrient release. Nutrients extracted with water are immediately available for plant uptake. Nitratenitrogen (NO_3-N), ammonium-nitrogen (NH_4-N), and phosphate-phosphorus (PO_4-P) were measured in all extracts. Although the fertilizer used to impregnate the EXSH with nutrients also contained urea-N, we did not measure that form of N in the water extracts. In general, the concentration of each constituent decreased with each sequential extract (Fig. 3). This demonstrates that a portion of soluble nutrients absorbed by EXSH can later redissolve and become available for plant uptake. If EXSH were present in a growing medium, individual aggregates would either absorb or release nutrients depending on the concentration gradient of the surrounding solution. In practical terms, if EXSH aggregates in the growing medium were surrounded by relatively pure water, then nutrients would diffuse from the interior pores towards the exterior pores due to a concentration gradient. In a similar but reverse process, nutrients from a fertilizer solution passed through the medium would possibly diffuse into the interior pores of the EXSH aggregates in response to the nutrient gradient. Phosphorus retention by expanded shale in constructed wetlands as reported by Forbes et al. (2004 & 2005) may have been related to this diffusion phenomenon.

Nitrate-N, NH_4-N and PO_4-P each have different binding mechanisms to solid particles in a growing medium. Nitrate-N has a negative charge and exists as a free (i.e., unbound) ion in typical soil conditions. It moves freely through the soil with soil water. Ammonium-N has a positive charge and is attracted to negatively charged exchange sites found in typical soils.

This allows it to be held more tightly in the soil and makes it less susceptible to loss through leaching. Orthophosphate-P is an oxyanion and tends to strongly bind to various soil components, primarily aluminum and iron minerals at low pH (7.6).

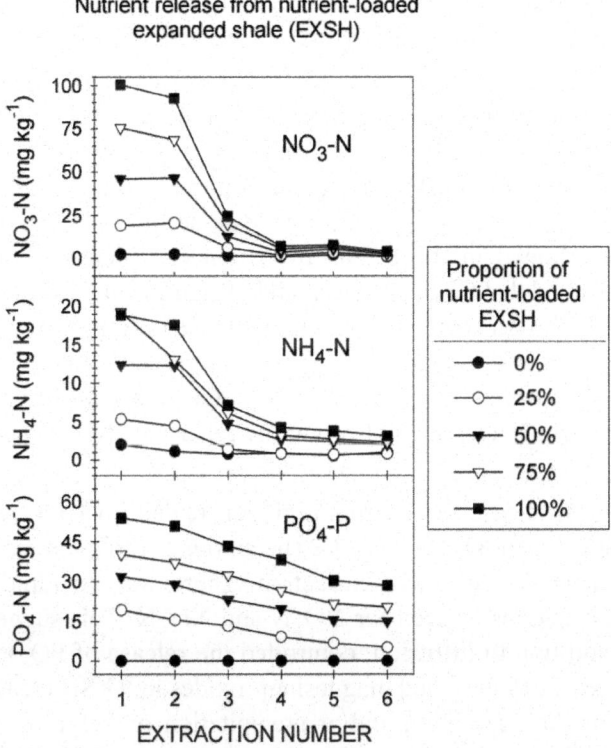

Nutrient release from nutrient-loaded
expanded shale (EXSH)

Proportion of
nutrient-loaded
EXSH

- ● 0%
- ○ 25%
- ▼ 50%
- ▽ 75%
- ■ 100%

Figure 3. Sequential extraction of NO3-N, NH4-N, and orthophosphate-P from expanded shale (EXSH) loaded with N, P, and K fertilizer

The different chemistries of NO_3-N, NH_4-N, and PO_4-P were apparent in the sequential extraction data presented in Fig. 3. Nitrate-N was easily extracted with water. This was best illustrated with the 100% fertilizer-treated EXSH treatment. The first extraction removed 100 mg NO_3-N per kg of EXSH. By the fourth extraction, the amount extracted had decreased to 7 mg kg^{-1}, but was still slightly greater than the concentration for the untreated 0% control (i.e., EXSH without fertilizer pre-treatment). The remaining extractions continued to remove small amounts of NO_3-N that were slightly greater than the control. The extraction of NO_3-N from fertilizer-treated EXSH was consistent with water uptake data that showed very rapid water uptake followed by a much slower uptake phase (Fig. 2). In the absence of any attractive force between NO_3-N and the EXSH aggregate, NO_3-N would diffuse out of the EXSH in a fashion similar to water. The rapid release of NO_3-N with the initial extractions corresponds to diffusion of NO_3-N from larger pores near the surface of the aggregate (Fig. 1). The much lower, yet significantly higher concentrations of NO_3-N released with the later extractions relative to the control suggest diffusion of NO_3-N from smaller pores deeper inside the EXSH aggregate.

Smaller amounts of NH_4-N were extracted from fertilizer-treated EXSH compared to NO_3-N because the fertilizer source used to pretreat the EXSH contained 1000 mg L^{-1} NO_3-N, but only 560 mg L^{-1} NH_4-N. However, the magnitudes of the difference between desorption of NO_3-N and NH4-N could not be explained by fertilizer composition alone. It is likely that some of the NH_4-N was retained in the EXSH by cation exchange capacity. The CEC analysis indicated that EXSH had a small but significant CEC of 2.75 cmol kg^{-1} (Table 1), which was adequate to retain up to 385 mg kg^{-1} NH_4-N. Other than the quantity extracted, release of NH_4-N from fertilizer-treated EXSH followed a trend similar to that shown by NO_3-N suggesting that diffusion processes were a significant factor in controlling the release of NH_4-N from the porous EXSH.

Phosphorus release curves from fertilizer-treated EXSH were distinct from those for NO_3-N and NH_4-N (Fig. 3). The amount of PO_4-P released decreased linearly with each successive extraction. There was no rapid release with the initial extractions as seen for NO_3-N and NH_4-N. This suggests that other factors, in addition to diffusion, controlled the release of PO_4-P from EXSH. The presence of calcium- and magnesium- oxides in EXSH created conditions favorable for the adsorption and/or precipitation of PO_4-P. Experience with calcareous soils has shown that dicalcium phosphate ($CaHPO_4$) is commonly a first precipitation product following application of phosphate fertilizer (Sposito, 1989). EXSH contained a substantial amount of extractable Ca as well as pH>8.3 (Table 1) suggesting the possible precipitation of the relatively soluble dicalcium phosphate ($CaHPO_4$) mineral. Phosphate adsorption and precipitation are reversible processes, particularly in an EXSH medium, which is much less complex than a soil medium. Dissolution and release of adsorbed and precipitated phosphorus from EXSH would occur at a slower and more constant rate than would be expected from simple diffusion due to a concentration gradient. The practical implication of these results is that phosphorus-impregnated EXSH could function as a relatively effective slow-release phosphorus fertilizer. Forbes et al. (2004 & 2005) also found that expanded shale was effective at removing soluble P from effluent water in a constructed wetland.

Lettuce Growth in Fertilizer-treated EXSH

The true test of nutrient bioavailability is whether plants are able to extract and utilize nutrients from the growing medium. For this experiment, Romaine lettuce (Lactuca sativa longifolia) was grown in the same fertilizer-treated,

EXSH media described in the preceding sequential extraction discussion. Briefly, the growing media consisted entirely of EXSH where 0, 25, 50, 75, or 100% (v/v) had been previously impregnated with fertilizer solution. Lettuce was grown for approximately 45 days with no additional fertilizer application and then harvested to determine shoot and root yields as well as nutrient content of the aboveground tissue (Fig. 4).

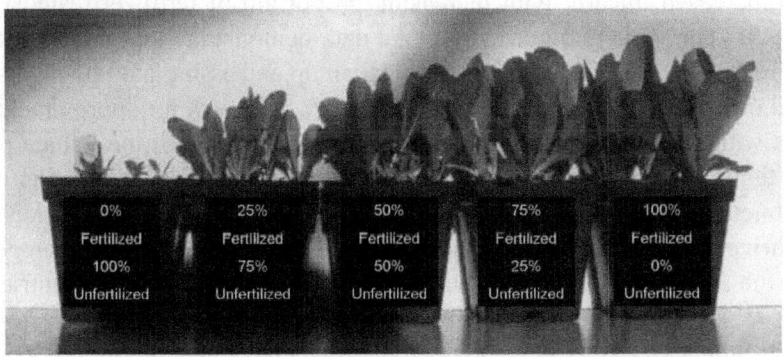

Figure 4. Increasing the proportion of nutrient-loaded expanded shale (EXSH) in the growing media from 0 to 100% resulted in significant increases in the size of Romaine lettuce.

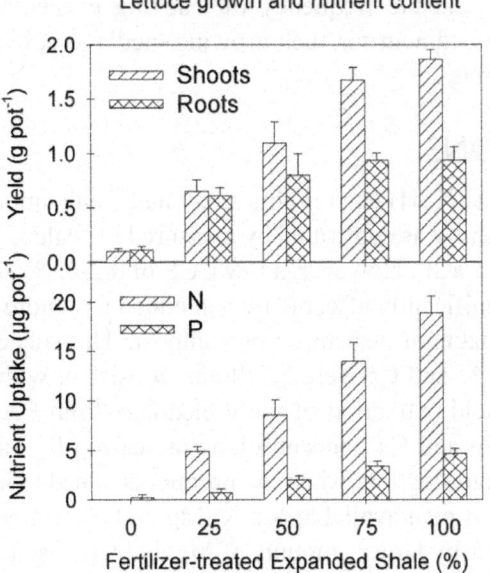

Figure 5. Lettuce yields and total nutrient uptake for lettuce grown in fertilizer-treated expanded shale.

Results for the shoot and root biomass yield as well as total nitrogen and phosphorus content of the aboveground tissue are shown in Fig. 5. The quantity of aboveground lettuce shoots increased linearly as the percentage of fertilizer-treated EXSH in the growing medium was increased (Fig. 5). Root mass also increased with increasing proportion of fertilizer- treated EXSH, but reached a maximum at the 75% fraction. Total uptake of nitrogen and phosphorus also increased linearly with increasing proportion of fertilizertreated EXSH (Fig. 5). The yield and nutrient uptake data demonstrate that nutrients in the fertilizer-treated EXSH medium were taken up and assimilated by plants. The level of fertility was sufficient to support lettuce growth for approximately 45 days with no additional fertilization. Following harvest of the lettuce plants, additional lettuce seeds were planted into these same EXSH growing media treatments, but no fertilizer was added. This second planting of lettuce failed to thrive in the same way as the first planting (data not shown) suggesting that the ability of EXSH to provide a reservoir of abundant plant nutrients is limited without additional fertilizer applications. Other studies (Nash et al., 1990; Smalley et al., 1993; Sloan et al., 2002) have also reported good plant growth in growing media containing EXSH as a major ingredient. Together these observations suggest that expanded shale can retain and release sufficient amounts of nutrients for healthy plant growth either alone or in a mixture, but additional fertilization will occasionally be needed depending on the intensity of plant growth and the frequency of leaching events. EXSH is therefore suitable as an amendment for such growing media as soil, potting mixes, and green roof substrates.

CONCLUSIONS

Expanded Shale (EXSH) used in this study had small amounts of Ca, Mg, K, and Na oxides. This was confirmed by measured pH values of >8 in both water and 0.01 mol L^{-1} $CaCl_2$. However, a low CCE of 0.2-0.3% suggests that EXSH is unlikely to significantly affect pH when added to a more buffered materials such as soil, sphagnum peat moss or compost. The concentrations of water soluble Na, K, Mg and Ca were small and consistent with the measured EC of 1.6 dS m^{-1}. Acid extraction of these elements from EXSH yielded higher amounts with Mg and Ca concentrations increasing by 20-fold and 10-fold, respectively, suggesting that when the product is added to a growing medium, small quantities of plantavailable Na, K, Mg and Ca can be provided initially and followed later by larger amounts of Mg and Ca in a slow release fashion. However, the concentration of these elements in EXSH is insufficient to affect plant nutrition positively or negatively. The release of NO_3N and NH_4N from nutrient-loaded EXSH was rapid and controlled by mainly by diffusion mechanisms, although cation exchange capacity was also responsible for retention of NH_4- N. The movement of PO_4-P on the other hand was influenced by both diffusion and the presence of calcium and magnesium oxides in the EXSH. The presence of these oxides enhanced the reversible processes of PO_4-P adsorption and precipitation in the EXSH medium. In practical terms, the ability of EXSH to reversibly adsorb fertilizer P suggests EXSH may be a suitable material for use as a slow-release phosphorus fertilizer. In this study, both the shoot and root mass of romaine lettuce increased with increasing proportions of fertilizer treated EXSH in the growing medium. This demonstrates that nutrients were released by the fertilizer treated EXSH and assimilated by the lettuce to support growth. The linear relationship between total nitrogen and phosphorus uptake and the proportion of fertilizer impregnated EXSH in the growing medium confirms the ability of EXSH to release nutrients for plant uptake.

REFERENCES

1.   Ampim, P.A.Y., J.J. Sloan, R.I. Cabrera, D.A. Harp, and F.H. Jaber. 2010. Green roof growing substrates: Types, ingredients, composition, and properties. J. Environ. Hort. 28:244-252.

2.   AOAC International. 2005. Agricultural Liming Materials. Official Methods of Analysis of AOAC International, 17th Edition. OAC International, 481 North Frederick Ave, Suite 500, Gaithersburg, MD 20877-2417, United States of America.

3.   ASTM D4972–01. 2007. Standard Test Method for pH of Soils. ASTM International, West Conshohocken, PA, 2003, DOI: 10.1520/D4972-01R07. www.astm.org.

4.   Bunt, A.C. 1988. Media and mixes for container-grown plants. Unwin Hyman. London.

5.   Chapman, H.D. 1965. Cation-Exchange Capacity. In C.A. Black et al. (eds.) Methods of Soil Analysis: Chemical and Microbiological Properties. Part 2. Agronomy No. 9. American Society of Agronomy. Madison, Wisconsin.

6.   Dunnett, N. and N. Kingbury. 2008. Planting green roofs and living walls. Timber Press Inc. Portland, OR.

7.   ESCSI (Expanded Shale, Clay and Slate Institute). 2002. Introducing the friendly material: Rotary kiln produced lightweight soil conditioner. Publication No. 8600.

8.   Evanylo, Greg and Mike Goatley Jr. 2011. Chapter 9. Organic and Inorganic Soil Amendments. In M. Goatley Jr. and K. Hensler (eds.) Urban Nutrient

Management Handbook. College of Agriculture and Life Sciences, Virginia Polytechnic Institute and State University. PUBLICATION 430-350. (Available online at: http://pubs.ext.vt.edu/430/430-350/430-350.html). Verified on 07-Sep-2011.

9.   Ferguson, B.K. 2005. Porous pavements. CRC Press. Boca Raton.

10.   Fischer, P. and E. Meinken. 1995. Expanded clay as a growing medium-Comparison of different products. Acta Hort. 401: 115-120.

11.   Fischer, P. and F. Penningsfeld. 1979. Hydrokultur – Beurteilung verschiedener Blähtonherkünfte. Gb + Gw 79:106-108.

12.   Forbes, M.G., K.R. Dickson, T.D. Golden, P. Hudak and R.D. Doyle. 2004. Dissolved phosphorus retention of light-weight expanded shale and masonry sand used in subsurface flow treatment wetlands. Environ. Sci. Technol. 38:892-898.

13.   Forbes, M.G., K.L. Dickson, F. Saleh and W.T. Waller. 2005. Recovery and fractionation of phosphorus retained by lightweight expanded shale and masonry sand used as media in subsurface flow treatment wetlands. Environ. Sci. Technol. 39: 4621-4627.

14.   Gajdos, R. 1997. Effects of two compost and seven commercial cultivation media on germination and yield. Compost Sci. and Util. 5:16-37.

15.   Gibson, M.J. and J.G. Farmer. 1986. Multi-step sequential chemical extraction of heavy metals from urban soils. Environ. Pollut. 11:117-135.

16.   Lemaire, F., A. Dartigues and L.-M. Rivière. 1985. Properties of substrate made with spent mushroom compost. Acta Hort. 172:13-29.

17.   Mengel, K. and E.A. Kirkby. 1982. Principles of Plant Nutrition (3rd ed.). International Potash Institute. P.O. Box CH-3048, Worblaufen-Bern/ Switzerland.

18.   Nash, M.A., T.P. Brubaker and B.W. Hipp. 1990. Expanded shale as a potting medium component for bedding plants. HortScience 25:1163.

19.   Rund, R.C. 1984. Agricultural Liming Material. p.1. In S. Williams (ed.) Official Methods of Analysis of the Association of Official Analytical Chemists, 14th ed.

20.   Sloan, J.J., R.I. Cabrera, P.A.Y. Ampim, S.A. George, and W.A. Mackay. 2010. Performance of ornamental plants in alternative organic growing media amended with increasing rates of expanded shale. HortTechnology 20:594-602.

21.   Sloan, J.J., S.W. George, W.A. Mackay, P. Colbaugh and S. Feagley. 2002. The suitability of expanded shale as amendment for clay soils. HortTechnology 12:646-651.

22. Smalley, T.J., G.L. Wade, W.L. Corley and P.A. Thomas. 1993. Soil amendment increases growth of Salvia splendens and Catharanthus roseus. HortScience 28:540

23. Sposito, G. 1989. The Chemistry of Soils. Oxford University Press. New York, NY.

24. (TXI) Texas Industries Inc. 2009. Lightweight aggregate as a soil amendment. Retrieved 2 July 20009. http://www.txi.com/TXI-products/ TXI-trugro.html.

25. Wright, R.D. 1986. Pour through nutrient extraction procedure. HortScience 21:227-229.

Chapter 5

BEHAVIOR OF CLAYEY SOIL STABILIZED WITH RICE HUSK ASH & LIME

B. Suneel Kumar[1] and T. V. Preethi[2]

[1]Graduate Student, Department of Civil Engineering, Geotechnical Engineering, SRM University, Kattankulathur-60320, Tamil Nadu, India

[2]Assistant professor, Department of Civil Engineering, SRM University, Kattankulathur-603203, Tamil Nadu, India

ABSTRACT

In India the soil mostly present is Clay, in which the construction of sub grade is problematic. In recent times the demands for sub grade materials has increased due to increased constructional activities in the road sector and due to paucity of available nearby lands to allow excavate fill materials for making sub grade. In this situation, a means to overcome this problem is to utilize the different alternative generated waste materials, which cause not only environmental hazards and also the depositional problems. Keeping this in view stabilization of weak soil in situ may be done with suitable admixtures to save the construction cost considerably. The present investigation has therefore been carried out with agricultural waste materials like Rice Husk Ash (RHA) which was mixed with soil to study improvement of weak sub grade in terms of compaction and strength characteristics. Silica produced from rice husk ashes have investigated successfully as a pozzolanic material in soil stabilization. However, rice husk ash cannot be used solely since the materials lack in calcium element. As a result, rice husk ash shall be mixed with other cementitious materials such as lime and cement to have a solid chemical reaction in stabilization process. Lime is calcium oxide or calcium hydroxide. It is the name of the natural mineral (native lime) CaO occurs as a product of coal seam fires and in altered lime stone xenoliths in volcanic ejection. In this study RHA and Lime is mixed in different percentage like (RHA as 5%, 10%, and 15%) and (Lime as 3%, 6%, 9%) and laboratory test CBR is done with a curing period of 4, 7 and 14 days with different percentages of RHA & Lime and Lime+ RHA.

INTRODUCTION

Soil improvement could either be by modification or stabilization or both. Soil modification is the addition of a modifier (cement, lime etc.) to a soil to change its index properties, while soil stabilization is the treatment of soils to enable their strength and durability to be improved such that they become totally suitable for construction beyond their original classification. Over the times, cement and lime are the two main materials used for stabilizing soils. These materials have rapidly increased in price due to the sharp increase in the cost of energy since 1970s (Neville, 2000). The over dependence on the utilization of industrially manufactured soil improving additives (cement, lime etc.), have kept the cost of construction of stabilized road financially high. This hitherto, has continued to deter the underdeveloped and poor nations of the world from providing accessible roads to their rural dwellers who constitute the higher percentage of their population and are mostly, agriculturally dependent. Thus the use of agricultural waste (such as rice husk ash) will considerably reduce the cost of construction and as well reducing the environmental hazards they causes. Therefore, replacing proportions of the Portland cement in soil stabilization with a secondary cementitious material like RHA will reduce the overall environmental impact of the stabilization process. Silica produced from rice husk ashes have investigated successfully as a pozzolanic material in soil stabilization. However, rice husk ash cannot be used solely since the materials lack in calcium element. As a result, rice husk ash shall be mixed with other cementitious materials such as lime and cement to have a solid chemical reaction in stabilization process.

Lime is a general term for calcium-containing inorganic materials in which carbonates, oxides and hydroxides predominate. Strictly speaking, lime is calcium oxide or calcium hydroxide. It is the name of the natural mineral (native lime) CaO occurs as a product of coal seam fires and in altered lime stone xenoliths in volcanic ejection. The word "lime" originates with its earliest use as building mortar and has a sense of "sticking and or adhering". "Burning" converts them into the highly caustic material quicklime (calcium oxide, Cao) and through subsequent addition of water, into less caustic (but still strongly alkaline) slaked lime or hydrated lime (calcium hydroxide, CA $(OH)_2$ =74.10), the process of which is called slaking of lime.

Rice husk is an agricultural waste obtained from milling of rice. About 108 tonnes of rice husk is generated annually in the world. Meanwhile, the ash has been categorized under pozzolana, with about 67-70% silica and about 4.9% and 0.95%, Alumina and iron oxides, respectively (Oyetola and Abdullahi, 2006). The silica is substantially contained in amorphous form, which can react with the CaOH librated during the hardening of cement to further form cementations compounds

Light compaction energy, the effect of Rice Husk Ash on the soil was investigated with respect to compaction characteristics, California Bearing Ratio (CBR) and unconfined compressive strength (UCS) tests. Results obtained, there was also a tremendous improvement in the CBR and UCS with increase in the RHA and lime at specified contents to their peak values at 6% lime and 10% RHA. The UCS values also improved with curing age and in the combination of lime + RHA, 6% lime+10% RHA showed good improvement in UCS and CBR value with increase in curing period.

Ario Muhammad, (2007) Silica produced from rice husk ashes has investigated successfully as a pozzolanic material in soil stabilization. However, rice husk ash cannot be used solely since the materials lack in calcium element. As a result, rice husk ash shall be mixed with other cementitious materials such as lime and cement to have a solid chemical reaction in stabilization process. For the stabilized soils, the admixture materials, i.e. lime, rice husk ash, were mixed in 12 % and 24 % of the dry weight of soil matrix respectively.

Biswas, (2007) The present investigation has therefore been carried out with agricultural waste materials like Rice Husk Ash (RHA) which was mixed with soil or lime-soil mixture to study improvement of weak sub grade in terms of compaction and strength characteristics. The laboratory test results show marked improvement of strength of soil on addition of admixtures in terms of California Bearing Ratio (CBR). In this study RHA is mixed with lime stabilized soil in which lime is 3%, 6% & 9%. Musa Alhassan, (2008) Soil sample collected from Maikunkele area of Minna, was stabilized with 2-12% rice husk ash (RHA) by weight of the dry soil. Performance of the soil-RHA was investigated with respect to compaction characteristics, California bearing ratio (CBR) and unconfined compressive strength (UCS) tests. The results obtained, indicates a general decrease in the maximum dry density (MDD) and increase in optimum moisture content (OMC) with increase in RHA content. There was also slight improvement in the CBR with increase in the RHA content. Chakraborty & Saibal, (2010) in recent times the need for suitable road materials has increased due to demand of construction activities in the road sector. Keeping this in view stabilization of weak soil in situ may be done with suitable admixtures. Investigation has therefore been carried out with agricultural waste materials like Rice Husk Ash (RHA) which was mixed with soil or lime-soil mixture to study improvement of weak sub grade in terms of compaction and strength characteristics. Abu Siddique (2011), the effects of lime stabilisation on plasticity, shrinkage, swelling, moisture-density relations and strength characteristics of an expansive soil have been investigated. The soil was stabilised with lime contents of 3%, 6%, 9%, 12% and 15%. With the increase in lime content, maximum dry density decreased while the optimum moisture content increased. California Bearing Raito (CBR) of the stabilised samples at all levels of compaction increased significantly with increasing lime content.

MATERIALS USED

Soil Used

Soil sample is collected from a proposed for the construction of road alignment in guduvanchery area, Chennai. Standard tests were conducted to determine the physical properties of the soil and the results are given in Table 1.

Table 1: Physical properties of soil

SL. no	Test Conducted	Result	
1	Wet sieve analysis	% passing 75 microns sieve is 56%, so it is fine grained soil	
2	Liquid limit	36.5 %	
3	Plastic limit	12.62 %	
4	Plasticity index on A-line	11.61 %	
5	Shrinkage limit	21.68 %	
6	Specific gravity	2.23	
7	Free swell index	18.18 %	
8	Standard proctor	OMC 15.15 %	MDD 1.797 kg/cm^3
9	UCS (kg/sq cm)	0.624	
10	CBR	5.48 %	

Classification of Soil Sample

Based upon the tests performed in laboratory for soil sample and according to the results obtained, the soil sample is classified as follows,

1. 56% of soil is passing through 75 microns sieve so it is fine grained soil.
2. According to A-line Chart , the soil can be classified as clay with Intermediate Compressibility –CI
3. According to free swell index value, the soil is classified as low compressible.

Rice Husk Ash (RHA)

Rice husk ash, basically a waste material, is produce by rice - mill industry while processing rice from paddy. Rice husk ash is a pozzolanic material that could be potentially used in soil stabilization, though it is moderately produced and readily available. About 20 – 22% rice husk is generated from paddy and about 25% of this total husk become ash when burn. It is non – plastic in nature. RHA has a good pozzolanic property. The chemical properties of RHA are shown in Table 2.

Table 2: Chemical properties of rice husk ash

Chemical	Percentage (%)
Silica(SIO_2)	83.60
Aluminium(Al_2O_3)	3.5
Iron(FEO_3)	1.10
Calcium (CAO)	1.80
Magnesium(MGO)	1.28
Sodium(NA_2O)	0.17
Potassium(K_2O)	0.29

Silica produced from rice husk ashes have investigated successfully as a pozzolanic material in soil stabilization. However, rice husk ash cannot be used solely since the materials lack in calcium element. Rice husk ash shall be mixed with other cementitious materials such as lime and cement to have a solid chemical reaction in stabilization process.

Lime

Lime is a general term for calcium-containing inorganic materials in which carbonates, oxides and hydroxides predominate. Strictly speaking, lime is calcium oxide or calcium hydroxide. It is the name of the natural mineral (native lime) CaO occurs as a product of coal seam fires and in altered lime stone xenoliths in volcanic ejects. The word "lime" originates with its earliest use as building mortar and has a sense of "sticking and or adhering".

These materials are still used in large quantities as building and engineering materials (including limestone products, concrete and mortar) and as chemical feedstock's, and sugar refining, among other uses. Lime industries and the use of many of the resulting products date from prehistoric periods in both the old & new worlds.

The rocks and minerals from which these materials are derived, typically limestone or chalk, are composed primarily of calcium carbonate. They may be cut, crush or pulverized and chemically altered. "Burning" converts them into the highly caustic material quicklime (calcium oxide, Cao) and through subsequent addition of water, into less caustic (but still strongly alkaline) slaked lime or hydrated lime (calcium hydroxide, CA (OH) $_2$=74.10), the process of which is called slaking of lime.

LABORATORY STUDIES

The testing program conducted on the clayey soil samples included determination of the physical and chemical properties of soils at their natural state. On the other hand, the testing program conducted on the clayey soil samples mixed with different percentages of rice husk ash and lime materials, included unconfined compression test and CBR test.

Unconfined Compression Test

UCS test is performed in accordance with IS:2720 part 10 (1973). The sample sizes were of 38 mm diameter and 76 mm length. At the optimum moisture content (OMC) and maximum dry unit weight, the tests were performed.

California Bearing Ratio (CBR)

California bearing ratio (CBR) is a penetration test for evaluation of the mechanical strength of road sub grades and base courses. It was developed by the California Department of Transportation before World War II.

The test is performed by measuring the pressure required to penetrate a soil sample with a plunger of standard area. The measured pressure is then divided by the pressure required to achieve an equal penetration on a standard crushed rock material. The CBR test is fully described in IS: 2720 part 16 (1987).

RESULTS AND DISCUSSIONS

UCC Test Results

UCC test was conducted in laboratory on soil sample with addition of different percentages of lime and RHA and the results obtained are shown in table no 3 and the figures 1 & 2 shows the graphs for UCC. Figure 1 shows the UCS value for different percentages of lime and figure 2 shows the UCS value for different percentages of RHA.

Table 3: UCC Test Results on soil sample

Additives	UCS Value (Kg/Sq Cm)		
	4 days	7 days	14 days
3% LIME	2.338	5.32	6.464
6% LIME	4.558	6.26	8.62
9% LIME	1.662	3.698	4.504
5% RHA	2.248	3.29	4.068
10% RHA	3.344	4.552	6.14
15% RHA	1.788	2.75	3.148

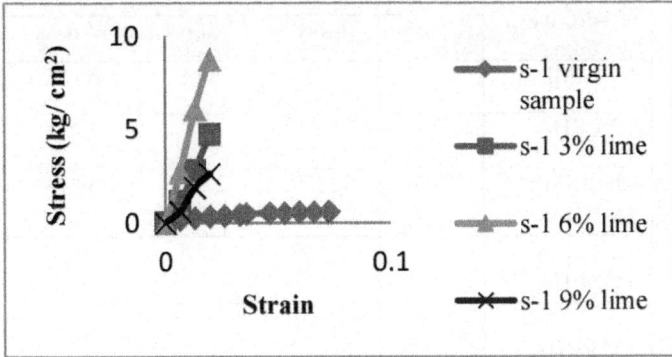

Figure 1: UCS value for lime.

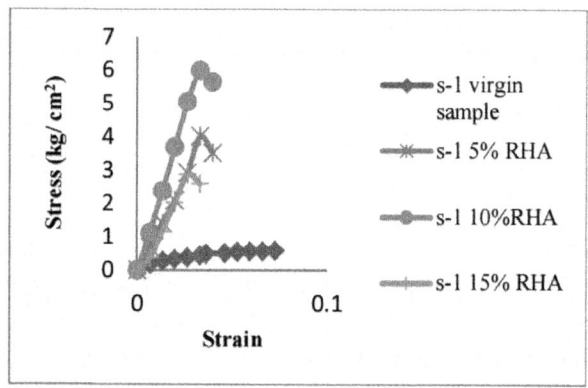

Figure 2: UCS value for RHA.

Discussion For Soil Sample

Based on the ultimate UCS value as shown in Figure 3 & 4, the CBR test was performed for 6% of lime and 10% of RHA individually and combination of lime + RHA is done for 6% lime with different percentages of RHA i.e. (5%, 10%, and 15%)

CBR Test Results

CBR test was conducted in laboratory on soil sample with addition of different percentages of lime and RHA and the results obtained are shown in Table 4.

Table 4: CBR Test Results for Soil Sample

Additives	CBR Value For Curing Period			
	0 day	4 days	7 days	14 days
6% LIME	13.95	33.11	50.8	66.75
10% RHA	7.9	8.21	9.8	13.77
6% LIME + 5% RHA	14.56	26.7	27.54	39.73
6% LIME + 10% RHA	17.21	29.66	45.8	56.68
6% LIME+ 15 % RHA	7.9	20.13	21.33	27.8

Figure 3 shows the graph drawn for CBR value for 6% of lime and 10 % of RHA which can be compared with virgin sample. Percentage of lime and RHA are taken from ultimate UCS value.

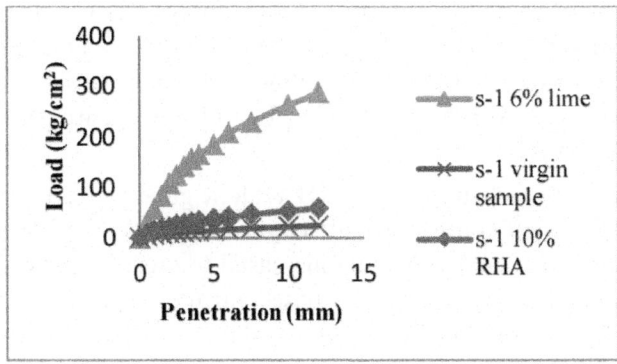

Figure 3: CBR value for lime and RHA.

Figure 4 shows the graphs drawn for combination of lime +RHA in which ultimate UCS value of 6% lime is taken and mixed with RHA in different percentages i.e. (5, 10, and 15%). For which 6%lime + 10% RHA is giving the ultimate value.

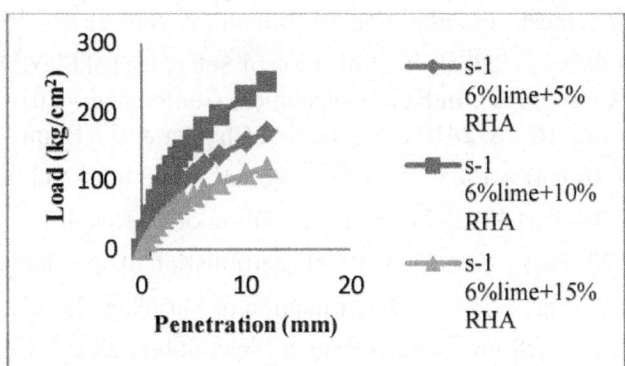

Figure 4: CBR value for lime+ and RHA.

CONCLUSION

- Based on the UCS value comparison, 6% addition of lime showed the good improvement of 92.74% in UCS value for 14 days curing compared to virgin soil and for 10% of RHA with 89.93% for a same period of curing.

- CBR test was performed based on addition of 6% of lime for curing period of 14 days which showed an improvement of 91.79% and for the 10% RHA, CBR value was increased up to 60.20% at 14th day. CBR test is also conducted for the combination of lime+RHA in which lime is taken as 6% and the combination of RHA is done in different percentages i.e (5%, 10%, & 15%) in which maximum improvement of CBR is observed in the combination of 6%lime+10%RHA for which % of CBR improvement are 90.65% for 14th day of curing comparing with virgin sample.

• Addition of industrial waste (RHA) alone gave an average improvement of 60% when compared with virgin sample. When the additive lime is added to it the CBR value increased to great extent which had been mentioned above (91%). So based on the respective results, quality of soil is increasing from poor condition to excellent condition based on CBR test values. As per the Pavement Design, when the California Bearing Ratio increases, the sub grade thickness can be reduced. So the RHA and Lime can be used to improve the CBR ratio respectively.

REFERENCE

1.   Agus Setyo Muntohar, "Uses of Lime -Rice Husk Ash And Plastic Fibers as Mixtures-material in High-plasticity Clayey Sub Grade", Journal Ilmiah Semesta Teknika, Vol. 10, 146 No. 2, 2007: 145 – 154.

2.   Chakraborty & Saibal, "Stabilization of Sub grades of Flexible Pavements with Admixtures", Indian Geotechnical Conference – 2010, GEOtrendz December 16–18, 2010 IGS Mumbai Chapter & IIT Bombay.

3.   IS: 2720- Part 5-1985 ,"Determination of liquid limit and Plastic limit".

4.   IS: 2720 - Part 40- 1977 ," Determination of free swell".

5.   IS: 2720- Part 3 –sect. 1-1980, "Determination of specific gravity".

6.   IS: 2720- Part 6-1972, "Determination of Shrinkage limit".

7.   Sudhira rath, "Lime Stabilization of Weak Sub-Grade for Construction of Rural Roads", International journal of earth sciences AND engineering Issn 0974-5904, vol. 05, no. 03 (01), june 2012, PP. 554-561.

8.   Koteswara Rao, D., "Stabilization Of Expansive Soil With Rice Husk Ash, Lime And Gypsum", International Journal of Engineering Science and Technology (IJEST) ISSN : 0975-5462 Vol. 3, No. 11 November 2011.

9.   " S.K.Khanna And C.E.G.Justo, "Highway engineering" khanna publications ninth edition (2011)

10.   Brooks, R. M., (2009), "Soil Stabilization with Fly ash and Rice Husk Ash", International Journal of Research and Reviews in Applied Sciences, Volume 1, Issue 3, pp. 209- 217.

11. Gidigasu, M.D., (1976), "Laterite Soil Engineering: Pedogenesis and Engineering Principles", Elsevier, Amsterdam, the Netherlands.

12. Ito, K. K, Senge, M., Adomako, J. T., and Afandi, (2008), "Amendment of Soil Physical and Biological Properties Using Rice Husk and Tapioca Wastes", Journal of Jpnanese Society of Soil Physics, No. 108, pp. 81-90.

13. Experimental Study", International Journal of Engineering Science and Technology, Vol. 3 No. 11, pp. 8076 – 8085.

14. Mtallib, M. O. A., and Bankole, G. M., (2011), "The Improvement of the Index Properties and Compaction Characteristics of Lime Stabilized Tropical Lateritic Clays with Rice Husk Ash (RHA) Admixtures", Electronic Journal of Geotechnical Engineering, Vol. 16, Bund. I, pp. 984-996.

15. Muntohar, S., and Hantoro, G., (2000), "Influence of Rice Husk Ash and Lime on Engineering Properties of a Clayey Sub-grade", Electronic Journal of Geotechnical Engineering, Vol. 5.

16. Neville A. M., (2000), "Properties of Concrete", 4th edition. Pearson Education Asia Ltd, Malaysia.

17. Ola, S.A., (1975), "Stabilization of Nigeria Lateritic Soils with Cement, Bitumen and Lime", Proc. 6th Reg. Conf. Africa on Soil Mechanics and Foundation Engineering. Durban, South Africa.

18. Osinubi K.J., (1999), "Evaluation of Admixture Stabilization of Nigeria Black Cotton Soil", Nigeria Soc. Engin. Tech. Trans., Vol. 34, No. 3, pp. 88-96.

19. Osinubi, K.J. and Katte, V.Y., (1997), "Effect of Elapsed Time after Mixing on Grain Size and Plasticity Characteristic, I: Soil-Lime Mixes", NSE Technical Transactions Vol. 32, No. 4.

20. Osula D. O. A., (1991), "Lime Modification of Problem Laterite", Engineering Geology, Vol. 30, pp. 141-149.

Chapter 6

PHYSIOLOGICAL AND BIOCHEMICAL MECHANISMS OF PLANT ADAPTATION TO LOW-FERTILITY ACID SOILS OF THE TROPICS: THE CASE OF BRACHIARIAGRASSES

T. Watanabe[1], M. S. H. Khan[2], I. M. Rao[3], J. Wasaki[4], T. Shinano[5], M. Ishitani[3], H. Koyama[6], S. Ishikawa[7], K. Tawaraya[8], M. Nanamori[1], N. Ueki[8], and T. Wagatsuma[8]

[1]Graduate School of Agriculture, Hokkaido University, Kita-ku, Sapporo, Japan

[2]Department of Soil Science, HMD Science and Technology University, Dinajipur, Bangladesh

[3]Centro Internacional de Agricultura Tropical (CIAT), A.A.6713, Cali, Colombia

[4]Graduate School of Biosphere Science, Hiroshima University, Higashi-Hiroshima, Japan

[5]National Agricultural Research Center for Hokkaido Region, Sapporo, Japan

[6]Faculty of Applied Biological Sciences, Gifu University, Gifu, Japan

[7]National Institute for Agro-Environmental Science, Tsukuba, Japan

[8]Faculty of Agriculture, Yamagata University, Tsuruoka, Japan

INTRODUCTION

Brachiaria species are the most widely planted tropical forage grasses in the world (Miles et al., 2004). For example, in Brazil alone, about 80 million hectares are planted to Brachiaria pastures (Macedo, 2005). They increase animal productivity by 5 to 10 times with respect to native savanna vegetation in the tropical areas of Latin America, thus representing a significant contribution to farmer's income (Rao et al., 1993). Although their origin is from the tropical areas of Africa, they are also used for livestock production in South-East Asia and Australia. Among them, Brachiaria decumbens cv. Basilisk, Brachiaria brizantha cv. Marandu, and Brachiaria ruziziensis cv. Kennedy have been more commonly utilized for livestock production in the tropics (Miles et al., 2004). Among the three grasses, B. decumbens is highly adapted to infertile acid soils, i. e., high level of tolerance to high aluminium (Al) saturation, low phosphorus (P) and low calcium (Ca) supply in soil (Louw-Gaume et al., 2010 a,b; Rao et al., 1995, 1996; Wenzl et al., 2001, 2003), but also highly sensitive to a major insect, spittlebugs (Miles et al., 2006) and produces mycotoxin after infection with Pithomyces chartarum (Andrade et al., 1978). B. brizantha cv. Marandu is highly resistant to spittlebugs, adapted to seasonal drought stress, highly responsive to fertilizer application but is not well adapted to low fertility acid soils (Miles et al., 2004, 2006). B. ruziziensis cv. Kennedy is sensitive to spittlebugs, performs better in well-drained fertile soils, has high forage quality but poorly adapted to low fertility acid soils (Ishigaki, 2010; Miles et al., 2004). B. decumbens and B. brizantha are generally tetraploid, apomicts while B. ruziziensis is diploid, sexual (Miles et al., 2004, 2006).

CIAT and its collaborators have an on-going breeding program to combine the desirable attributes from the three grasses (Miles et al, 2004, 2006). B. hybrid cv. Mulato is the product of three generations of crosses between B. ruziziensis, B. decumbens and B. brizantha. This grow well in low P, low fertility acid soils in both wet and dry seasons (Rao et al., 1998), and produces a large numbers of panicles with well synchronized flowering and good caryopsis formation, which leads to good-quality seed. Defining the specific physiological and biochemical mechanisms that are associated with greater adaptation to low fertility acid soils will contribute to developing rapid and reliable methods to select the phenotypes and to develop molecular markers for marker assisted breeding of brachiariagrasses. Developing superior Brachiaria hybrids from the on-going breeding programs that combine the desirable attributes including adaptation to major biotic and abiotic constraints, forage quality, and seed production will facilitate sustainable intensification of crop-livestock systems in the tropics (Miles et al., 2004, 2006; Rao, 2001 a,b). This chapter reviews the progress made in defining the physiological and biochemical mechanisms of adaptation of brachiariagrasses to low fertility acid soils.

There is limited knowledge on the comparative differences in Al resistance among B. decumbens, B. brizantha, B. ruziziensis and B. hybrids (Mulato and Mulato ⊇) grown in hydroponic system. Identification of plant attributes that contribute to greater ability to acquire nutrients under low pH, low P and high Al conditions is critical to develop brachiariagrases that are productive and persistant under infertile acid soil conditions.

There have been some discussions on the validity of short-term screening technique that uses simple solution of Al and Ca to test the effect of Al on relative root elongation of young seedlings: whether the results obtained by this short-term screening technique can apply to the behaviour of older plants is under discussion (Ryan et al., 2011). However, significant positive correlations were observed on Al resistance of 15 cultivars of sorghum (Sorghum bicolor Moench) and 10 cultivars of maize (Zea mays L.) with data obtained using short-term (1 day) screening and long-term screening technique using hydroponic system (Akhter et al., 2009). Similar results were also observed with 8 rice cultivars (unpublished data). In this chapter, we consider short-term vs long-term responses of several plant species including brachiariagrasses that differed widely in their level of Al resistance.

B. decumbens is known for very high level of Al resistance, however, the mechanisms responsible for this high level of Al resistance was not associated with exudation capacity of organic acid anions from root tips (Wenzl et al., 2001). It is important to define the specific mechanism(s) contributing to the high level of Al resistance in B. decumbens.

Higher level of Al exclusion was found in B. decumbens, however, the specific mechanisms related to Al exclusion are not known. Several mechanisms other than organic acid anion exudation were found and the related Al resistance genes have been reported (Huang et al., 2009; Yamaji et al., 2009; J.L. Yang et al., 2008; Z-B Yang et al., 2010). We found a new mechanism for higher level of Al tolerance of an Al-resistant rice cultivar based on higher abundance of sterols in plasma membrane (PM) lipids of root-tip cells (Khan et al., 2009). Rice is also known for its greater level of Al resistance than other cereal crops and its level of resistance was also not related to organic acid anion exudation (Ishikawa et al., 2000; Ma et al., 2002). Therefore, it is crucial to test the role of sterols in PM lipids of root-tip cells of brachiariagrasses.

High concentrations of phenolic compounds have been reported as one of the promising mechanisms to explain higher level of Al resistance of a forage legume, Lotus pedunculatus Cav. (Stoutjesdijk et al., 2001) and several common woody plants (Ofei-Manu et al., 2001).

This can be explained by higher complexing abilities of phenolic compounds with Al ions (Cornard & Merlin, 2002; Yoneda & Nakatsubo, 1998). Phenolic compounds have also been reported to be solubilized into lipid layer (Boija et al., 2006, 2007). Existence of phenolic compounds in lipid layer was found to make PM less fluid (Arora et al., 2000). Lipid layer with less fluidity will make PM less permeable even in the presence of Al ions (Khan et al., 2009). The contribution of phenolic compounds in root-tip portion to high level of Al resistance in B. decumbens is not known.

A major constraint to agriculture on tropical and subtropical soils is P deficiency (Fairhust et al., 1999). Applying large amounts of fertilizer to correct P deficiency is not feasible for most resource-poor farmers in developing countries. Thus the agricultural productivity becomes limited in the near future. Moreover, P fertilizer is receiving more attention as a nonrenewable resourse (Cordell et al., 2009; Steen, 1998). For sustainable P management in agriculture on tropical and subtropical soils, it is essential to define the mechanisms involved in making plants more efficient in P acquisition and use (Lynch, 2011; Ramaekers et al., 2010; Rao et al., 1999).

P is a constituent of phospholipids (PL), nucleic acids, nucleosides, coenzymes, and phosphate esters in plants. P helps regulate plant metabolisms by controlling enzymatic activity through phosphorylation and/or dephosphorylation. To overcome P deficiency, plants develop several strategies, including the well-known one of secreting acid phosphatase (APase), ribonuclease (RNase), and organic acids into the rhizosphere to improve P availability in the soil (Duff et al., 1994; Green, 1994; Jones, 1998; Rao et al., 1999; Tadano et al., 1993; Wasaki et al., 2003a). Besides from acquiring P from the outside of the plant, APase is also recognized for its role in efficient utilization of absorbed P for metabolism (Duff et al., 1989, 1991). One of the mechanisms to increase P recycling ability is dependent on the activity of APase (Duff et al., 1991, 1994) and RNase (Howard et al., 1998) in the cell. The increase of both enzymes is reported and these enzymes are considered to utilize those P compounds that are stored in vacuole. The other mechanism is bypassing several metabolic pathways to reduce the usage of P molecule. Theodorou and Plaxton (1993) showed that P deficiency induces some glycolytic enzymes, such as phosphoenolpyruvate carboxylase (PEPC) and phosphoenolpyruvate phosphatase (PEPP). These catalyze the bypass reaction of pyruvate kinase (PK), which is responsible for regulating carbon flow from glycolysis to the TCA cycle. PEPC replenishes intermediates of the TCA cycle, and may help regulate both carbon and nitrogen metabolism, and P recycling under P deficiency. Kondracka and Rychter (1997) observed that, in P-deficient bean leaves, the rate of malate synthesis increases, and the accumulation of aspartate and alanine (products of PEP metabolism) is also enhanced. In early stages of P deficiency, the increased activity of PEPC and use of PEP in amino acid synthesis are probably the most important reactions for P recycling in bean leaves during photosynthesis. Thus, PEP metabolism by PEPC and PEPP, or PEP transport via PPT (which transports PEP from the cytosol into chloroplasts in leaves) may affect carbon distribution within the plant under P deficiency.

Under P-deficient conditions, the brachariagrasses improve their P acquisition by enhancing root growth, uptake efficiency, and ability to use poorly available plant P (Louw-Gaume et al., 2010b; Rao et al., 1999, 2001a). Although they have much lower internal requirements for P than do other grasses, they also show interspecific differences (Rao et al., 1996).

Recent advances in post-genomic studies have indicated that transcriptomic analysis is a useful tool for understanding gene expression networks. Some transcriptomic studies of low-P adaptation strategies have also been carried out using cDNA arrays (Hammond et al., 2003; Misson et al., 2005; Ramaekers et al., 2010; Uhde-Stone et al., 2003; Wang et al., 2002; Wasaki et al., 2003b, 2006; Wu et al., 2003). Although some aspects of plant strategies for coping with P-deficient conditions are understood, the majority are not; a broad view of gene expression is necessary to fully elucidate all of the mechanisms involved (Ramaekers et al., 2010). In this chapter, we review the progress in understanding the transcriptomic changes by P deficiency in rice plants, which is a model of Gramineae plants and also relatively tolerant of low P and low pH conditions. We use qRT-PCR for quantification of the effects of P deficiency on phosphohydrolases and carbon metabolism in rice leaves. We also review the progress in defining the physiological and biochemical bases of improved P use efficiency in B. hybrid (cv. Mulato).

MATERIALS AND METHODS

We randomly selected seven plant species, i.e., rice (Oryza sativa L. cv. Sasanishiki), maize (Zea mays L. cv. Pioneer 3352), pea (Pisum sativum L. cv. Kinusaya), barley (Hordeum vulgare L. cv. Manriki), tea (Camellia sinensis L. cv. Yabukita), siclepod (Cassia tora L.) and B. brizantha. Seeds of rice, maize, pea were soaked in tap water under aeration for 3 to 24 h depending on plant species. Seeds of siclepod were notched by a razor to facilitate germination and soaked in tap water for 12 h. Seeds of barley and B. brizantha were not soaked. These seeds were germinated on a nylon screen that was put on a polypropylene container filled with tap water under aeration at 27°C in a growth room. Seeds of tea were germinated in quarts sand wetted with deionized water, and then seedlings with roots approximately 1 cm long were transferred to the container described above. All the seedlings with roots approximately 5 cm long were used in the following experiments. In the short-term experiment, ten seedlings of each plant species were pretreated in 600 mL volume of a solution with (Al treatment) or without (control) 50 μ M $AlCl_3$ containing 0.2 mM $CaCl_2$ at pH 4.7 for 1 h. After measurement of the root length with a ruler, all the seedlings were transferred into a Al-free 0.2 mM $CaCl_2$ solution at pH 4.7. Root length was measured again after 24 h. For the tea plant, the root length was measured after 3 d because of the slower root elongation than in the others. Relative growth (%) in the short-term treatment with Al was calculated as the ratio of net root re-elongation of the primary root in the Al treatment to that in the control. In the long-term experiment, twenty seedlings of each plant species were precultured in a 54-L volume of nutrient solution at pH 5.2. The nutrient solution was composed of 1.43 mM NH_4NO_3, 0.7 mM $NaNO_3$, 0.13 mM NaH_2PO_4, 0.78 mM K_2SO_4, 1 mM $CaCl_2$, 0.6 mM $MgSO_4$, 36 μM $FeSO_4$, 9 μ M $MnSO_4$, 0.08 μ M $CuSO_4$, 0.03 μ M $(NH_4)_6Mo_7O_{24}$, 18.5 μ M H_3BO_3, 1.5 μ M $ZnCl_2$. After 1 week, the seedlings were transferred into the nutrient solution with same composition as above (control) or the nutrient solution containing 100 μ M Al and 10 μ M P in soluble form (Al treatment) at pH 5.2 or 4.5, respectively. The solutions were renewed weekly. The concentrations of Al and P, and pH of the solution were monitored every day and adjusted if required. Two months after the Al treatment, the plants were harvested and dried in a draft oven (60°C). Relative growth (%) in the long-term treatment with Al was calculated as the ratio of the dry weight of whole plant in the Al treatment to that in the control. The concentrations of K, P, Mg, and Ca were determined by ICP-AES (inductively coupled plasma-atomic emission spectrometry) after digestion of the plant samples using an acid mixture (HNO_3: $HClO_4$ = 5: 3, v/v). Extraction and determination of the phenolic compounds were carried out as described by Ofei-Manu et al. (2001). Seedlings of Brachiaria hybrid (B. ruziziensis Ger. & Ev. clone 44-06 × B. brizantha (A. Rich.) Stapf CIAT 36061, also known as cv. Mulato), Andropogon gayanus Kunth (CIAT 621), barley (cv. Ryofu) were transferred to 36-L containers containing nutreint solution with or without 0.37 mM Al (as $Al_2[SO_4]_3$). At the end of 10 days of treatment, the roots of seedlings from each treatment were sampled, dried in a forced-air oven at 80°C for 72 h, and then weighed. The dried samples were ground and digested with H_2SO_4-H_2O_2 for Al analysis by ICP-AES. Brachiaria seedlings were prepared as described above, and transferred to 36-L containers carrying the standard nutrient solution, but with 2.8 mM Al at pH 3.7 added, and left to grow for 1 month. The much higher Al concentration was used to ensure clear peaks in the 27Al NMR spectrum. Even so, the Brachiaria seedlings grew well (data not shown). After treatment, roots were removed from the seedlings and washed, first with tap water, then with deionized water. The roots were grouped into three: Fraction (a), roots given the water washings only, and used to determine total amounts of Al and organic acid anions; and Fraction (b), roots were also washed with 0.1 M HCl for 5 min to remove apoplastic, soluble or loosely bound, components. Each fraction of Brachiaria roots was placed in a 10-mm-diameter NMR tube. $AlCl_3$ (0.1 M) solution was used as an external reference to calibrate the chemical shift (0 ppm). 27Al NMR spectra were recorded, using a Bruker MSL400 spectrometer at 104.262 MHz. The spectra were obtained by using a frequency range of 62.5 kHz, a pulse width of 12 μs, a delay time of 0.16 ms, a cycle time of 0.5 s, and 4000 scans. For estimation of low P tolerance, we have selected Brachiaria hybrid (cv. Mulato) and rice (Oryza sativa L. cv. Nipponbare). For the seedling growth nutrient solution was prepared with 2.12 mM N (NH_4NO_3), 0.77 mM K (K_2SO_4:KCl = 1:1), 1.25 mM Ca ($CaCl_2 \cdot 2H_2O$), 0.82 mM Mg ($MgSO_4 \cdot 7H_2O$), 35.8 μM Fe ($FeSO_4 \cdot 7H_2O$), 9.1 μM Mn ($MnSO_4 \cdot 4H_2O$), 46.3 μM B (H_3BO_3), 3.1 μM Zn ($ZnSO_4 \cdot 7H_2O$), 0.16 μM Cu ($CuSO_4 \cdot 5H_2O$), 0.05 μM Mo (($NH_4)_6Mo_7O_{24} \cdot 4H_2O$), with 6 μM P ($NaH_2PO_4 \cdot 2H_2O$) and pH was maintained at 5.2 for 1 week preculture. Then phosphorus level was changed as 0, 6 and 32 µM for 2 weeks. After sampling, samples were freeze-dried, then P fractionation was carried out based on Schmidt-Thannhauer-Schneider method. Samples were also used for measurements of APase and RNase activities. $14CO_2$ was generated by adding 30% PCA into $NaH_{14}CO_2$ (18.5 kBq), then applied to plants for 5 min in a vinyl package under natural light condition. After sampling, samples were fractionized by using column method to obtain organic acids, amino acids, and sugars fraction, then 14C content in each sample was determined by using scintillation counter. In transcriptomic analyzes and subsequent molecular analysis, we have selected rice (Oryza sativa L. cv. Michikogane) and the growth condition was same as shown before except for the P treatment was done using 0 or 32 µM levels. Total RNA was extracted from frozen samples using a sodium dodecyl sulfate (SDS)- phenol method. The real-time PCR was performed by the LightCyclerTM system (Roche) with the LightCycler DNA Master SYBR Green I kit for PCR (Roche), and the TaqStartTM antibody (Clontech) was used for the repression of unspecific amplification.

Mechanisms of Al Resistance in Brachiariagrasses

Differences in Al Resistance and Nutrient Acquisition among Crops and Brachiariagrasses

Differences in Al resistance among crops and B. brizantha are shown in Fig.1 (the left panel). Al resisance was ranked as follows: B. brizantha > rice > tea > maize > pea, siclepod > barley. The order of Al resistance in the short-term experiment was well correlated with that of the long-term experiment in spite of marked difference in the treatment conditions, i.e., duration of Al treatment, Al concentration, composition of co-existing nutrients, pH, etc. ($R^2 = 0.785$ [p< 0.01], right side figure of Fig.1) (Ishikawa et al., 2000). Although B. brizantha is relatively less adapted than B.decumbens to infertile acid soils, B. brizantha was found to be superior in its level of Al resistance to the other crops tested.

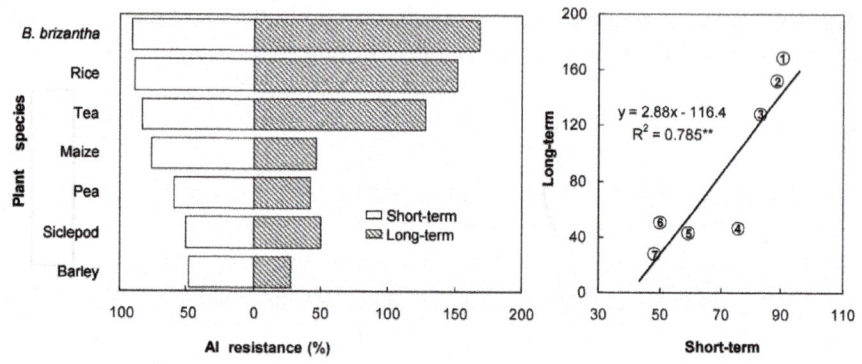

Figure 1. Differences in Al resistance among crops and Brachiaria brizantha. The left panel shows Al resistance values in long-term experiment and short-term experiment. The right figure shows the relationship between Al resistance in long-term experiment and that in short-term experiment. ① *B. brizantha,* ② rice, ③ tea, ④ maize, ⑤ pea, ⑥ siclepod, ⑦ barley.

We also compared nutrient acquisition ability of B. brizantha with that of crops. Nutrient acquisition ability was compared by quantifying relative nutrient status in shoots in Al treatment to that in low pH · low P treatment. Among the 5 species compared, B. brizantha was found to have superior nutrient acquisition abilities for K,P and Mg (Fig. 2). However, B. brizantha showed the lowest acquisition ability for Ca. B. brizantha can acquire highest amounts of K, P and Mg even under high Al, low P and low pH conditions. B. brizantha may have the lowest requirement of Ca for normal growth even in acid soil conditions. This greater ability of B. brizantha to acquire nutrients under simulated acid soil conditions could be responsible for its greater vigour in acid soil conditions during the pasture establishment phase (Rao et al., 1996).

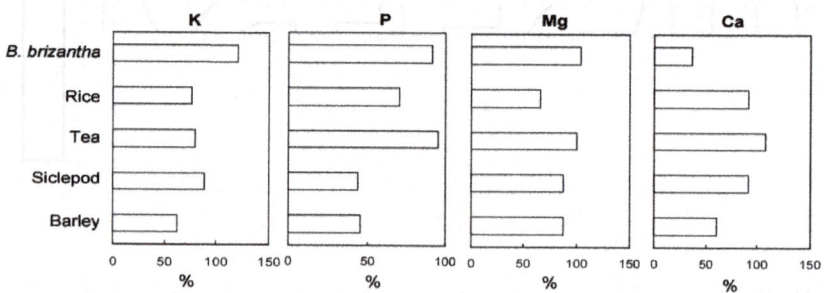

Figure 2. Relative nutrient status of each nutrient concentration in Al treatment to that in control treatment.

Differences in Al Resistance, Al Accumulation and Plasma Membrane Permeability among Brachiariagrasses

Adaptive responses of several brachiariagrasses to infertile acid soils have been identified and described by previous research (Louw-Gaume et al., 2010 a,b; Rao et al., 1995, 1996; Wenzl et al., 2001, 2003). We compared Al resistance, Al accumulation (hematoxylin staining method [Wagatsuma et al., 1995] and PM permeability (FDA-PI fluorescence staining method [Ishikawa et al., 2001]) among 4 brachiariagrasses in short-term experimental conditions that were described in the former section, together with most Al-resistant rice cultivar Rikuu-132 (Khan et al., 2009) as a reference plant species. Al resistance was ranked as follows: B. decumbens, B. hybrid, B. brizantha > B. ruziziensis > Rikuu-132 (Fig. 3). Al resistance of B. hybrid (cv. Mulato) and B. brizantha was found to be comparable to that of B.decumbens which has been ranked as the most Al-resistant brachiariagrass (Wenzl et al., 2001). Although Al resistance of B. ruziziensis was found to be markedly lower than B. decumbens (Wenzl et al., 2001, 2003), its resistance level was higher than that of the most Alresistant rice cultivar. It was suggested that the highest Al resistant phenotype of B. hybrid may be ascribed to the Al resistance genes from B. decumbens or B. brizantha and not from B. ruziziensis. Al accumulation was localized mainly within 1- mm root-tip portion and its concentration corresponds reversely to Al resistance order: the least Al accumulation was recognized for the most Al-resistant B. decumbens. PM lipid layer was less permeable to Al in brachiariagrasses than in the most Al-resistant rice cultivar and its less permeable PM characteristic was localized mainly within 1- mm root-tip portion (Fig.4).

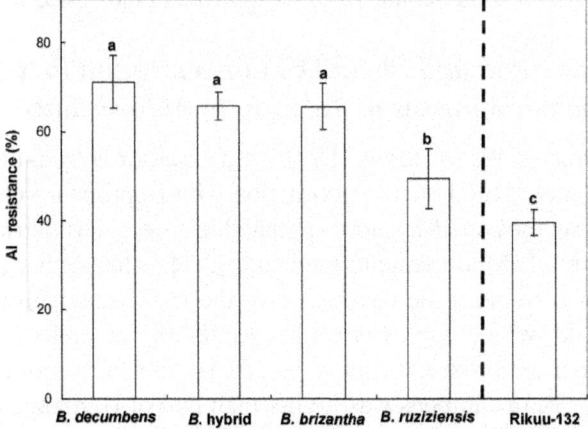

Figure 3. Differences in Al resistance among brachiariagrasses and Al-resistant rice cultivar Rikuu-132.

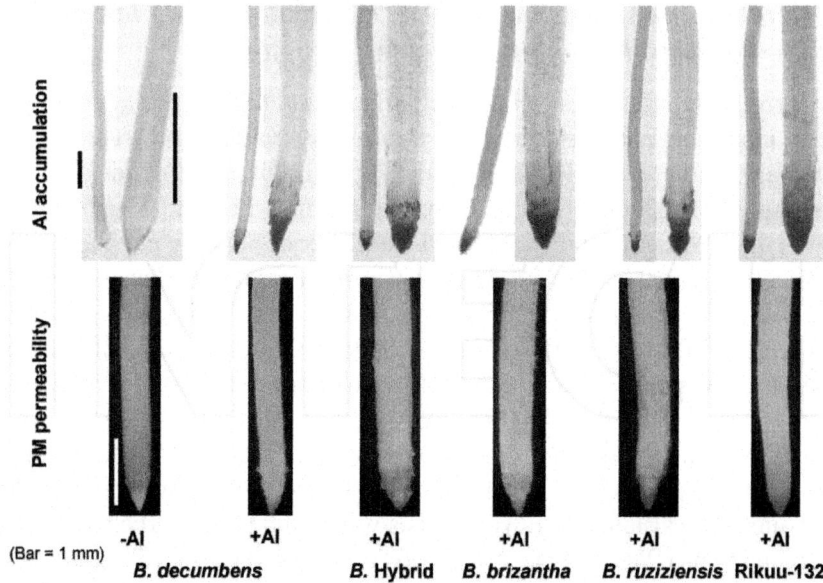

Figure 4. Differences in Al accumulation and PM permeability among brachiaria-grasses and Al-resistant rice cultivar (Rikuu-132).

Permeability of PM was negatively associated with Al resistance: the least PM permeabilization was observed with the most Al-resistant B. decumbens. The less permeability of PM and the lower Al accumulation in root-tip portion in Al-resistant brachiariagrass agree well with the former results which have been recognized in Alresistant plant species, cultivars, or lines (Ishikawa et al., 2001; Ishikawa & Wagatsuma, 1998; Wagatsuma et al., 2005).

Lipid Composition and Phenolics Concentration in Root-tip Portion of Brachiariagrasses in Relation to Al Resistance

The lower ratio of PL to sterols (S) (PL was measured by molybdenum blue spectrophotometric method after extraction with isopropanol-chloroform-H_2O [2:2:1] ; S was measured by ortho-phthalaldehyde colorimetric method after extraction with dichloromethane-methanol [2:1] [Khan et al., 2009]) in root-tip portion was found to be beneficial for the less permeability of PM in the presence of Al, which agrees with the results of rice cultivars (Khan et al., 2009). In the more proximal root region (0-10 mm from root apex), the ratio of PL to S for B. decumbens was higher than that of B. ruziziensis, but on the contrary, it was lower in root-tip portion (0-2 mm from root apex) under Al treatment conditions (Fig.5).

The lower negativity of PM surface that was associated with the lower ratio of PL to S in root-tip portion could contribute to lower permeability of PM to Al. This is highly consistent with Gouy-Chapman-Stern model of Al rhyzotoxicity (Kinraide, 1999). In case of rice cultivars (Khan et al., 2009), wheat lines, triticale lines, maize cultivars (unpublished data), lipid compositional difference in connection with Al resistance were recognized in root-tip portion of 0-10 from root apex. However, in brachiariagrasses, lipid compositional difference was related with Al resistance only in root-tip portion of 0-2 mm from root apex. We suggest that the high level of Al resistance in brachiariagrass is extremely localized at the root tip.

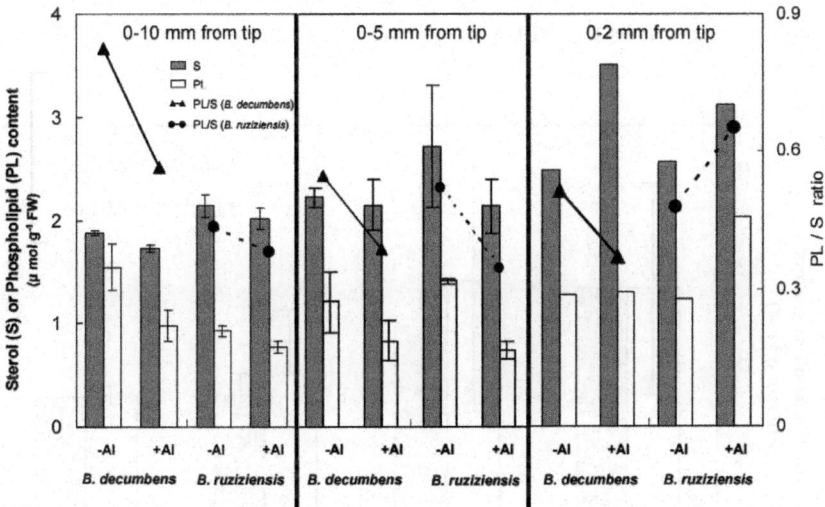

Figure 5. Sterol or phospholipid content in the different segment of root of brachiaria-grasses treated with or without Al in solution.

It is known that phenolic compounds can be solubilized into lipid layer, and the lipid layer solubilized with phenolic compounds is transformed into the less fluid layer (Arora et al., 2000; Boija & Johansson, 2006). Higher concentration of phenolic compounds was detected in root-tip portion (0-5 mm and 0-2 mm from root apex) of B. decumbens than in B. ruziziensis (Fig.6).

The concentration of phenolic compounds was lower in the portion of 0-10 mm from root apex than that of the shorter part from root apex (data not shown). Phenolic compounds have been detected basically in the cell wall, vacuole, and to a small extent in the cytoplasm and nucleus (Hutzler et al., 1998). At around neutral pH of cytosol, the binding affinity to Al ions was significantly higher for phenolic compounds than for organic acids (Ofei-Manu et al., 2001). Higher concentration of phenolic compounds is considered to be more effective for greater detoxification of Al ions in cytosol of B. decumbens. Additionally, higher inclusion of phenolic compounds into PM lipid layer may be more favourable for making the PM less permeable in the presence of Al ions, although there are no reports on the inclusion of phenolic compounds in plant lipid layer. Several quantitative and qualitative changes in PM may contribute to superior level of Al resistance in B. decumbens. These include: higher proportion of S relative to PL, higher concentration of phenolic compounds in cytosol, and higher inclusion of phenolic compounds in PM lipid layer in root-tip portion. These changes may contribute to an extremely strong PM lipid layer which plays a key role in exclusion of Al and high level of Al resistance in B. decumbens. Direct demonstration of the existence of phenolic compounds in PM lipid layer will be an important task for the future research.

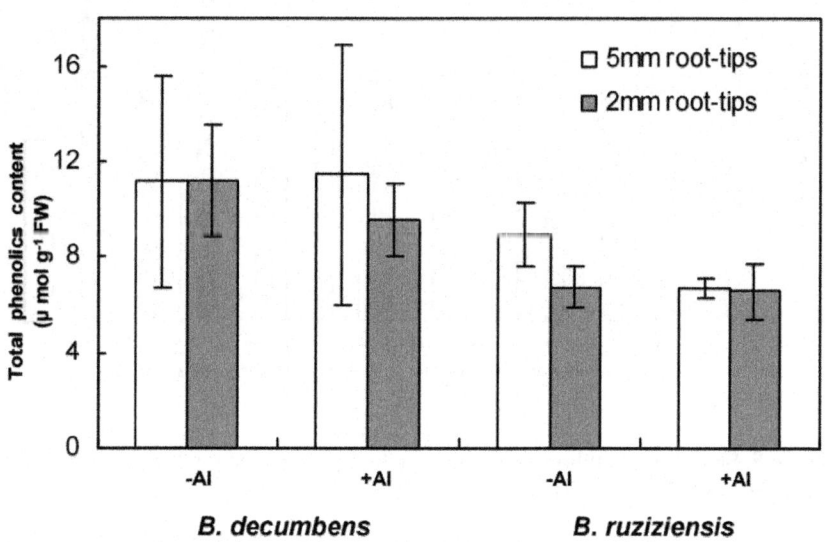

Figure 6. Total phenolic compounds in root-tips of two brachiariagrasses treated with or without Al in solution.

Mechanisms of High Level of Al Resistance in B. Hybrid (cv. Mulato)

B. hybrid showed higher resistance to Al similar to B. decumbens. When B. hybrid seedlings were grown with an extremely high concentration of Al (0.37 mM) for 10 days, no growth inhibition was observed (Fig. 7). Moreover, Al application did not inhibit the uptake of nitrogen (N), P and K in B. hybrid. Andropogon gayanus, a poaceous pasture grass, is also very resistant to Al and Al application significantly increased Al concentration in both leaf and root of this species (Fig. 8). In B. hybrid, by contrast, significant increase in Al accumulation was also observed in root but not in leaf. This indicates that some mechanisms restricting Al translocation from roots to shoots should exist in B. hybrid. The 27Al NMR spectrum obtained from intact roots showed several peaks downfield at 10-20 ppm (Fig.9a), suggesting that most of the soluble Al in roots makes complexes presumably with organic acid anions (Fatemi et al., 1992; Kerven et al., 1995). Since the 27Al NMR spectrum did not change after removing soluble and/or loosely bound apoplastic Al, these Al complexes in roots were likely to be localized in the symplast of cells. In many Al-accumulator species, leaves and roots with high concentration of Al are detoxified by organic ligands, such as Al oxalate in Melastoma malabathricum (Watanabe et al., 1998, 2005). The same mechanisms are considered possible in roots of B. hybrid. It has been reported that Cd translocation from roots to shoots is restricted by Cd isolation in root vacuoles (Miyadate et al., 2010). Al in the B. hybrid may also compartmentalize in root vacuoles and, thus, may not be translocated to shoots.

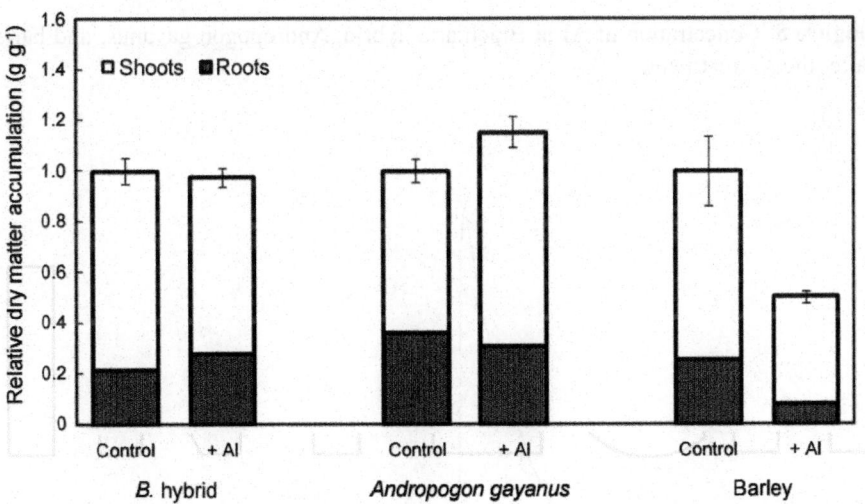

Figure 7. Effects of Al toxicity on growth of Brachiaria hybrid, Andropogon gaya-nus, and barley. Growth was expressed as the relative dry matter accumulation (i.e. [dry weight after treatment – initial dry weight in each treatment]/[dry weight after treatment-initial dry weight in control treatment]).

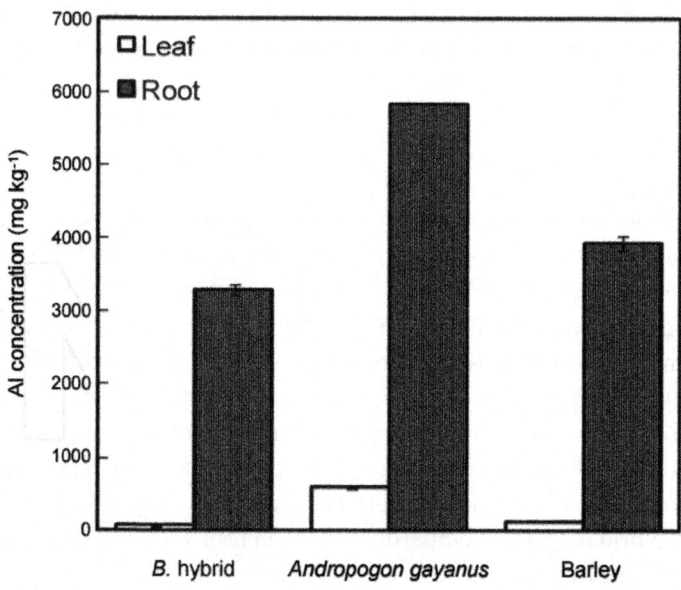

Figure 8. Concetration of Al in Brachiaria hybrid, Andropogon gayanus, and barley after the Al treatment.

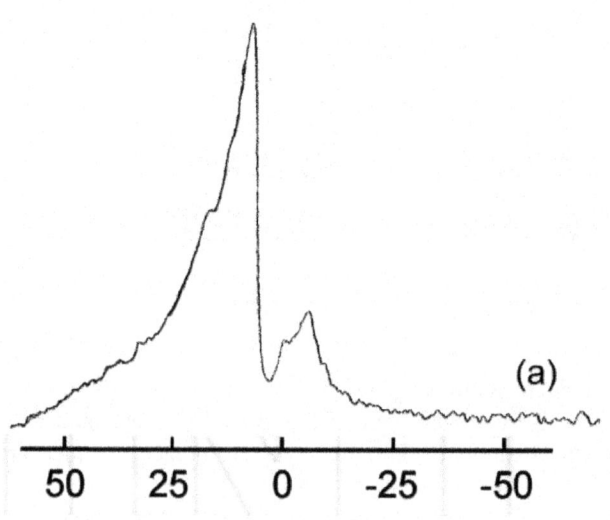

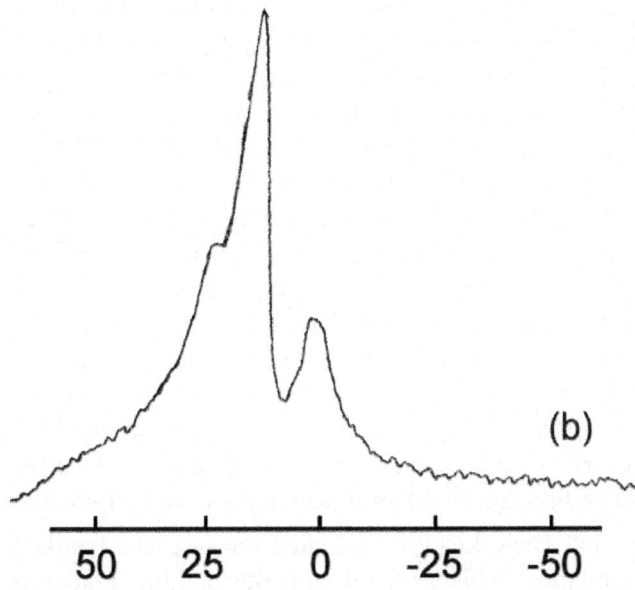

Figure 9. 27Al NMR spectra in intact roots (a) and roots after removing soluble and/or loosely bound apoplastic Al by 0.1 M HCl (b). AlCl$_3$ (0.1 mM) was used as an external reference to calibrate the chemical shift (0 ppm).

Mechanisms of Low P Tolerance in Brachiariagrass Comparing with Those in Rice

Low P tolerance of B. Hybrid

As indicated above, Brachiaria species are well adapted to the low-fertility acid soils of the tropics because they are highly tolerant of high Al and low supplies of P and Ca (LouwGaume et al., 2010b; Rao et al., 1995, 1996, 2001b; Wenzl et al., 2001). They have lower internal requirements for P than other grasses because they are able not only to acquire P with their extensive root systems but also to use the acquired P more efficiently for growth and metabolism (Rao et al., 1996, 1999). However, mechanisms of P-use efficiency are relatively less known in plants, including B. hybrid. Because carbon metabolism is well known to be affected by the P status in plant tissue (Rao, 1996), we studied low-P-tolerance mechanisms, in terms of P recycling and carbon metabolism, in the B. hybrid comparing them with those of rice (Nanamori et al., 2004).

B. hybrid and rice plants were cultivated in nutrient solutions with or without 32 µM P. The data obtained on growth parameters and nutrient status are shown in Fig. 10. When P supply in the nutrient solution was low, root:shoot ratio increased, especially in the B. hybrid. We found that, for the B. hybrid, vigorous root growth is a mechanism for acquiring larger amounts of P from low P conditions. This finding was supported by the high levels of N concentration found in B. hybrid roots, while P concentration in B. hybrid leaves was significantly lower than that of rice leaves. Lower P concentration in B. hybrid leaves may indicate that the B. hybrid uses P more efficiently to sustain active metabolism for dry matter production. The P concentration of B. hybrid was quite low (0.44 and 0.56 mgP/gDW in roots and leaves, respectively) and less than rice, which is also known as a low P tolerant plant. Results on the fractionation of P compounds indicated that acid-soluble Pi accounted for about half of the total P in the B. hybrid (Fig. 11). Results on the Pi:total P ratio in B. hybrid leaves under P deficiency indicate that the B. hybrid can survive with extreme low intracellular Pi concentration. This may be due to rapid turnover of other organic P pools under P-deficient conditions.

Chapin and Bieleski (1982) studied the impact of mild P stress on P fractions in relation to plant growth in barley and low-P-adapted barleygrass. They found that barleygrass had a higher proportion of Pi at each level of P supply. They explained this as a consequence of slower growth in barleygrass and higher P status rather than any inherent difference in mechanism. However, in our study, the higher Pi proportion in the B. hybrid, compared with that of rice, coincided with lower P concentrations, as explained above. We, therefore, speculate that recycling of internal organic P compounds could be an important mechanism of P-use efficiency in the B. hybrid.

Bosse and Köck (1998) have shown activities of APase and RNase were induced during P deficiency, and that this induction is associated with P turnover in plants. In our study, APase and RNase activities were both strongly induced in both rice and B. hybrid by P deficiency (Fig. 12). Induction of APase activity was markedly higher in roots under Pdeficient conditions. Duff et al. (1994) reported the existence of extracellular APase in roots, where it is localized mainly in apical meristems and outer and surface cells. It is involved in hydrolyzing and mobilizing Pi from organic phosphates in the soil for plant nutrition. The induction of APase in roots may also be associated with excretion. Bosse and Köck (1998) suggested that the increase in activity of phosphohydrolases was a specific response to the decline of cellular available Pi in Pi-starved tomato seedlings. Although Pi in roots was lower than in shoots of both test crops, it was impossible to account for the difference of APase induction between roots and shoots only by the difference in intracellular Pi concentration. Thus, we suggest that some other signal transduction pathway must be operating between roots and shoots against P starvation in the cell.

APase activity in shoots was greater in the B. hybrid than in rice, suggesting the possibility of rapid P turnover in the B. hybrid. This may enable the B. hybrid to survive under low P conditions. APase may not be a major mechanism for scavenging or acquiring P because differences in APase induction could not sufficiently account for the diverse growth response of genotypes of both common bean and maize plants under P deficiency (Yan et al., 2001; Yun & Kaeppler, 2001).

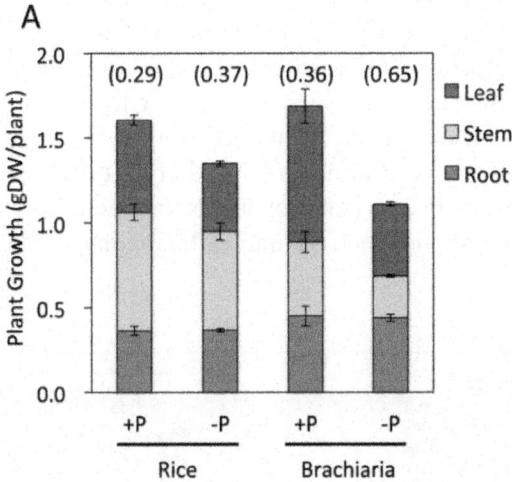

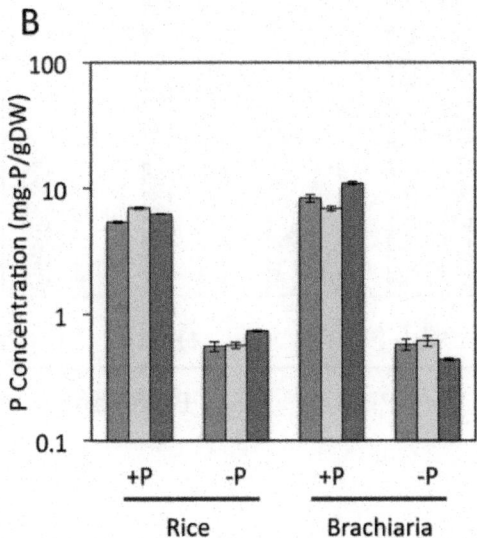

Figure 10. Growth (A) and P concentration (B) of rice and Brachiaria hybrid plants grown under P sufficient and deficient conditions. Error bars indicate S.E. (n = 3).

However, we observed in our study that APase activity was induced by P deficiency and the activity seems to be correlated well with P-use efficiency, as indicated by the lower value of total P concentration, so that the function of APase in adaptation to low-P conditions should not be underestimated. RNase activity was also high in roots under P-deficient conditions (Fig. 12). Nürnberger et al. (1990) and Löffler et al. (1992) showed that both extracellular and intracellular RNase activities were induced in tomato-cell culture under P deficiency. Extracellular RNase could help degrade the RNA from senescing cells that have been either damaged or lysed, and also help degrade any RNA that might be present in the rhizosphere. Thus, the high RNase activity in roots may be associated with secretion similar to APase. RNase activity in shoots was also greater in the B. hybrid than in rice, indicating that RNase also contributes to rapid P turnover. Glund et al. (1990) showed that, in the relationship between Pi concentration and RNase activity, induction of RNase under P starvation occurs when the intracellular content of P is very high.

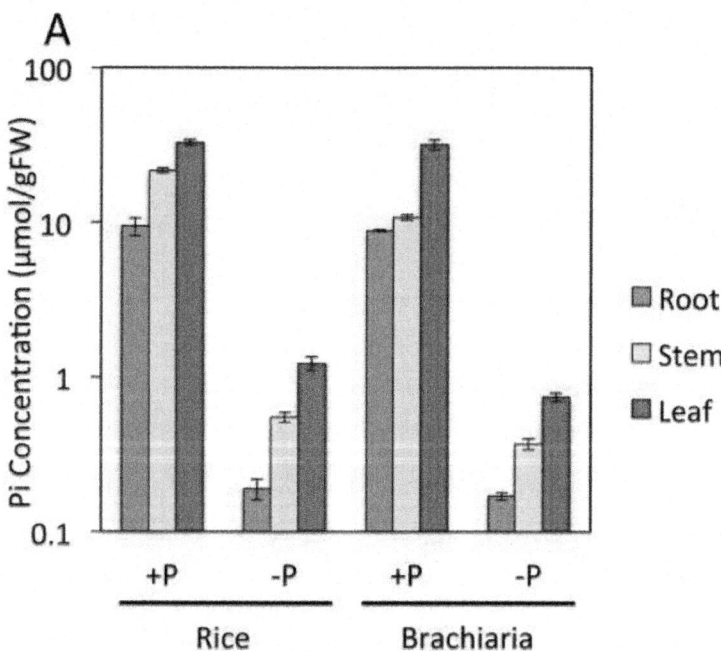

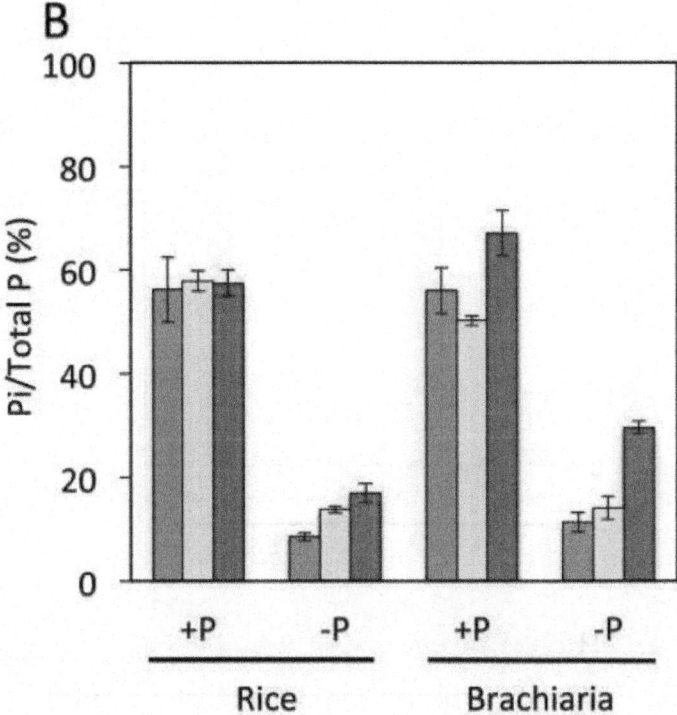

Figure 11. Pi concentration (A) and ratio of Pi to total P (B) of rice and Brachiaria hybrid plants grown under P sufficient and deficient conditions. Error bars indicate S.E. (n= 3).

The above studies indicate that phosphohydrolases, such as APase and RNase, were induced by P deficiency as a P-recycling system. Coinciding with such a mechanism, it is possible that carbon metabolism could also be altered under P deficiency. We therefore studied photosynthate partitioning under P deficiency, tracing photosynthetically fixed 14C in leaves. In rice, photosynthates mainly distributed to sugars, which consist of sucrose, indicating that rice enhanced the sucrose synthesis pathway (Fig. 13). The mRNA accumulation of sucrose phosphate synthase (SPS) also increased as mentioned previously. Hence, sucrose concentration in rice leaves was remarkably high (Fig. 13). The 14C distribution proportion to sugars increased with P deficiency. Enhanced sucrose synthesis in rice leaves through P deficiency may contribute to P recycling because P is liberated during sucrose synthesis (Rao, 1996). However, sucrose catabolism was restricted because the 14C distribution ratio to amino acids and organic acids decreased with P deficiency and with carbohydrate accumulation (Fig. 13).

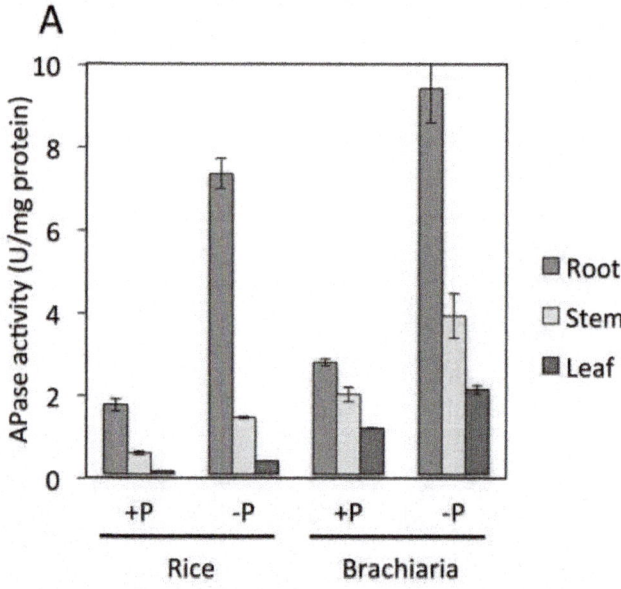

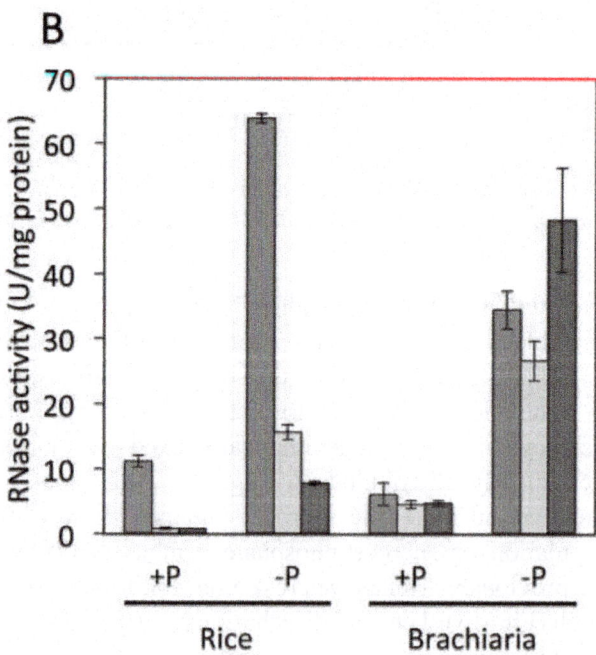

Figure 12. Acid phosphatase (A) and ribonuclease (B) activities in leaves of rice and Brachiaria hybrid grown under P sufficient and deficient conditions. Error bars indicate S.E. (n = 3).

Sucrose synthesis may, therefore, not contribute efficiently to P recycling. However, the 14C distribution proportion to sugars in the B. hybrid was not as marked as in rice (Fig. 13), and the effect of P deficiency was smaller. The 14C distribution ratio to amino acids and organic acids in the B. hybrid was greater than in rice, and slightly affected by P deficiency. The decrease of total organic acids and carbohydrates in B. hybrid leaves under P deficiency suggests that the B. hybrid can sustain active amino acid and organic acid pathways with enhanced sugar catabolism, using P efficiently under P deficiency. PK and its bypassing enzymes catalyze the PEP-consuming reaction in leaves, with PEPP activity increasing by a factor of 5.6 to 6.0 with P deficiency. This induction of PEPP is likely to be associated with P recycling, as Duff et al. (1989) suggested. PK was also induced by P deficiency, but not significantly in the B. hybrid. PEPC activity was slightly induced by P deficiency in rice but not in the B. hybrid. The decrease of PEPC activity in B. hybrid leaves would result from reduced net photosynthesis under P deficiency. Kondracka and Rychter (1997) suggest that facilitating the PEP metabolism may be important in view of the P recycle. PEPC and PEPP are considered to function in P recycling as PK-bypass pathways. If these enzyme activities are induced in P recycling, then the carbon flow to the TCA cycle is expected to increase. The 14C distribution ratio to amino acids and organic acids increased slightly in the B. hybrid with P deficiency (Fig. 13), indicating that these bypassing enzymes may function to facilitate carbon flow to the TCA cycle. However, in rice, the 14C distribution ratio to amino acids and organic acids decreased with P deficiency. Therefore, the PK bypassing mechanism under P deficiency may not contribute to facilitating the carbon flow to the TCA cycle in rice. In addition to the PK-bypassing mechanism, carbon export from chloroplast to cytosol via the triose-phosphate translocator (TPT) may be a process that significantly affects carbon partitioning under P deficiency (Rao, 1996). When plants are starved for P, triose-P exports from chloroplast to cytosol via TPT, and subsequent sucrose synthesis in the cytosol is likely to be restricted (Rao, 1996). The 14C distribution ratio to sugars and to residue, which mainly consists of sucrose and starch respectively, increased with P deficiency in both rice and B. hybrid (Fig. 13), indicating that restriction of triose-P exports from chloroplast to cytosol via TPT may not occur.

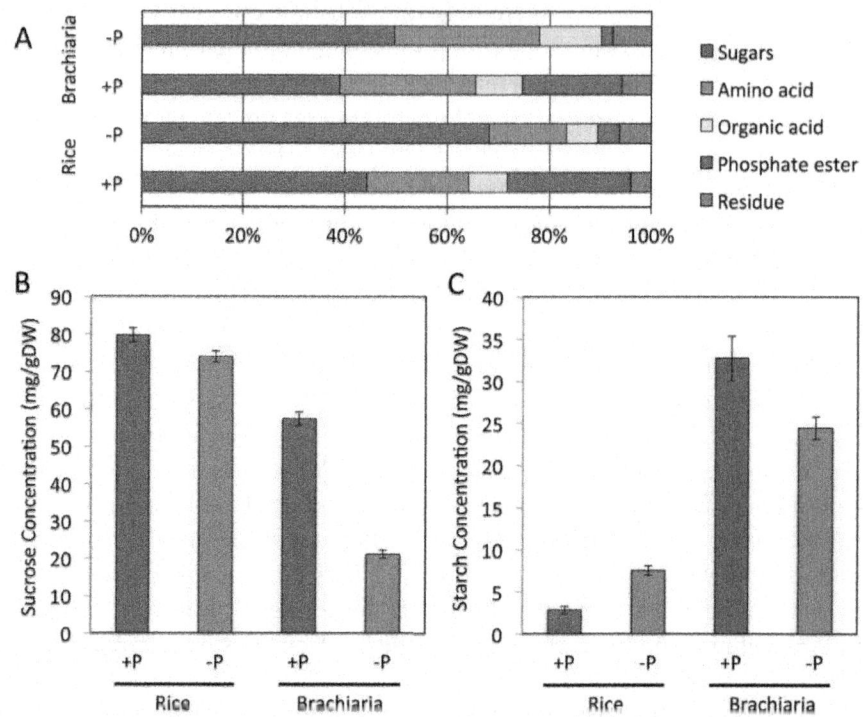

Figure 13. Photosynthetically assimilated 14C distribution (A), sucrose and starch concentration(B and C, respectively) in leaves of rice and Brachiaria hybrid. Error bars indicate S.E. (n = 3).

Transcriptomic Analysis of P Deficient Rice Plants

Rice (Oryza sativa L. ssp. japonica) plants were germinated and cultured in nutrient solutions containing 0 and 32 μM NaH_2PO_4 for –P and +P treatments, respectively. The seedlings were cultivated for 9 days after transplanting. Total RNA of leaves and roots was used for transcriptomic analyzes by using cDNA arrays (Wasaki et al., 2003b, 2006). As the response of rice roots, there were 15 up-regulated genes in the short-term (24 h) and 86 in the longterm (9 d) treatment with –P, whereas there were 23 and 97 down-regulated genes in the two treatments, respectively. The number of genes regulated (especially down-regulated genes) by the P deficiency was lower in leaves than in roots. There was one up-regulated gene in the short-term (24 h) and 48 genes in the long-term (9 days) –P treatments, whereas there were eight and four down-regulated genes in these two treatments, respectively. None of the genes were regulated in a similar manner between the short and long-term –P leaves. This result suggests that the responses in P-deficient rice leaves are different between shortand long-term treatments, whereas those of roots are relatively similar. OsPI1 (Oryza sativa phosphate-limitation inducible gene 1; Wasaki et al. 2003c), showed the most significant increase in its transcription in the long-term –P treatment, in both the roots and leaves. This gene was classified as a member of TPSI1/Mt4 family, which is the P-deficient responsive non-coding RNA. The SqdX-like gene, a homolog of sulfoquinovosyl diacylglycerol (SQDG) synthesis related genes, was up-regulated significantly in the –P roots. P deficiency enhances dynamic lipid reconstruction and causes SQDG or galactolipids accumulation and expression of a related gene in leaves (Essigmann et al., 1998; Nakamura et al., 2009). Because SQDG has the ability to substitute for PL, it was suggested that the increase of SQDG synthesis is available for the efficient use of P in the membrane (Essigmann et al., 1998). Four genes related to P metabolism were induced in leaves by long-term –P treatment. Inorganic pyrophosphatase and a phosphatase probably contributed to the maintenance of Pi concentration in the tissue by the direct production of Pi from organic phosphate compounds. It was concluded that the function of inorganic pyrophosphatase was common in both roots and leaves, because expression was induced in both organs by long-term –P treatment. Both bi-functional nuclease and S-like RNase expression levels were increased by the –P conditions; their contribution is to produce monomeric nucleotides as substrates for phosphatases (Duff et al., 1994; Green, 1994; Palma et al., 2000).

Many genes involved in polysaccharide metabolisms were up- and down-regulated in leaves by long-term –P and P re-supply treatments, respectively. It is probable that the upregulation of ADP-glucose pyrophosphorylase, which is a key enzyme of starch synthesis, and starch synthetic enzymes such as starch branching enzyme and starch synthases, induces the accumulation of starch in leaves under –P conditions. In fact, there are many reports of the accumulation of starch in the chloroplasts of P-deficient rice and other plants (Ciereszko & Barbachowska, 2000; Fredeen et al., 1989; Nomura et al., 1995; Qui and Israel, 1992; Rao et al., 1993; Usuda & Shimogawara, 1991). We concluded that the starch accumulation in leaves grown under P-deficient conditions was caused by the disruption of the export of triose phosphate from the stroma by the Pi translocator (Nátr, 1992). Nátr (1992) also noticed the liberation of Pi by the enhancement of starch synthesis. Because starch synthesis and the induction of Pi utilizing enzymes are synchronized, it is a reasonable speculation that the starch accumulation in the P-deficient leaves is a result of the maintenance of the internal Pi concentration.

Fig. 14 shows a summary of metabolic changes based on the regulation of gene expression in the leaves and roots of rice exposed to –P stress. Some important metabolic changes in roots by –P are suggested, namely: (1) acceleration of carbon supply for organic acid synthesis through glycolysis; (2) alteration of lipid metabolism; (3) rearrangement of compounds for cell wall; and (4) changes of gene expression related to the response for metallic elements such as Al, Fe and Zn. The major responses in leaves were involved in internal P utilization. The response in leaves seems to be less dramatic than that in roots; however, it is probable that an important function is regulated in shoots, such as the regulation of the novel TPSI1/Mt4 gene family (Burleigh & Harrison, 1999), which contains rice OsPI1 (Wasaki et al., 2003c).

Bypass Pathways in rice for P Use Efficiency in Plant

From our previous study using microarray on P deficient rice, we found that several genes relating C and P metabolism in chloroplast changed their expression level.

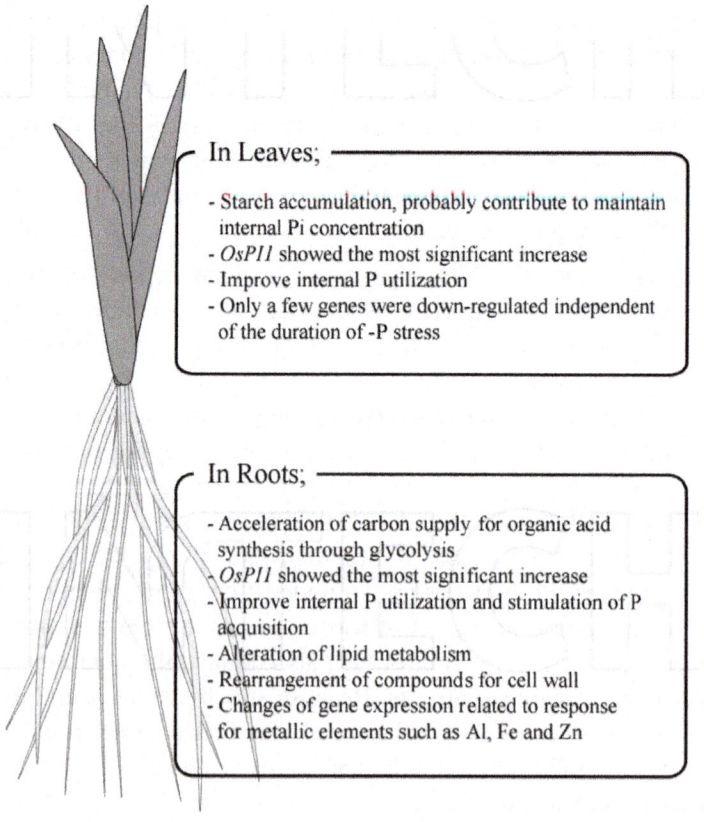

In Leaves;
- Starch accumulation, probably contribute to maintain internal Pi concentration
- *OsPI1* showed the most significant increase
- Improve internal P utilization
- Only a few genes were down-regulated independent of the duration of -P stress

In Roots;
- Acceleration of carbon supply for organic acid synthesis through glycolysis
- *OsPI1* showed the most significant increase
- Improve internal P utilization and stimulation of P acquisition
- Alteration of lipid metabolism
- Rearrangement of compounds for cell wall
- Changes of gene expression related to response for metallic elements such as Al, Fe and Zn

Figure 14. Summary of plant responses to phosphorus deficiency in shoot and root tissue.

One of them is phosphoenolpyruvate/phosphate translocator (PPT), and it showed enhancement under phosphorus deficient condition. PPT transports PEP into the chloroplast and antiports Pi to cytosol (Hausler et al., 2000), the role of PPT under P deficient condition is considered to supply substrate for the shikimate pathway. There exit another phosphate transporter; triose phosphate translocater (TPT) on chloroplast membrane which loads triose phosphate into cytoplasm and antiports phosphate into chloroplast. These two phosphate translocators are considered to regulate the phosphate level in the chloroplast.

As comprehensive analysis of each pathway using intact plant has not reported, we evaluated it by using quantitative real time PCR (qRT-PCR) to determine the expression level of each gene under P deficient condition (Shinano et al., 2005). While the expression level of mRNA is not simply representing the activity of those enzymes corresponding to the gene, obtained information will be very useful to consider plant response to P deficiency.

Gene Coding Key Enzyme of Sucrose Synthesis

The synthesis of sucrose will liberate phosphate from intermediate compounds, thus it is expected that the level of mRNA for SPS increased with -P treatment. SPS exist in cytosol of mesophyll cell and the combined reaction of SPS and sucrose phosphate phosphatase is main route for sucrose synthesis. That is, during sucrose synthesis, one molecule of Pi is liberated from sucrose phosphate. The –P treated plants first uses Pi stored in vacuole but after they used up all Pi in vacuole, cytosolic Pi content became lower. Then the plants with lower Pi concentration may facilitate sucrose synthesis and excrete Pi from sugar phosphate to keep up the Pi concentration in cytosol of mesophyll cell. Sucrose content in phosphate starved plant varies with species. In common bean and sugar beet, leaf sucrose content increased by P deficiency (Ciereszko & Barbachowska, 2000; Rao et al., 1990), although in leaves of Arabidopsis it decreased. Our results indicate that rice increases sucrose synthesis with P deficiency.

Genes Coding Candidates for Glycolytic Bypass Enzymes

NADP dependent glyceraldehyde 3-phosphate dehydrogenase (NADP-G3PDH) instead of NAD dependent G3PDH (NAD-G3PDH), and PEPC instead of PK are expected to play alternative pathways to regulate carbon flow under P deficient condition.

In rice leaves under P deprivation, we did not see any increase in relative expression of NADP-G3PDH, which is known as P starvation inducible bypassing enzyme for NADG3PDH in Brassica nigra (Duff et al., 1989). On the contrary, NAD-G3PDH relative expression was significantly high in –P plants at 21 days. In the level of gene expression argument, this result may suggest NADP-G3PDH is not working as glycolytic bypass in rice plant. The lack of induction of nonphosphorylating pathway was also seen in other plant species, such as S. minutum (Theodorou et al., 1991) and A. brevipes (Guerrini et al., 2000). Also PEPC and PK have the relationships of glycolytic bypass induced under P deficiency (Li & Ashiharam, 1990). Even though PEPC was thought to be catalyzed with the alternative pathway of PK under P deficiency (Li & Ashiharam, 1990), relative expression of both genes was increased by –P treatment. Increase of PEPC expression by P deficiency is also known in lupin.

Genes Encoding Chloroplast Membrane Transporters

Precise value of Pi concentration in cytosol and chloroplast is not known while it is suggested that the value is between 10 to 15 mM in cytosol (Mimura, 1999) and 20 to 35 mM in chloroplast (Diez & Heber, 1984). This indicates that higher requirement for maintaining Pi level in chloroplast rather than in cytosol, and low P condition is expected to increase the level of TPT and in versa in PPT. While the expression level of TPT was not changed by P deficient condition, the expression level of PPT increased dramatically. When one molecule of P is transported into chloroplast as PEP while exporting one Pi, the incorporated PEP is decomposed in the chloroplast thereby having no net change in the P level of the chloroplast. We assumed that the role of PPT is increasing the PEP metabolism and makes a cycle from primary photosynthate synthesized in chloroplast and metabolized in cytosol with glycolysis then re-enter chloroplast with PEP then decomposed to release Pi in the chloroplast. From the analysis of rice microarray, PKp (plastid type PK) and shikimate kinase expression were enhanced under P deficiency. These results indicate physiological adaptation to incorporate PEP into chloroplast to support photosynthetic carbon flow and synthesis of secondary metabolic compounds. Recently, another type of phosphate transporter (PHT2; 1, which has high homology with Na+/Pi symporter of fungi) was reported (Versaw & Harrison, 2002). There is need to evaluate how these transporters are operated to regulate phosphate flux within these subcellular organs.

CONCLUSIONS

Brachiariagrasses are highly adapted to infertile acid soils, however, the physiological and biochemical mechanisms responsible for their superior adaptation have not yet been fully defined. This chapter summarizes the recent progress towards this objective. Comparative differences in Al resistance among 4 brachiariagrasses and 6 reference plant species were analyzed, and the following order of Al resistance was observed: B. decumbens, B. hybrid, B. brizantha > B. ruziziensis > rice (the most Al-resistant cultivar Rikuu-132) > tea (cv. Yabukita) > maize (cv. Pioneer 3352) > pea (cv. Kinusaya) > siclepod > barley (cv. Manriki). The order of Al tolerance in the short-term experiment with exposure to Al (1-h of 50 μ M $AlCl_3$ in 0.2 mM $CaCl_2$ followed by 24-h of Al-free 0.2 mM $CaCl_2$) was well correlated with that in the long-term exposure experiment (2 months of Al treatment with full nutrients) in spite of the differences in the treatment conditions, i.e., duration of Al treatment, Al concentration, composition of co-existing nutrients, and pH. Short-term Al resistance screening technique is accepted to be useful for the evaluation of Al resistance in spite of the simple composition of the treatment solution, considering the positive correlation data obtained formerly among 15 cultivars of sorghum, 10 cultivars of maize, and 8 cultivars of rice. Brachiariagrass showed greater abilities to acquire K, P and Mg, and to tolerate to lower concentration of Ca in shoots in the presence of high concentration of Al in the growing medium including low P at low pH conditions. The level of Al resistance of B. hybrid was ranked to be high as comparable to the most Al-resistant B. decumbens. It was suggested that the highest Al resistance phenotype of B. hybrid may be ascribed to the Al resistance genes from B. decumbens or B. brizantha but not from B. ruziziensis. Extremely high level of Al resistance found in B. decumbens was attributed to localized tip portion of less than 2 mm from root apex due to low amount of Al accumulation, low permeability of PM to Al, lower ratio of PL to S, and higher concentration of phenolic compounds in the tip portion of root as compared with other brachiariagrasses. Thus B. decumbens is considered to possess multiple physiological and biochemical mechanisms to resist high level of Al in soil solution, and its strategy may be extremely localized in the tip portion of the root apex. B. hybrid also exhibited good level of Al resistance. When an extremely high concentration of Al (0.37 mM) was included into the culture solution, significant increase in Al accumulation was observed only in root part. 27Al NMR analysis suggested that the most part of Al in roots was likely to be localized in the cytosol of cells in organically complexed forms and this complexation may inhibit greater upward translocation of Al to shoots, which is beneficial to reduce Al toxicity in shoots.

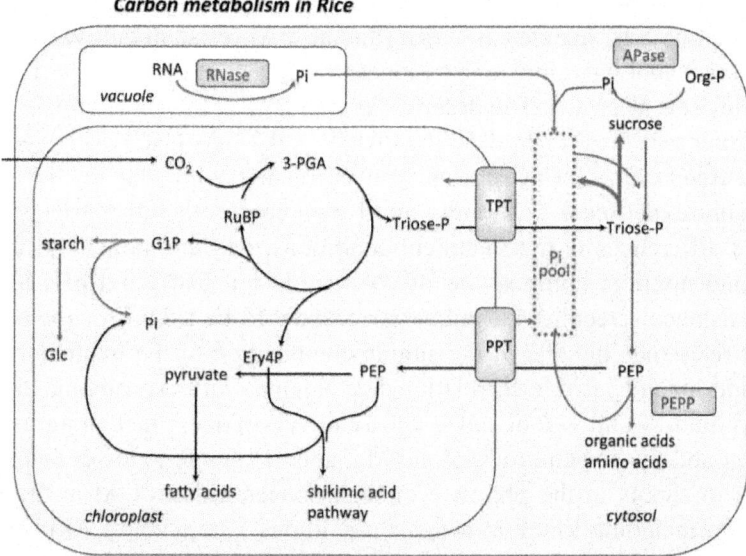

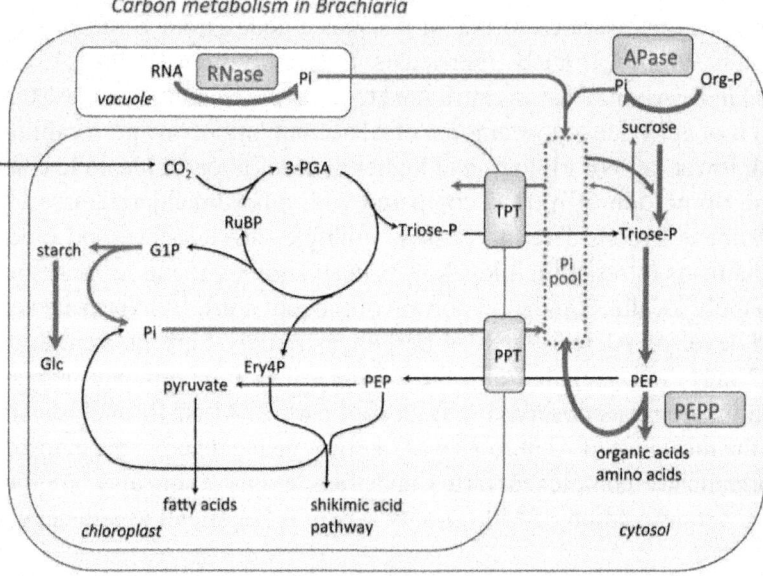

Figure 15. Summary of metabolic changes in leaves and roots with P deficiency: A. Rice; B Brachiaria hybrid. Red arrows indicate P deficiency inducible pathways.

Our study shows that tolerance of low P in both rice and B. hybrid involved marked differences in P recycling and carbon metabolism. We summarized the proposed P recycling mechanisms involved in carbon metabolism of rice and B. hybrid in Fig. 15. For rice, strategies for low-P tolerance include (1) decreased carbon flow to amino acids and organic acids, and decreased N concentration; and (2) improved partitioning of photosynthates to sucrose, combined with restricted sugar catabolism. For the B. hybrid, low-P tolerance involved two major strategies under P deficiency: (1) increasing the ability to use P efficiently by inducing APase and RNase in shoots; and (2) enhancing sugar catabolism and subsequent synthesis of amino acids and organic acids in leaves. Brachiariagrasses also showed greater abilities to acquire K, P and Mg, and to tolerate low concentration of Ca in shoots in the presence of high Al concentration in the growing medium.

In summary, studies on physiological and biochemical mechanism of adaptation of brachiariagrasses grown under simulated conditions of low fertility acid soils indicated their higher level of resistance to Al and tolerance to low supply of P and Ca. This was mainly attributed to their greater ability of Al complexation and Al localization in roots, less upward translocation of Al to shoot tissue, improved P utilization efficiency due to high PPT, and greater acquisition efficiency of K, P and Mg. Identifying the genes responsible for these superior traits of brachiariagrasses is a major objective for future research.

ACKNOWLEDGMENT

This work was supported by a Grant-in-Aid for Scientific Research to Wagatsuma T (18208008 and 23380041) from the Japan Society for the Promotion of Science.

REFERENCES

1. Akhter, A., Khan, M. S. H., Egashira, H., Tawaraya, K., Rao, I. M., Wenzl, P., Ishikawa, S & Wagatsuma, T. (2009). Greater Contribution of Low-nutrient Tolerance to Sorghum and Maize Growth under Combined Stress Conditions with High Aluminum and Low Nutrients in Solution Culture Simulating the Nutrient Status of Tropical Acid Soils. Soil Science and Plant Nutrition, Vol. 55, No. 3, (May 2009), pp. 394-406, ISSN 1747-0765

2. Andrade, S. O., da Sila Lopes, H. O., de Almeida Barros, M., Leite, G. G., Dias, S. M., Saueressig, M., Nobre, D & Temperini, J. A. (1978). Photosensitization in Cattle Grazing on Pasture of Brachiaria decumbens Stapf Infested with Pithomyces chartarum (Berk. & Curt.) M. B. Ellis. Arquivos do Instituto Biologico, Vol.45, No. 2, (April-June 1978), pp. 117-136, ISSN 0020-3653 (In Portuguese with English abstract)

3.   Arora, A., Byrem, T. M., Nair, M. G & Strasburg, G. M. (2000). Modulation of Liposomal Membrane Fluidity by Flavonoids and Isoflavonoids. Archives of Biochemistry and Biophysics, Vol.373, No. 1, (Januray 2000), pp. 102-109, ISSN 0003-9861

4.   Boija, E & Johansson, G. (2006). Interactions Between Model Membranes and Lignin-related Compounds Studied by Immobilized Liposome Chromatography. Biochimica et Biophysica Acta, Vol. 1758, No. 5, (May 2006), pp. 620-626, ISSN 0005-2736

5.   Boija, E., Lundquist, A., Edwards, K. & Johansson, G. (2007). Evaluation of Bilayer Disks as Plant Cell Membrane Models in Partition Studies. Analytical Biochemistry, Vol. 364, No. 2 (May 2007), pp. 145-152, ISSN 0003-2697

6.   Bosse, D. & Köck, M. (1998). Influence of Phosphate Starvation on Phosphohydrolases during Development of Tomato Seedlings. Plant, Cell and Environment, Vol.21, No.3, (March 1998), pp. 325-332, ISSN 0140-7791

7.   Burleigh, S. H. & Harrison, M. J. (1999). The Down-regulation of Mt4-like Genes by Phosphate Fertilization Occurs Systemically and Involves Phosphate Translocation to the Shoots. Plant Physiology, Vol.119, No.1 (January 1999), pp. 241-248, ISSN 0032- 0889

8.   Chapin, F. S. III & Bieleski, R. L. (1982). Mild Phosphorus Stress in Barley and Related Lowphosphorus-adapted Barleygrass: Phosphorus Fractions and Phosphate Absorption in Relation to Growth. Physiologia Plantarum, Vol.54, No.3, (March 1982), pp. 309- 317, ISSN 0031-9317

9.   Ciereszko, I. & Barbachowska, A. (2000). Sucrose Metabolism in Leaves and Roots of Bean (Phaseolus vulgaris L.) during Phosphate Deficiency. Journal of Plant Physiology, Vol.156, No. 5/6, pp. 640-644, ISSN 0176-1617

10.   Cordell, D., Drangert, J. O. & White, S. (2009). The Story of Phosphorus: Global Food Security and Food for Thought. Global Environmental Change, Vol.19, No.2, (May 2009), pp. 292-305, ISSN 0959-3780

11.   Cornard, J. P. & Merlin, J. C. (2002). Specroscopic and Structural Study of Complexes of Quercetin with Al($\sqsubset$). Journal of Inorganic Biochemistry, Vol. 92, No. 1, (September 2002), pp. 19-27, ISSN 0162-0134

12.   Dietz, K. & Heber, U. (1984). Rate-limiting Factors in Leaf Photosynthesis. I. Carbon Fluxes in the Calvin Cycle. Biochimica et Biophysica Acta, Vol.767, No. 3, (December 1984), pp. 432-443, ISSN 0005-2728

13.   Duff, S. M., Lefebvre, D. D. & Plaxton, W. C. (1989). Purification and Characterization of a Phosphoenolpyruvate Phosphatase from Brassica

nigra Suspension Cells. Plant Physiology, Vol.90, No. 2, (June 1989), pp. 734-741, ISSN 0032-0889

14. Duff, S. M. G., Plaxton, W. C. & Lefebvre, D. D. (1991). Phosphate-starvation Response in Plant Cells: De novo Synthesis and Degradation of Acid Phosphatases. Proceedings of the National Academy of Science of the United States of America, Vol.88, No.21, (November 1991), pp. 9538-9542, ISSN 0027-8424

15. Duff, S. M. G., Sarath, G. & Plaxton, W. C. (1994). The Role of Acid Phosphatase in Plant Phosphorus Metabolism. Physiologia Plantarum, Vol.90, No.4, (April 1994) pp. 791- 800, ISSN 0031-9317

16. Essigmann, B., Güler, S., Narang, R. A., Linke D. & Benning C. (1998). Phosphate Availability Affects the Thylakoid Lipid Composition and the Expression of SQD1, a Gene Required for Sulfolipid Biosynthesis in Arabidopsis thaliana. Proceedings of the National Academy of Sciences of United States of America, Vol.95, No.5, (March 1998), pp. 1950-1955, ISSN 0027-8424

17. Fairhust, T., Lefrovy, R., Mutert, E. & Batjes, N. (1999). The Importance, Distribution and Causes of Phosphorus Deficiency as a Constraint to Crop Production in the Tropics. Agroforestry Forum, Vol.9, No.1, pp. 2-8, ISSN 0966-8616

18. Fatemi, S. J. A., Williamson, D. J. & Moore, G., R. (1992). A 27Al NMR Investigation of Al3 + Binding to Small Carboxylic Acids and the Proteins Albumin and Transferrin. Journal of Inorganic Biochemistry, Vol. 46, No. 1, (April 1992) pp. 35-40, ISSN 0162- 0134

19. Fredeen, A. L., Rao, I. M. & Terry, N. (1989). Influence of Phosphorus Nutrition on Growth and Carbon Partitioning in Glycine max. Plant Physiology, Vol. 89, No. 1, (January 1989), pp. 225-230, ISSN 0032-0889

20. Glund, K., Nürnberger, T., Abel, S., Jost, W., Preisser, J. & Komor, E. (1990). Intracellular Picompartmentation during Phosphate Starvation–triggered Induction of an Extracellular Ribonuclease in Tomato Cell Culture. In: Progress in Plant Cellular and Molecular Biology. H.J.J. Nijkamp, L.W.H. Van Der Plas, & J. Van Aartrijk, (Eds.), 338–342, Kluwer Academic Publishers, 978-0-7923-0873-7, Boston

21. Green, P. J. (1994). The Ribonuclease of Higher Plants. Annual Review of Plant Physiology, Vol.45, (Jun 1994), pp. 421-425, ISSN 0066-4294

22. Guerrini, F., Cangini, M. & Boni, L. (2000). Metabolic Responces of the Diatom Achanthes brevipes (Bacillariophyceae) to Nutrient Limitation.

Journal of Phycology, Vol.36, Nol. 5, (October 2000), pp. 882-890, ISSN 0022-3646

23. Hammond, J. P., Bennett, M.J., Bowen, H.C., Broadley, M.R., Eastwood, D.C., May, S.T., Rahn, C., Swarup, R., Woolaway, K. E. & White , P. J. (2003). Changes in Gene Expression in Arabidopsis Shoots during Phosphate Starvation and the Potential for Developing Smart Plants. Plant Physiology, Vol. 132, No. 2, (June 2003), pp. 578-596, ISSN 0032-0889

24. Hausler, R. E., Baur, B., Scharte, J., Teichmann, T., Eicks, M., Fischer, K. L., Flugge, U. I., Schubert, S., Weber, A. & Fischer, K. (2000). Plastidic Metabooolite Transporters and Their Physiological Functios in the Inducible Crassulacean Acid Metabolism Plant Messembryanthemum crystallinum. The Plant Journal, Vol.24, No.3, (November 2000), pp. 285-296, ISSN 0960-7412.

25. Howard, C. J., LeBrasseur, N. D., Bariola, P. A. & Pamela, J. G. (1998). Control of Ribonuclease in Response to Phosphate Limitation: Induction of RNS1 in Arabidopsis. In: Phosphorus in Plant Biology: Regulatory Roles in Molecular, Cellular, Organic, and Ecosystem Processes, J.P. Lynch & J. Deikman (Eds.), American Society of Plant Physiology, ISBN 094-3088-38-0, Rockville, MD, U.S.A.

26. Huang, C. F., Yamaji, N., Mitani, N., Yano, M., Nagamura, Y. & Ma, J. F. (2009). A Bacterialtype ABC Transporter is Involved in Aluminum Tolerance in Rice. The Plant Cell, Vol.21, No. 2, (Feburuary 2009), pp. 655-667, ISSN 1040-4651

27. Hutzler, P., Fischbach, R., Heller, W., Jungblut, T. P., Reuber, S., Schmitz, R., Veit, M., Weissenbock, G & Schnitzler, J-P. (1998). Tissue Localization of Phenolic Compounds in Plants by Confocal Laser Scanning Microscopy. Journal of Experimental Botany, Vol. 49, No. 323, pp. 953-965, ISSN 0022-0957

28. Ishigaki, G. (2010). Studies on Establishment of Tissue Culture System and its Utilization for Breeding in Ruzigrass (Brachiaria ruziziensis). (in Japanese with English Tables, Figures, and Summary). Dissertation thesis, University of Miyazaki, Miyazaki, Japan, Retrieved from http:// hdl.handle.net/10458/2755

29. Ishikawa, S. & Wagatsuma, T. (1998). Plasma Membrane Permeability of Root-tip Cells Following Temporary Exposure to Al Ions is a Rapid Measure of Al Tolerance among Plant Species. Plant and Cell Physiology, Vol. 39, No. 5, (May 1998), pp. 516- 525, ISSN 0032-0781

30. Ishikawa, S., Wagatsuma, T., Sasaki, R. & Ofei-Manu, P. (2000). Comparison of the Amount of Citric and Malic Acids in Al Media of Seven Plant Species and Two Cultivars Each in Five Plant Species. Soil Science and Plant Nutrition, Vol. 46, No. 3, (September 2000), pp. 751-758, ISSN 1747-0765

31. Ishikawa, S., Wagatsuma, T., Takano, T., Tawaraya, K. & Oomata, K. (2001). The Plasma Membrane Intactness of Root-tip Cells is a Primary Factor for Al Tolerance in Cultivars of Five Plant Species. Soil Science and Plant Nutrition, Vol. 47, No. 3, (September 2001), pp. 489-501, ISSN 1747-0765

32. Jones, D. L. (1998). Organic Acids in the Rhizosphere – A Critical Review. Plant and Soil, Vol. 205, No. 1, (August 1998), pp. 25-44, ISSN 0032-079X

33. Kerven, G. L., Larsen, P. L., Bell, L. C. & Edwards, D. G. (1995). Quantitative 27Al NMR Spectroscopic Studies of Al($\subset$) Complexes with Organic Acid Ligand and Their Comparison with GEOCHEM Predicted Values. Plant and Soil, Vol. 171, No. 1, pp. 35-39, ISSN 0032-079X

34. Khan, M. S. H., Tawaraya, K., Sekimoto, H., Koyama, H., Kobayashi, Y., Murayama, T., Chuba, M., Kambayashi, M., Shiono, Y., Uemura, M., Ishikawa, S. & Wagatsuma, T. (2009). Relative Abundance of Δ5-sterols in Plasma Membrane Lipids of Root-tip Cells Correlates with Aluminum Tolerance of Rice. Physiologia Plantarum, Vol. 135, No. 1, (January 2009), pp. 73-83, ISSN 0031-9317

35. Kinraide, T. (1999). Interactions Among Ca2 $+$, Na$+$ and K$+$ in Salinity Toxicity : Quantitative Resolution of Multiple Toxic and Ameliorative Effects. Journal of Experimental Botany, Vol. 50, No. 338, (September 1999), pp. 1495-1505, ISSN 0022-0957

36. Kondracka, A. & Rychter, A. M. (1997). The Role of Pi Recycling Processes during Photosynthesis in Phosphate-deficient Bean Plants. Journal of Experimental Botany, Vol. 48, No. 7, (July 1997), pp. 1461-1468, ISSN 0022-0957

37. Li, X. N. & Ashiharam H. (1990). Effects of Inorganic Phosphate on Sugar Catabolism by Suspension-cultured Catharanthus roseus. Phytochemistry, Vol. 29, No. 2, (February 1990), pp. 497-500, ISSN 0031-9422

38. Löffler, A., Abel, S., Jost, W., Beintema, JJ. & Glund, K. (1992). Phosphate-regulated Induction of Intracellular Ribonulease in Cultured Tomato (Lycopersicon esculentum) Cells. Plant Physiology, Vol.98, No.4, (April 1992), pp. 1472-1478, ISSN 0032-0889

39. Louw-Gaume, A. E., Rao, I. M., Gaume, A. J. & Frossard, E. (2010a).
     A Comparative Study on Plant Growth and Root Plasticity Responses
     of Two Brachiaria Forage Grasses Grown in Nutrient Solution at Low
     and High Phosphorus Supply. Plant and Soil, Vol. 328, No. 1-2, (March
     2010), pp.155-164, ISSN 0032-079X

40. Louw-Gaume, A., Rao, I. M., Frossard, E. & Gaume, A. (2010b).
     Adaptive Strategies of Tropical Forage Grasses to Low Phosphorus
     Stress: The Case of Brachiariagrasses. In: Handbook of Plant and Crop
     Stress. Third Edition, M. Pessarakli, Ed., Taylor & Francis Group, pp.
     1111-1144, ISBN 1439813965, USA, Boca Raton

41. Lynch, J.P. (2011). Root Phenes for Enhanced Soil Exploration and
     Phosphorus Acquisition: Tools for Future Crops. Plant Physiology, Vol.
     156, No. 3, (July 2011), pp. 1041-1049, ISSN 0032-0889

42. Ma, J. F., Shen, R., Zhao, Z., Mathias, W., Takeuchi, Y., Ebitani, T &
     Yano, M. (2002). Response of Rice to Al Stress and Identification of
     Quantitative Trait Loci for Al Tolerance. Plant and Cell Physiology, Vol.
     43, No. 6, (Jun 2002), pp. 652-659, ISSN 0032-0781

43. Macedo, M. A. M. (2005). Pastagens no Ecosistema Cerrados: Evalucao
     das Pesquisas Para o Desenvolvimento Sustentavel. Reuniao Anual da
     Siciedade Brasileira de Zootecnia, Vol. 41, pp. 56-84. UFGO, SBZ,
     Goiania.

44. Miles, J.W.; do Valle, C.B.; Rao, I.M. & Euclides, V.P.B. (2004).
     Brachiariagrasses. In: Warmseason (C4) grasses (L. Moser, B. Burson
     & L.E. Sollenberger, Eds.). ASA-CSSA-SSSA, Madison, WI, USA, pp.
     745-783, ISSN 0011-183X

45. Miles, J. W., Cardona, C. & Sotelo, G. (2006). Recurrent Selection in
     a Synthetic Brachiariagrass Population Improves Resistance to Three
     Spittlebug Species. Crop Science, Vol. 46, No. 3, (June 2006), pp. 1088-
     1093, ISSN 0011-183X

46. Mimura, T. (1999). Regulation of Phosphate Transport and Homeostasis
     in Plant Cells. In: International Review of Cytology, K.W. Jeon (Ed.),
     Vol. 191, pp. 149-200. ISBN 978- 0123-645-95-1

47. Misson, J., Raghothama, K. G., Jain, A., Jouhet, J., Block, M. A., Bligny,
     R., Ortet, P., Creff, A., Somerville, S., Rolland, N., Doumas, P., Nacry, P.,
     Herrerra-Estrella, L., Nussaume, L. & Thibaud, M. C. (2005). Arabidopsis
     thaliana Affymetrix Gene Chips Determined Plant Responses to Phosphate
     Deprivation. Proceedings of the National Academy of Sciences of USA,
     Vol.102, No.33, (August 2005), pp. 11934-11939, ISSN 0027-8424

48. Miyadate, H., Adachi, S., Hiraizumi, A., Tezuka, K., Nakazawa, N., Kawamoto, T., Katou, K., Kodama, I., Sakurai, K., Takahashi, H., Satoh-Nagasawa, N., Watanabe, A.,Fujimura, T. & Akagi, H. (2010). OsHMA3, P1B-type of ATPase Affects Root-toshoot Cadmium Translocation in Rice by Mediating Efflux into Vacuoles. New Phytologist, Vol. 189, No. 1, (January 2010), pp. 190-199, ISSN 0028-646X

49. Nakamura, Y., Koizumi, R., Shui, G., Shimojima, M., Wenk, M.R., Ito, T. & Ohta, H. (2009). Arabidopsis Lipins Mediate Eukaryotic Pathway of Lipid Metabolism and Cope Critically with Phosphate Starvation. Proceedings of the National Academy of Sciences of USA, Vol. 106, No. 49, (December 2009), pp. 20978-20983, ISSN 0027-8424

50. Nanamori, M., Shinano, T., Wasaki, J., Yamamura, T., Rao, I.M. & Osaki, M. (2004). Low Phosphorus Tolerance Mechanisms: Phosphorus Recycling and Photosynthate Partitioning in the Tropical Forage Grass, Brachiaria Hybrid cultivar Mulato Compared with Rice. Plant and Cell Physiology, Vol. 45, No. 4, (April 2004), pp. 460- 469, ISSN 0032-0781

51. Nátr, L. (1992). Mineral Nutrients – A Ubiquitous Stress Factor for Photosynthesis. Photosynthetica, Vol.27, No. 3, (September 1992), pp. 271-294, ISSN 0300-3604

52. Nomura, M., Imai, K. & Matsuda, T. (1995). Effects of Atmospheric Partial Pressure of Carbon Dioxide and Phosphorus Nutrition on the Ultrastructure of Rice (Oryza sativa L.) Chloroplasts. Japanese Journal of Crop Science, Vol.64, No. 4, pp. 784-793, ISSN 0011-1848

53. Nürnberger, T., Abel, S., Jost, W. & Glund, K. (1990). Induction of An Extracellular Ribonulease in Cultured Tomato Cells upon Phosphate Starvation. Plant Physiology, Vol. 92, No. 4, (April 1990), pp. 970-976, , ISSN 0032-0889

54. Ofei-Manu, P., Wagatsuma, T., Ishikawa, S. & Tawaraya, K. (2001). The Plasma Membrane Strength of the Root-tip Cells and Root Phenolic Compounds are Correlated with Al Tolerance in Several Common Woody Plants. Soil Science and Plant Nutrition, Vol. 47, No. 2, (June 2001), pp. 359-375, ISSN 1747-0765

55. Palma, D. A., Blumwald, E. & Plaxton, W. C. (2000). Upregulation of Vacuolar H+- translocating Pyrophosphatase by Phosphate Starvation of Brassica napus (Rapeseed) Suspension Cell Cultures. FEBS Letter, Vol. 486, No. 2, (December 2000), pp. 155-158, ISSN 0014-5793

56. Qui, J. & Israel, D. W. (1992). Diurnal Starch Accumulation and Utilization in Phosphorusdeficient Soybean Plants. Plant Physiology, Vol.98, No.1, (January 1992), pp. 316- 323., ISSN 0032-0889

57. Ramaekers, L., Remans, R., Rao, I. M., Blair, M. W. & Vanderleyden, J. (2010). Strategies for Improving Phosphorus Acquisition Efficiency of Crop Plants. Field Crops Research, Vol. 117, No. 2-3, (June 2010), pp. 169-175.

58. Rao, I. M., Fredeen, A. L. & Terry, N. (1990). Leaf Phosphate Status, Photosynthesis and Carbon Partitioning in Sugar Beet. III. Diurnal Changes in Carbon Partitioning and Export. Plant Physiology, Vol. 92, No.1, (January 1990), pp. 29-36 , ISSN 0032-0889

59. Rao, I. M. (1996). The Role of Phosphorus in Photosynthesis. In : Handbook of Photosynthesis, M. Pessarakli, (Ed.), pp. 173-194, Marcel Dekker, New York

60. Rao, I. M. (2001a). Adapting Tropical Forages to Low-fertility Soils. In: Proceedings of the XIX International Grassland Congress. (J.A. Gomide.; W.R.S. Mattos. & S.C. da Silva. Eds.), Brazilian Society of Animal Husbandry, Piracicaba, Brazil, pp. 247-254, ISBN 857- 1330-1-7

61. Rao, I. M. (2001b). Role of Physiology in Improving Crop Adaptation to Abiotic Stresses in the Tropics: The Case of Common Bean and Tropical Forages. In: Handbook of Plant and Crop Physiology, M. Pessarakli Ed., pp. 583-613, Marcel Dekker, ISBN 0824705467, Inc., New York, USA

62. Rao, I. M., Ayarza, M. A. & Garcia, R. (1995). Adaptive Attributes of Tropical Forage Species to Acid Soils. ⊆. Differences in Plant Growth, Nutrient Acquisition and Nutrient Utilization among C4 Grasses and C3 Legumes. Journal of Plant Nutrition, Vol. 18, No. 10, (October 1995), pp. 2135-2155, ISSN 0176-1617

63. Rao, I. M., Friesen, D. K. & Osaki, M. (1999). Plant Adaptation to Phosphorus-limited Tropical Soils. In: Handbook of Plant and Crop Stress. (M. Pessarakli , Ed.), pp. 61-96, Marcel Dekker, Inc., ISBN 0824719484, New York, USA.

64. Rao, I. M.; Kerridge, P. C. & Macedo, M. C. M. (1996). Nutritional Requirements of Brachiaria and Adaptation to Acid Soils. In: Brachiaria: Biology, Agronomy, and Improvement (J.W. Miles., B. L. Maass. & C. B. Valle, Eds.), Centro Internacional de Agricultura Tropical, Cali, Colombia, pp. 53-71, ISBN 958-9439-57-8

65. Rao, I. M., Miles, J. W & Granobles, J. C. (1998). Differences in Tolerance to Infertile Acid Soil Stress Among Germplasm Accessions and Genetic Recombinants of the Tropical Forage Grass Genus, Brachiaria. Field Crops Reseach, Vol. 59, No. 1, (October 1998), pp. 43-52, ISSN 0378-4290

66. Rao, I. M., Zeigler, R. S., Vera, R. & Sarkarung, S. (1993). Selection and Breeding for Acid-soil Tolerance in Crops: Upland Rice and Tropical Forages as Case Studies. BioScience, Vol. 43, No. 7, (July 1993), pp. 454-465, ISSN 0006-3658

67. Ryan, P. R., Tyerman, S. D., Sasaki, T., Furuichi, T., Yamamoto, Y., Zhang, W. H. & Delhaize, E. (2011). The Identification of Aluminium-resistance Genes Provides Oppotunities for Enhancing Crop Production on Acid Soils. Journal of Experimental Botany, Vol. 62, No. 1, (January 2011), pp. 9-20, ISSN 0022-0957

68. Shinano, T., Nanamori, M., Dohi, M., Wasaki, J. & Osaki, M. (2005). Evaluation of Phosphorus Starvation Inducible Genes Relating to Efficient Phosphorus Utilization in Rice. Plant and Soil, Vol. 269, No. 1-2, (February 2005), pp. 81-87, ISSN 1573-5036

69. Steen, I. (1998). Phosphorus Availability in the 21st Centrury. Phosphorus &Potassium, Vol. 217, (September-October 1998), ISSN 0031-8426

70. Stoutjesdijk, P. A., Sale, P. W. & Larkin, P. J. (2001). Possible Involvement of Condensed Tannins in Aluminium tolerance of Lotus pedunculatus. Australian Journal of Plant Physiology, Vol. 28, No. 11, (November 2001), pp. 1063-1074, ISSN 0310-7841

71. Tadano, T., Ozawa, K., Sakai, H., Osaki, M. & Matsui, H. (1993). Secretion of Acid Phosphatase by the Roots of Crop Plants under Phosphorus-deficient Conditions and Some Properties of the Enzyme Secreted by Lupin Roots. Plant and Soil, Vol. 155/156, No. 1, (October 1993), pp. 95-98, ISSN 1573-5036

72. Theodrou, M.E., Elrifi, I.R., Turpin, D.H. & Plaxton, W.C. (1991). Effect of Phosphorus Limitation on Respiratory Metabolism in the Green Alga Selenastrum minutum. Plant Physiology, Vol. 95, No. 4, (April 1991), pp. 1089-1095. , ISSN 0032-0889

73. Theodorou, M. E. & Plaxton, W. C. (1993). Metabolic Adaptations of Plant Respiration to Nutritional Phosphate Deprivation. Plant Physiology, Vol. 101, No.2, (February 1993), pp. 339-344, ISSN 0032-0889

74. Uhde-Stone, C., Zinn, K. E., Ramirez-Yáñez, M., Li, A., Vance, C. P. & Allan, D. L. (2003b). Nylon Filter Arrays Reveal Differential Gene Expression in Proteoid Roots of White Lupin in Response to Phosphorus Deficiency. Plant Physiology, Vol. 131, No. 3, (March 2003), pp. 1064-1079, ISSN 0032-0889

75. Usuda, H. & Shimogawara, K. (1991). Phosphate Deficiency in Maize. ⊇. Enzyme Activities. Plant and Cell Physiology, Vol. 32, No. 8, (December 1991), pp.1313-1317, ISSN 0032- 0781

76. Versaw, W. K. & Harrison, M. I. (2002) A Chloroplast Phosphate Transporter, PHT2; 1, Influences Allocation of Phosphate within the Plant and Phosphate-starvation Responses. The Plant Cell, Vol. 14, No. 8, (August 2002), pp. 1751-1766, ISSN 1040- 4651

77. Wagatsuma, T., Ishikawa, S., Obata, H., Tawaraya, K. & Katohda, S. (1995). Plasma Membrane of Younger and Outer Cells is the Primary Specific Site for Aluminum Toxicity in Roots. Plant and Soil, Vol. 171, No. 1, (April 1995), pp. 105-112, ISSN 0032-079X

78. Wagatsuma, T., Ishikawa, S., Uemura, M., Mitsuhashi, W., Kawamura, T., Khan, M. S. H. & Tawaraya, K. (2005). Plasma Membrane Lipids are the Powerful Components for Early Stage Aluminum Tolerance in Triticale. Soil Science and Plant Nutrition, Vol. 51, No. 5, pp. 701-704, (September 2005), ISSN 1747-0765

79. Wang, Y. -H., Gravin, D. F. & Kochian, L. V. (2002). Rapid Induction of Regulatory and Transporter Genes in Response to Phosphorus, Potassium, and Iron Deficiencies in Tomato Roots. Evidence for Cross Talk and Root/Rhizosphere-mediated Signals. Plant Physiology, Vol. 130, No. 3, (November 2002), pp. 1361-1370, ISSN 0032-0889

80. Wasaki, J., Shinano, T., Onishi, K., Yonetani, R., Yazaki, J., Fujii, F., Shimbo, K., Ishikawa, M., Shimatani, Z., Nagata, Y., Hashimoto, A., Ohta, T., Sato, Y., Miyamoto, C., Honda, S., Kojima, K., Sasaki, T., Kishimoto, N., Kikuchi, S. & Osaki, M. (2006). Transcriptomic Analysis Indicates Putative Metabolic Changes Caused by Manipulation of Phosphorus Availability in Rice Leaves. Journal of Experimental Botany, Vol. 57, No. 9, (September 2006), pp. 2049-2059, ISSN 0022-0957

81. Wasaki, J., Yamamura, T., Shinano, T. & Osaki, M. (2003a). Secreted Acid Phosphatase is Expressed in Cluster Roots of Lupin in Response to Phosphorus Deficiency. Plant and Soil, Vol. 248, No. 1-2, (January 2003), pp. 129-136, ISSN 1573-5036

82. Wasaki, J., Yonetani, R., Kuroda, S., Shinano, T., Yazaki, j., Fujii, F., Shimbo, K., Yamamoto, K., Sakata, K., Sasaki, T., Kishimoto, N., Kikuchi, S., Yamagishi, M. & Osaki, M. (2003b). Transcriptomic Analysis of Metabolic Changes by Phosphorus Stress in Rice Plant Roots. Plant, Cell and Environment, Vol. 26, No. 9, (September 2003), pp. 1515-1523, ISSN 0140-7791

83. Wsaki, J., Yonetani, R., Shinano, T., Kai, M. & Osaki, M. (2003c). Expression of the OSPI1 Gene, Cloned from Rice Roots Using cDNA Microarray, Rapidly Responds to Phosphorus Status. New Phytologist, Vol. 158, No. 2, (May 2003), pp. 239-248, ISSN 0028-646x

84. Watanabe, T., Misawa, S. & Osaki, M. (2005). Aluminum Accumulation in the Roots of Melastoma malabathricum L., an Aluminum-accumulating Plant. Canadian Journal of Botany, Vol. 83, No. 11, (Nobember 2005), pp. 1518-1522, ISSN 0008-4026

85. Watanabe, T., Osaki, M., Yoshihara, T. & Tadano, T. (1998). Distribution and Chemical Speciation of Aluminum in the Al Accumulator Plant, Melastoma marabathricum L. Plant and Soil, Vol. 201, No. 2, pp. 165-173, ISSN 0032-079X

86. Wenzl, P., Mancilla, L. I., Mayer, J. E., Albert, R. & Rao, I. M. (2003). Simulating Infertile Acid Soils with Nutrient Solutions : The Effects on Brachiaria species. Soil Science Society of America Journal, Vol. 67, No. 5, (September-October 2003), pp. 1457-1469, ISSN 1747-0765

87. Wenzl, P., Patino, G. M., Chaves, A. L., Mayer, J. E. & Rao, I. M. (2001). The High Level of Aluminum Resistance in Signalgrass is not Associated with Known Mechanisms of External Detoxification in Root Apices. Plant Physiology, Vol. 125, No. 3, (March 2001), pp. 1473-1484, ISSN 0032-0889

88. Wu, P., Ma, L., Hou, X., Wang, M., Wu, Y., Liu, F. & Deng, X. W. (2003). Phosphate Starvation Triggers Distinct Alterations of Genome Expression in Arabidopsis Roots and Leaves. Plant Physiology, Vol. 132, No. 3, (July 2003), pp. 1260-1271, ISSN 0032- 0889

89. Yamaji, N., Huang, C. F., Nagao, S., Yano, M., Sato, Y., Nagamura, y. & Ma, J. F. (2009). A Zinc Finger Transcription Factor ART 1 Regulates Multiple Genes Implicated in Aluminum Tolerance in Rice. The Plant Cell, Vol. 21, No. 10, (October 2009), pp. 3339-3349, ISSN 1040-4651

90. Yan, X., Liao, H., Trull, M. C., Beebe, S. E. & Lynch, J. P. (2001). Induction of a Major Leaf Acid Phosphatase Does not Confer Adaptation to Low Phosphorus Availability in Common Bean. Plant Physiology, Vol. 125, No. 4, (April 2001), pp. 1901-1911, ISSN 0032-0889

91. Yang, J. L., Li, Y. Y., Zhang, Y. J., Zhang, S. S., Wu, Y. R., Wu, P., Zheng, S. J. (2008). Cell Wall Polysaccharides Are Specifically Involved in the Exclusion of Aluminum from the Rice Root Apex. Plant Physiology, Vol. 146, No. 2, (February 2008), pp. 602-611, ISSN 0032-0889

92. Yang, Z-B., Eticha, D., Rao, I. M. & Horst, W. J. (2010). Alteration of Cell-wall Porosity is Involved in Osmotic Stress-induced Enhancement of Aluminium Resistance in Common Bean (Phaseorus velgaris L.). Journal of Experimental Botany, Vol. 61, (July 2010), pp. 3245-3258, ISSN 0022-0957

93. Yoneda, S. & Nakatsubo, F. (1998). Effect of the Hydroxylation of Patterns and Degrees of Polymerization of Condensed Tannins on Their Metal-chelating Capacity. Journal of Wood Chemistry and Technology, Vol. 92, No. 2, (April 2008), pp. 193-205, ISSN 0277- 3813

94. Yun, S. J. & Kaeppler, S. M. (2001). Induction of Maize Acid Phosphatase Activities under Phosphorus Starvetion. Plant and Soil, Vol. 237, No. 1, (November 2001), pp. 109- 115, ISSN 1573-5036

Chapter 7

LONG-TERM EFFECTS OF RESIDUE MANAGEMENT ON SOIL FERTILITY IN MEDITERRANEAN OLIVE GROVE: SIMULATING CARBON SEQUESTRATION WITH ROTHC MODEL

O.M. Nieto[1,2], J. Castro[1], and E. Fernández[2]

[1]IFAPA Centro Camino de Purchil, Junta de Andalucía, Granada, Spain

[2]Dpto. Edafología y Química Agrícola, Facultad de Ciencias, Universidad de Granada, Granada, Spain

INTRODUCTION

Olive orchards are widely cultivated throughout the semiarid Mediterranean region. During olive growth a large quantity of vegetable residues are produced, either from the biannual pruning or from the olive-fruit cleaning in the oil mill, where the olive fruit is separated from the leaves, twigs and soil. These residues are generally discarded and the pruning debris is usually burned in situ or used for energy. Such practices not only release a large quantity of CO_2 into the atmosphere, but also fail to return to the soil the elements taken up by the tree. The use of crop residues is being widely debated today because of its impact on the soil degradation (Lal, 2008).

In Andalusian olive orchards, conventional agricultural practices such as tillage or non-tillage with bare soil also reduces the incorporation of plant remains into the soil, thus changing the quantity of soil organic carbon (SOC) in a variable way and accelerating erosive processes (Pastor, 2004). Nevertheless in recent years, the technique of shredding pruning debris and spreading the material over the orchard is becoming generalized as an alternative to burning. The residues left from fruit cleaning in the oil mill prior to extracting the oil, composed of leaves, green twigs, and superficial soil, can also be spread on the soil surface, returning to the soil the elements previously taken up by the tree. These new soil-management systems are an alternative for improving the soil quality and fertility in sustainable agricultural system (Ordóñez et al., 2001; Rodríguez-Lizana et al., 2008; Nieto et al., 2010).

Many studies on agricultural ecosystems, as reviewed by Jarecki & Lal (2003), have documented the changes in soil properties when the soil management shifts from tillage to cover crop, mainly the increase of SOC and nitrogen (N). The management of crop residues is an important aspect of conservation systems (Six et al., 1999; Paustian, 2000; Lal, 2008), since proper distribution on the ground surface reduces water losses and thus discourages soil erosion (Schomberg et al., 1999). Water and erosion constitute especially serious issues in zones that have a Mediterranean climate and can be only partially solved by recycling the crop debris (Rodríguez-Lizana et al., 2008).

When the soil includes great quantities of fresh plant material, it is necessary to separate the soil organic matter in order to quantify the SOC which is truly fixed. Some authors recommend methods to separate the water-floatable organic matter (FOM) from soil fractions according to size, using physical procedures such as ultrasound, together with a mixture of physical and chemical methods (Buyanovsky et al., 1994; Hevia et al., 2003). These latter procedures of fractionation appear to be more adequate when the residues added provide certain quantities of soil together with the plant debris.

According to Ingram & Fernandes (2001), the factors determining the current level of carbon in agricultural soils are the losses of soil and clay by erosion, the decline in plant debris, and the elimination of this. Franzluebbers (2002) proposed that soil quality is correlated with the stratification of the SOC. High SOC levels on the soil surface mitigate the direct impact of raindrops, protecting against sealing and the disruption of the soil structure (Hernanz et al., 2002).

Cultivation practices that improve soil quality and fertility, such as the use of crop residues, progressively change the physical and chemical properties of the soil (Rhoton et al., 1993; Ordóñez et al., 2001; Hernández et al., 2005). In addition to SOC and N, other nutrients have been made evaluated in this sense such as K^+ (Thomas et al., 2007), as has soil properties such as cation-exchange capacity (Oorts et al., 2003), or soil-water content (Rawls et al., 2003; Bescansa et al., 2006). The impact of different soil-management systems on soil properties have been studied for olive orchards (e.g. Hernández et al., 2005; Soria et al., 2005; Castro et al., 2008; Gómez et al., 2009) but only a few works have evaluated the effect of shredded olive-pruning debris (Ordóñez et al., 2001, Sofo et al., 2005; Rodríguez-Lizana et al., 2008).

Recently, agricultural soils have been identified as the major carbon pool in the context of its global cycle. Some authors (Jarecki & Lal, 2003;

Hernández et al. 2005; Smith et al., 2008) have reported that strategies based upon changes in soil management in agricultural soils are potentially important in increasing carbon sequestration by the soil and in reducing the atmospheric CO_2 concentration. The main processes responsible for lowering current carbon levels in agricultural soils include erosion, tillage, and low inputs of agricultural residues (Lal, 2008; Álvaro-Fuentes et al., 2009). Some authors have emphasized that intensive tilling accelerates the decomposition of organic matter as result of the break-up of soil aggregates (Balesdent et al., 2000; Paustian et al., 2000) and contributes considerably to soil loss through erosion (Rodríguez-Lizana et al., 2008). Soil-management techniques that combine a restriction on tillage and the addition of organic residues are considered to be one potential way for improving soil properties and diminishing atmospheric CO_2 concentrations by storing carbon in the form of organic matter (IPCC, 2000; Jarecki & Lal, 2003).

The present work describes the effect on the soil after the spreading of olive-pruning debris together with the residues of the olive-fruit cleaning in two predominant soils in Andalusian olive orchards. In specific, study was made of the content, distribution, and stabilization of SOC (floatable and non-floatable in water) and N, the soil potential for carbon sequestration, as well as the effect of SOC in the K^+ content, bulk density (ρb), pH, cation-exchange capacity (CEC), and soil-water content (SWC) at -33 and -1500 kPa.

MATERIAL AND METHODS

Field Description

The study plot was located in the Cortijo El Empalme (Villacarrillo, Jaén), south-eastern Spain (38.175°N, 3.15°W) and is 812 m a.s.l. The climatic characteristics of the area are given in Table 1 (MAPA, 1989). The average annual rainfall was 550 mm, and average annual maximum and minimum temperatures of 37.0°C and 2.8°C, respectively. The natural vegetation is a perennial, sclerophyllous Holm oak (Quercus ilex L.) forest typical of the Mediterranean basin.

The orchard was comprised of adult olives (cv. picual) with 2-3 trunks and planting density of 82 trees ha^{-1}. The average slope is 3%. The orchard has underground drip irrigation and no fertilizers are applied to the soil. Following the WRBSR (FAO, 2006), the soils studied are classified as Chromic Calcisols (CLcr) and Calcic Vertisols (VRcc). The parent material is limestone in the CLcr and marls in the VRcc. The colour of the dry bare soils was dull brown (7.5 YR 5/4) in the CLcr and light grey (2.5 Y 7/1) in the VRcc.

Before the experiment, the soil-management system of the orchard was conventional tillage (T), consisting of two or three passes (0.20 m deep) with a disc harrow and cultivator, twice a year to control weeds. This tillage was applied only to the open gaps between the trees (50% of the total area of the grove). Under the tree canopy (UC), the soil was completely cleared every year using pre- and post-emergence herbicides. Dead leaves, dried fruit and twigs were removed by manual blowers and a mechanical sweeper without breaking the surface crust. The trees were pruned every 2 yr and the debris was burned.

Table 1. Maximum, minimum and average monthly air temperature, monthly rainfall and potential evapotranspiration (ETo) for the study area (MAPA, 1989)

Month	Temperature (ºC)			Rainfall	ETo
	Maximum	Minimum	Mean	(mm/month)	(mm/month)
January	10.9	3.7	7.3	69	33
February	13.3	4.5	8.9	74	46
March	17.9	5.6	11.8	68	87
April	20.3	8.6	14.5	59	110
May	25.8	13.4	19.6	52	154
June	30.1	18.6	24.4	25	170
July	35.5	20.6	28.1	6	211
August	34.1	20.3	27.2	8	182
September	28.6	17.0	22.8	24	122
October	22.2	11.6	16.9	53	81
November	14.8	6.4	10.6	47	43
December	11.3	4.0	7.7	67	30

The soil-management system was changed in 1996 on the CLcr and in 2000 on the VRcc to cover crop, whereby shredded olive-pruning and the residues from the olive-fruit cleaning (PD+CR) were spread between the trees. The ground was not tilled and all these residues remained on the surface. The biomass input was quantified by the use of a 30 x 30 cm metal frame tossed at random 40 times between trees. The mean annual input was 23.9 ± 14.3 Mg C ha[-1] yr[-1]. This area was studied by Soria (2002) under traditional tillage in 1997, just before to the experiment was started. The clay types are given in Table 2.

Table 2. Clay type (%) for 0-30 cm depth, in each soil type according to Soria (2002)

Soil type	Illite	Montmorillonite	Kaolinite
Chromic Calcisols	77	13	10
Calcic Vertisols	45	36	19

Sampling and Analytical Methods

Random soil samples were taken in two different areas: (i) between trees in PD+CR soil after removing the superficial plant-residue layer and (ii) UC area where the soil was completely bare. In addition, to establish the time-zero conditions for the experiment, two neighbouring tilled olive groves were sampled. A trench of 50 x 100 x 50 cm was opened and the samples were taken at depth intervals of 0-2, 2-5, 5-10, 10-15, and 15-30 cm. Three replicate plots per type of soil and area were sampled. Soil samples were also taken from the pits to determine ρb, following the method of Blake & Hartge (1986), using a set of cylinders of 2, 3, and 5 cm high specifically manufactured for this purpose.

The soil samples were dried and sieved (2-mm grid size). In the fine-earth fraction, the following analyses were performed: the textural analysis was made by the pipette method of Robinson (Soil Conservation Service, 1972); the SWC at field capacity was extracted in a pressure plate at -33 kPa, and the moisture at the wilting point was measured at -1500kPa (Cassel & Nielsen, 1986); the assimilable K^+ was extracted with NH4OAc 1M; and the CEC was determined by saturation in sodium and, prior to washing with alcohol, extraction by sodium adsorbed with NH4OAc 1M (Soil Conservation Service, 1972); the pH was measured in a soil suspension in distilled water (1:2.5).

For the determination of the SOC, N, and $CaCO_3$ equivalent, the sample was ground again (0.125 mm). The content of the SOC was determined using the method of Tyurin (1951); water-floatable organic matter (FOM) and non-water-floatable organic matter (NFOM) was separated following the method described by Hevia et al. (2003); for the total nitrogen, the Kjeldahl method was used (Bremner, 1965); and the $CaCO_3$ equivalent was determined by a manometric method (Williams, 1948). For plant remains, carbon and N were determined by the same methods. The SOC, N, and clay contents per hectare were computed by multiplying the soil mass (i.e. bulk-density) by the depth and the SOC, N, and clay concentrations, respectively. The CO_2 emissions from burning residues were determined from the values of carbon concentration in pruning debris using a molecular-weight ratio (1.00 g C = 3.67 g CO_2) (IPCC, 2000).

RothC Model

A detailed description of the model is given in Coleman & Jenkinson (1996). In brief, the RothC model separates the SOC into four active compartments and a small amount of inert organic matter (IOM). Plant residues reintroduced to the soil (Figure 1) are divided into decomposable plant materials (DPM) and resistant plant materials (RPM), both undergoing decomposition to produce microbial biomass (BIO), humified organic matter (HUM) and CO_2 (lost from the system). The clay content of the soil determines the proportions that go to CO_2 or to BIO + HUM. Each compartment, except for IOM, undergoes decomposition by first-order kinetics at its own characteristic rate, which is determined by using modifiers for soil moisture, temperature and plant cover.

The climatic input parameters include monthly average air temperature, monthly precipitation and monthly open-pan evaporation. Other input parameters are soil clay content, monthly carbon input from plant residues or farmyard manure and monthly information on soil cover, whether the soil is bare or covered by plants. As no data for openpan evaporation were available, the average values for monthly potential evaporation (converted to open-pan evaporation) were used (Coleman & Jenkinson, 1996). The IOM was calculated using the equation proposed by Falloon et al. (1998). The turnover time was calculated as the total organic carbon content except IOM divided by the annual input of carbon into the soil (Jenkinson & Rayner, 1977).

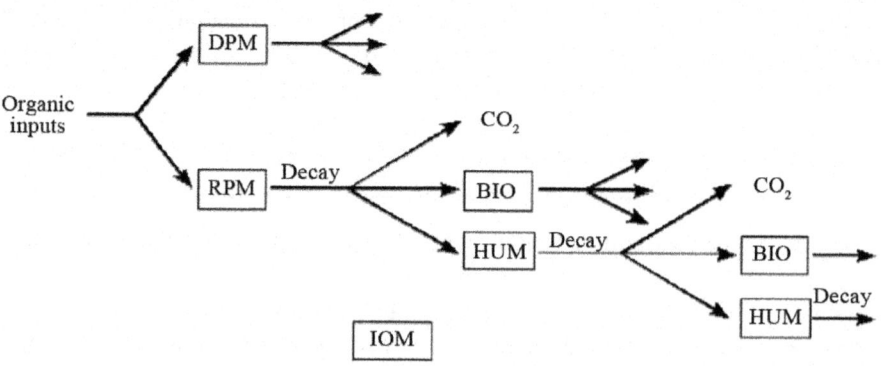

Figure 1. Structure of the Rothamsted carbon model (from Coleman & Jenkinson, 1996).

RothC was designed to run in two modes: 'forward', in which known inputs are used to calculate changes in soil organic matter; and 'inverse', when inputs are calculated from known changes in soil organic matter. This model performed well for changing the carbon inputs to fit measured SOC values in other studies (Falloon & Smith, 2002). In our work, we used this model to assess the changes in the soil carbon content when soil management was changed from tillage to cover crop with pruning debris and the residues from the olive-fruit cleaning.

To run RothC at equilibrium (Coleman & Jenkinson, 1996), we needed to assume that the soils were in equilibrium (more than 30 yr with the same management). To determine the SOC value for the soil equilibrium under tillage, we sampled two neighbouring areas (one for each soil type) that ploughed the soils over the last 30 years. For the tilled olive grove, the soil was assumed to be in equilibrium. In this case, the RothC model was run iteratively in reverse to calculate how much organic carbon needs to enter a soil annually to give the measured amount of SOC. This value for the annual input of organic carbon was chosen to optimise the fitting between the modelled and measured data.

Statistical Analysis

Data were analysed using SPSS v.10.0. Effects of the location and soil type for each variable were determined by one and two-way analysis of variance (ANOVA) at a confidence level of 95%. Tukey tests was performed for post-hoc comparisons between levels within each factor considered. Bartlett and Shapiro–Wilk tests were applied to check homoscedasticity and normality, respectively, to ensure that assumptions of the model were met. Spearman correlation was used to determine the degree of dependence between the SOC and other variables.

RESULTS

Textural analysis, ρd, CaCO$_3$ and pH

Table 3 presents the results from the textural analysis, the percentage of gravel in the soils studied, ρb, CaCO$_3$ content, and pH. The gravel content increased in depth, with values greater in the CLcr. Despite differences in clay and sand percentages, both soils presented a loamy-clayey texture. The ρb diminished significantly in the uppermost 5 cm of the PD+CR with respect to the UC soils, with values equal to or lower than 1; beyond this depth, the values rose and tended to be equal. The CaCO$_3$ content was greater than 200 g kg-1 in both soils, with higher values in VRcc. However, the concentration of this element increased in depth until reaching a maximum of 706 g kg^{-1} in the last depth under the tree canopy in the CLcr. Both soils studied had basic pH values, which were significantly lower at the first two depths of the cover crop.

SOC, N and C:N

The spreading of the pruning debris and the cleaning residues significantly increased the SOC and N content in both soils, with maximum values in the uppermost 10 cm (Table 4). The differences were significant for depth, location and their interaction. No differences were registered in the SOC and N for each soil type. Under the canopy, both soils presented similar SOC and N contents, with values slightly higher in the upper 5-10 cm.

The C:N ratio reached maximum values in the uppermost 5 cm of PD+CR, being higher than 20 (Table 4). The differences were significant also for the location and type of soil, with higher values in CLcr than in VRcc. Under the canopy and in the last cm of the cover crop, all the C:N values were close to or lower than 10. In all cases, this value diminished in depth. The percentages of FOM and NFOM are represented in Figure 2. The presence of fresh plant residues increased the FOM in the uppermost cm of the soil; under the canopy and in depth, the FOM decreased.

The addition of debris in PD+CR increased the SOC and N content in the bulk soil, from 26.4 ± 1.2 and 27.1 ± 1.0 Mg C ha^{-1} to 158.0 ± 11.6 and 113.6 ± 17.8 Mg C ha^{-1} in CLcr and VRcc, respectively (Table 5). Under the tree, these values were intermediate for both soils. The stratification was similar in both soil types, higher than 10 for the cover crop and minimum ($\sim$1) in conventional tilled soil.

K+ , CEC and SWC

The CEC and K^+ contents were high for the PD+CR cover of both soils, diminishing over the profile (Table 6). The strongest differences were presented in the first layers of VRcc, with 30 cmolc kg^{-1} for CEC and K^+ values greater than 2 cmolc kg^{-1}. The presence of the plant residues on the soil surface also changed the SWC, with values significantly higher in the uppermost 2 or 5 cm of PD+CR and at all depths for the VRcc.

Table 3. Values for contents in gravel (>2 mm), sand (2-0.05 mm), silt (0.05-0.002 mm) and clay (

Depth (cm)	Gravel (g kg⁻¹)	Texture			ρb (Mg m⁻³)	CaCO₃ (g kg⁻¹)	pH (1:2.5)
		Sand (g kg⁻¹)	Silt (g kg⁻¹)	Clay (g kg⁻¹)			
Chromic Calcisols – Pruning debris + cleaning-residues cover							
0-2	188 (35)	405 (31)	256 (3)	339 (34)	0.9 (0.1)	234 (36)	7.8 (0.1)
2-5	184 (24)	384 (34)	276 (26)	340 (37)	1.0 (0.1)	228 (21)	7.8 (0.1)
5-10	317 (176)	464 (29)	235 (30)	301 (18)	1.7 (0.1)	340 (112)	8.3 (0.2)
10-15	238 (155)	496 (35)	215 (51)	288 (19)	1.8 (0.1)	344 (136)	8.3 (0.2)
15-30	173 (102)	433 (72)	246 (31)	322 (47)	1.7 (0.1)	399 (117)	8.4 (0.2)
Chromic Calcisols – Under canopy							
0-2	147 (94)	458 (59)	203 (52)	339 (30)	1.6 (0.2)	353 (73)	8.3 (0.1)
2-5	167 (97)	480 (41)	181 (28)	339 (18)	1.6 (0.3)	374 (76)	8.3 (0.1)
5-10	160 (49)	483 (25)	184 (23)	333 (14)	1.6 (0.1)	422 (3)	8.3 (0.1)
10-15	163 (124)	500 (9)	177 (84)	322 (28)	1.6 (0.1)	524 (57)	8.3 (0.1)
15-30	276 (104)	521 (40)	201 (28)	278 (16)	1.6 (0.1)	706 (118)	8.4 (0.1)
Calcic Vertisols – Pruning debris + cleaning-residues cover							
0-2	81 (36)	345 (58)	330 (13)	325 (45)	0.9 (0.2)	325 (40)	7.7 (0.1)
2-5	70 (41)	304 (88)	365 (24)	331 (81)	0.9 (0.1)	413 (91)	7.8 (0.1)
5-10	138 (47)	243 (18)	395 (51)	362 (21)	1.3 (0.1)	523 (49)	8.1 (0.1)
10-15	112 (37)	251 (21)	385 (17)	365 (16)	1.4 (0.1)	597 (16)	8.3 (0.1)
15-30	120 (35)	249 (19)	393 (20)	358 (11)	1.4 (0.1)	585 (17)	8.3 (0.1)
Calcic Vertisols – Under canopy							
0-2	87 (17)	225 (13)	407 (15)	368 (3)	1.4 (0.1)	588 (6)	8.3 (0.1)
2-5	56 (11)	226 (7)	401 (15)	372 (17)	1.4 (0.1)	586 (11)	8.4 (0.1)
5-10	64 (39)	231 (25)	400 (14)	369 (14)	1.3 (0.1)	576 (26)	8.4 (0.1)
10-15	85 (18)	250 (59)	407 (28)	344 (34)	1.4 (0.1)	616 (74)	8.4 (0.1)
15-30	115 (105)	304 (53)	393 (45)	302 (14)	1.3 (0.1)	645 (54)	8.4 (0.1)
FACTOR			ANOVA *p*-value				
Depth	0.160	0.349	0.961	0.522	**<0.001**	**<0.001**	**<0.001**
Location	**<0.001**	**<0.001**	**<0.001**	0.822	**<0.001**	**<0.001**	**<0.001**
Soil type	0.002	**<0.001**	**<0.001**	**<0.001**	**<0.001**	**<0.001**	0.009
L x D	0.224	0.115	0.327	0.150	**<0.001**	0.058	**<0.001**
L x S	**<0.001**	**<0.001**	**<0.001**	0.985	**<0.001**	0.166	0.001

Table 4. Values of soil organic carbon (SOC), nitrogen (N) concentration, and C:N ratio, in each soil type and location at the different depths. For each value, the standard error is shown in parenthesis. Factors L: location, D: depth, S: soil type

Depth (cm)	SOC (g kg⁻¹)	N (g kg⁻¹)	C:N
	Chromic Calcisols – Pruning debris + cleaning-residues cover		
0-2	124.7 (9.0)	5.8 (1.0)	21.8 (3.1)
2-5	122.3 (15.5)	5.6 (1.0)	21.9 (1.3)
5-10	60.0 (13.9)	3.9 (0.5)	15.5 (3.1)
10-15	13.3 (2.1)	1.2 (0.3)	11.7 (3.9)
15-30	9.7 (1.5)	1.2 (0.3)	8.1 (0.9)
	Chromic Calcisols – Under canopy		
0-2	18.3 (10.1)	1.9 (1.0)	9.4 (1.9)
2-5	14.3 (4.9)	1.4 (0.5)	10.0 (0.1)
5-10	14.3 (4.0)	1.5 (0.5)	10.0 (1.8)
10-15	12.0 (1.0)	1.5 (0.5)	8.2 (1.8)
15-30	8.7 (1.2)	1.0 (0.2)	8.7 (0.3)
	Calcic Vertisols – Pruning debris + cleaning-.residues cover		
0-2	118.7 (1.5)	6.2 (0.7)	19.4 (1.9)
2-5	90.7 (0.6)	4.2 (0.9)	22.3 (5.1)
5-10	33.7 (7.6)	2.7 (0.3)	12.4 (1.3)
10-15	14.7 (1.2)	1.9 (0.4)	8.0 (1.2)
15-30	11.7 (2.5)	1.6 (0.2)	7.3 (2.0)
	Calcic Vertisols – Under canopy		
0-2	15.0 (1.0)	2.4 (0.2)	6.2 (0.7)
2-5	15.0 (3.0)	1.9 (0.3)	7.9 (1.9)
5-10	12.7 (3.8)	1.7 (0.6)	7.7 (2.6)
10-15	8.7 (1.5)	1.5 (0.6)	7.1 (4.7)
15-30	6.7 (1.5)	1.1 (0.3)	6.4 (0.9)
FACTOR	ANOVA *p*-value		
Depth	<0.001	<0.001	<0.001
Location	<0.001	<0.001	<0.001
Soil type	0.226	0.101	<0.001
L x D	<0.001	<0.001	<0.001
L x S	0.366	0.086	<0.001

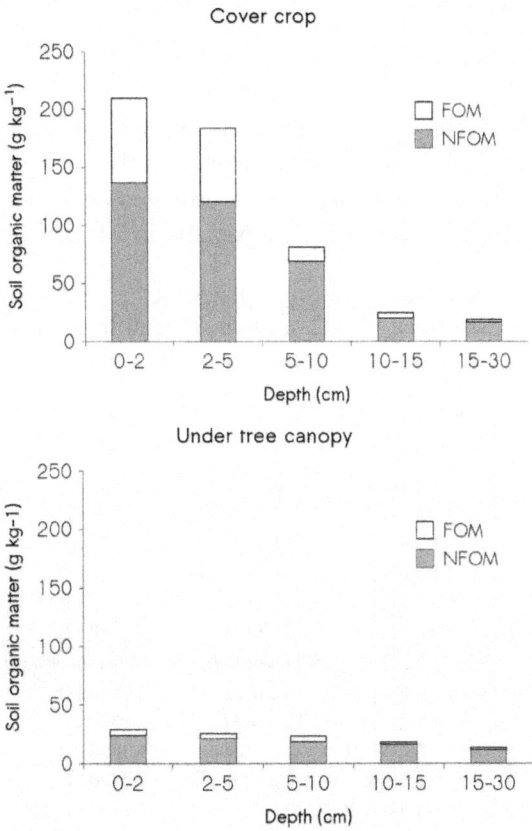

Figure 2. Contents in floatable (FOM) and non-floatable (NFOM) organic matter in water for each location at the different depths.

Table 5. Soil organic carbon (SOC), nitrogen (N), and stratification ratio of SOC for 0-30 cm depth, in each soil type and location. For each value, the standard error is shown in parenthesis

Soil type		SOC (Mg ha⁻¹)	N (Mg ha⁻¹)	Stratification ratio of SOC
Chromic Calcisols	Cover crop	158.0 (11.6)a	11.2 (1.1)a	13.1 (1.6)a
	Under canopy	55.4 (11.8)b	6.2 (1.1)b	2.0 (0.9)b
	Tillage	26.4 (1.2)c	3.4 (0.6)c	0.9 (0.1)b
	ANOVA (*p*-value)	**<0.001**	**<0.001**	**<0.001**
Calcic Vertisols	Cover crop	113.6 (17.8)a	10.1 (0.6)a	10.5 (2.2)a
	Under canopy	37.9 (4.5)b	5.4 (2.1)b	2.3 (0.6)b
	Tillage	27.1 (1.0)b	3.8 (0.3)b	1.2 (0.1)b
	ANOVA (*p*-value)	**0.001**	**0.006**	**<0.001**

Correlations with SOC

The correlations between SOC and other soil parameters are summarized in Table 7. The SOC was correlated negatively with the silt content in all cases, and positively with the clay content only in soils without carbon addition (UC). As expected, the SOC correlated negatively with ρb, $CaCO_3$ and pH, and positively with N. The K^+ and CEC correlated positively with the organic fractions (SOC, NFOM, and FOM), especially in VRcc.

Table 6. Potassium (K^+), cation-exchange capacity (CEC) and soil-water content at -33 and - 1500 kPa in each soil type and location at the different depths. For each value, the standard error is shown in parenthesis. Factors L: location, D: depth, S: soil type

Depth (cm)	K^+ (cmol$_c$ kg^{-1})	CEC (cmol$_c$ kg^{-1})	Soil-water content (m³ m⁻³) -33 kPa	-1500 kPa
		Chromic Calcisols – Pruning debris + cleaning-residues cover		
0-2	1.5 (0.8)	22.5 (8.6)	0.34 (0.02)	0.22 (0.01)
2-5	1.5 (0.7)	25.0 (7.0)	0.35 (0.02)	0.25 (0.02)
5-10	1.5 (0.4)	19.6 (3.5)	0.24 (0.03)	0.15 (0.03)
10-15	0.9 (0.4)	17.2 (3.3)	0.17 (0.01)	0.09 (0.01)
15-30	0.6 (0.2)	15.9 (2.2)	0.19 (0.02)	0.11 (0.02)
		Chromic Calcisols – Under canopy		
0-2	1.4 (0.5)	15.1 (0.6)	0.19 (0.04)	0.10 (0.01)
2-5	1.1 (0.5)	17.5 (6.3)	0.17 (0.03)	0.09 (0.01)
5-10	0.8 (0.2)	13.0 (0.6)	0.17 (0.01)	0.10 (0.01)
10-15	0.4 (0.1)	12.0 (1.4)	0.17 (0.01)	0.10 (0.01)
15-30	0.4 (0.1)	10.5 (2.1)	0.17 (0.02)	0.10 (0.01)
		Calcic Vertisols – Pruning debris + cleaning-residues cover		
0-2	2.6 (0.8)	30.0 (7.2)	0.35 (0.03)	0.30 (0.06)
2-5	1.9 (0.5)	25.1 (8.7)	0.31 (0.03)	0.23 (0.05)
5-10	1.3 (0.3)	19.6 (4.8)	0.30 (0.01)	0.18 (0.02)
10-15	0.5 (0.3)	15.9 (2.9)	0.29 (0.02)	0.17 (0.02)
15-30	0.4 (0.1)	15.4 (2.6)	0.29 (0.02)	0.17 (0.02)
		Calcic Vertisols – Under canopy		
0-2	0.8 (0.3)	16.7 (2.9)	0.30 (0.01)	0.18 (0.02)
2-5	0.6 (0.2)	18.0 (1.5)	0.29 (0.01)	0.19 (0.03)
5-10	0.3 (0.1)	16.4 (0.7)	0.29 (0.01)	0.17 (0.02)
10-15	0.2 (0.2)	16.7 (4.8)	0.29 (0.01)	0.17 (0.03)
15-30	0.1 (0.1)	12.5 (2.6)	0.28 (0.02)	0.16 (0.03)
FACTOR		ANOVA *p*-value		
Depth	<0.001	<0.001	<0.001	<0.001
Location	0.008	<0.001	<0.001	<0.001
Soil type	0.470	<0.001	<0.001	<0.001
L x D	0.013	0.003	<0.001	<0.001
L x S	0.053	<0.001	<0.001	<0.001

Under canopy, the correlation coefficients lowered the significance value, and were not significant for FOM. The correlations between the SWC and the organic fractions were positive and significant in the cover crop of the CLcr; the degree of significance diminishing in the VRcc. No firm correlations were found for the soil fine fraction and SWC.

Table 7. Correlation coefficients of soil organic fractions with the properties studied, for the entire dataset (all data) and for each soil type and location. CLcr: Chromic Calcisols; VRcc: Calcic Vertisols; PD+CR: cover crop with pruning debris and the fruit-cleaning residues; UC: under the tree canopy. * Significant at P < 0.05; ** Significant at P < 0.01 according to Spearman's test

	All data	CLcr		VRcc	
	(n=60)	PD+CR	UC	PD+CR	UC
			SOC (Mg ha^{-1})		
Silt	-0.94 **	-0.77 **	-0.64 *	-0.68 **	-0.69 **
Clay	0.20	-0.04	0.94 **	0.19	0.89 **
			SOC (%)		
ρb	-0.43 **	-0.82 **	0.07	-0.78 **	0.27
CaCO$_3$	-0.50 **	-0.63 *	-0.36	-0.84 **	-0.44
pH	-0.64 **	-0.80 **	-0.13	-0.90 **	-0.22
N	0.86 **	0.92 **	0.87 **	0.97 **	0.63
K$^+$	0.80 **	0.58 *	0.65 **	0.94 **	0.77 **
CEC	0.63 **	0.61 *	0.49	0.83 **	0.64 *
SWC -33 kPa	0.53 **	0.84 **	0.61 *	0.64 **	0.04
SWC -1500 kPa	0.56 **	0.86 **	0.51	0.81 **	0.23
			NFOM		
K$^+$	0.79 **	0.54 *	0.74 **	0.90 **	0.72 **
CEC	0.63 **	0.62 *	0.45	0.80 **	0.70 **
SWC -33 kPa	0.50 **	0.87 **	0.53 *	0.68 **	0.07
SWC -1500 kPa	0.53 **	0.87 **	0.30	0.74 **	0.28
			FOM		
K$^+$	0.68 **	0.67 **	0.42	0.81 **	0.46
CEC	0.52 **	0.56 *	0.31	0.65 **	0.04
SWC -33 kPa	0.52 **	0.81 **	0.57 *	0.52 *	0.09
SWC -1500 kPa	0.55 **	0.81 **	0.51	0.71 **	-0.17

RothC

The results for changing soil management from conventional tillage to cover crop are summarized in Table 8. At first, we calculated the annual carbon input for the tillage of olive trees from the SOC concentration measured in neighbouring areas, assuming a steady state. For both soils, the annual input modelled was 1.0 Mg C ha^{-1} yr^{-1}. During the experiment, we measured a carbon input at the soil-surface in PD+CR cover of 23.9 ± 14.3 Mg C ha^{-1} yr^{-1}. After 10 and 6 yr of change in the soil-management system, the annual carbon inputs needed to reach the SOC values between trees as estimated by the model were very similar to those measured in both soils. The turnover time decreased from 26 yr in T soils to 6 and 5 yr for CLcr and VRcc, respectively.

Table 8. Measured and modelled data for the turnover of organic carbon in olive-grove soils under conventional tillage (T) and mulched with residues from pruning debris and olivefruit cleaning (PD+CR) (from Nieto et al., 2010)

Scenario	Soil type	SOC measured (Mg C ha^{-1})	IOM (Mg C ha^{-1})	Input C modelled (Mg C ha^{-1} yr^{-1})	Input C measured (Mg C ha^{-1} yr^{-1})	Turnover time (yr)
T	CLcr (equ)	26.4	2.0	1.0	-	26
	VRcc (equ)	27.1	2.1	1.0	-	26
PD+CR	CLcr (10 yr)	158.0	2.0	25.3	23.9	6
	VRcc (6 yr)	113.6	2.1	23.6	23.9	5

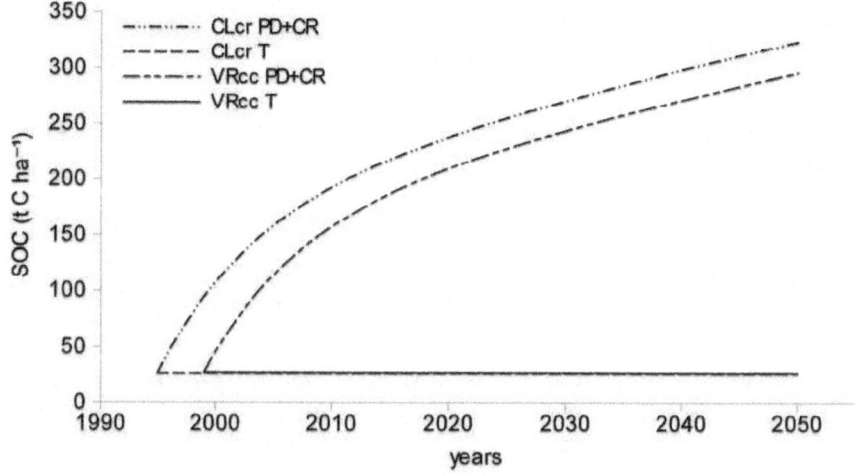

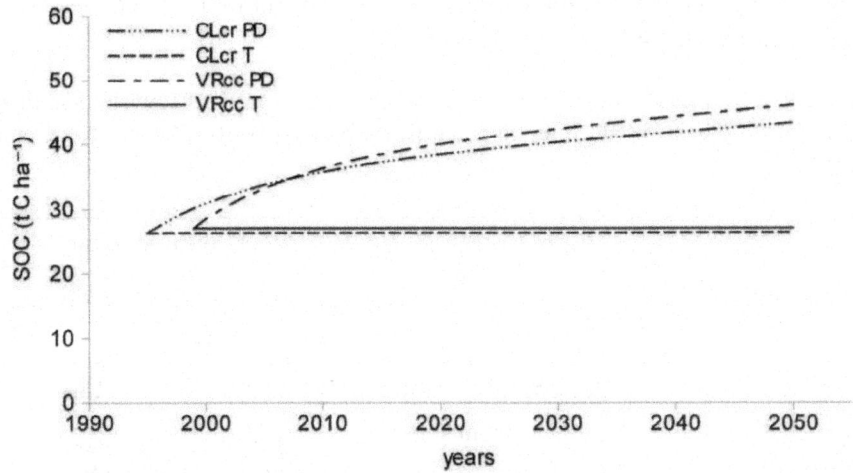

Figure 3. SOC modelled for CLcr and VRcc under conventional tillage (T), shredded olivepruning debris cover (PD) and mulching with residues from olive-fruit cleaning and pruning debris (PD+CR) (from Nieto et al., 2010).

The C sequestration was calculated as the difference between CO_2 emissions in each type of management. Under tillage, the cleaning residues were stored in areas close to the oil mill, so that we could not include them in the CO_2 emission balance. Although the pruning debris had a different use in the two cases, under tillage they were burnt and in the cover crop they were spread on the soil.

Figure 3 shows SOC modelled by RothC for 50 yr in both soils under three management systems: cover with PD+CR, only pruning debris (PD), and tillage (Nieto et al., 2010). When mulched with PD+CR the model predicted a continuous increase in SOC, but without reaching a state of equilibrium. No differences could be discerned between either type of soil, due mainly to the large quantities of organic carbon added. When the input was that of PD alone, the model gave a lower SOC content than in the previous case.

CO_2 emissions are given in Table 9 when the PD were burnt in tillage, and when shredded and spread on the ground. As a result of the addition of PD to the soil, the CO_2 that was previously released into the atmosphere during the years of tilled management was reduced more than 55% for both soils. For the decrease in CO_2 emission, the RothC model estimated a potential carbon sequestration of 0.5 and 0.6 Mg C ha^{-1} yr^{-1} for CLcr and VRcc (Table 10). We did not model carbon sequestration after mulching with CRs because this waste matter is normally discarded. Nevertheless, the total carbon content in the soil registered an increase during the experiment of 13.2 and 14.4 Mg C ha^{-1} yr^{-1} for each soil type.

Table 9. CO_2 released into the atmosphere after 10 and 6 years of conventional tillage and mulching with shredded pruning debris, for both types of soil (from Nieto et al., 2010)

Scenario		CO_2 released to the atmosphere (Mg CO_2 ha^{-1})	
		CLcr (10 yr)	VRcc (6 yr)
Tillage	Burn pruning debris	24.0	14.4
	CO_2 lost from soil	12.0	7.2
	Total	36.0	21.6
Pruning debris	Burn pruning debris	-	-
	CO_2 lost from soil	15.9	9.2
	Total	15.9	9.2

Table 10. SOC increase as result of changes in soil management from conventional tillage to mulching with residues from pruning debris and olive-fruit cleaning (PD+CR). CO_2 reduction and carbon sequestration after the addition of only pruning debris (from Nieto et al., 2010)

Soil type	SOC increase (Mg C ha^{-1} yr^{-1})	CO_2 reduction (Mg CO_2 ha^{-1} yr^{-1})	C sequestration (Mg C ha^{-1} yr^{-1})
CLcr	13.2	2.0	0.5
VRcc	14.4	2.1	0.6

DISCUSSION

Soil Organic Carbon and Related Soil Properties

The two types of soils studied were close together and are widely represented in the areas dedicated to olive cultivation in the Mediterranean region. The parent material, limestone in the CLcr and marls in the VRcc, conditioned the development of different profiles with respect to such characteristics as color, and content, as well as type of clay, sand, and silt. Large quantities of soil from nearby zones were applied with the plant residues from olivefruit cleaning. For this reason, such properties as texture and CaCO$_3$ content presented similar values for the first cm of PD+CR in both soils types. The pH values changed with the SOC content (Thomas et al., 2007).

As opposed to the findings of Fontaine et al. (2004), the addition of fresh organic matter did not cause a negative balance in the SOC, since we began with degraded soils of very low SOC contents. Many authors have also found a rise in SOC values in agricultural soils within at least the uppermost 10 cm in depth, after using conservation practices (Angers et al., 1997; Hernanz et al., 2002; Jarecki & Lal, 2003). In olive orchards, some authors as Hernández et al. (2005), Sofo et al. (2005), Castro et al. (2008) and Gómez et al. (2009), have reported increases in the SOC and N content after applying plant residues. Our values were higher than those reported by these authors because, together with the shredded pruning debris, a major quantity of residues composed of soil and plant debris were brought from the olive-cleaning processes, reaching a biomass accumulation greater than that indicated by these authors. In addition, the soil from the cleaning process originated from the soil under the tree canopy, with small aggregates, fine material and a high SOC content.

Under the tree canopy, despite that the soil was maintained free of weeds with herbicides and free of plant debris by sweeping, SOC values reached 64.7 and 43.3 Mg C ha^{-1}, higher than tilled soils between rows (Table 5). This was due to the greater presence of roots in this zone and the continual dropping of olive leaves, which fell under the canopy, where they remained until the annual cleanup before harvest (Ordóñez et al., 2001; Soria et al., 2005). The high values of the SOC found in PD+CR indicate the effectiveness of the treatment in terms of storage, reaching values of up to 158.0 Mg ha^{-1} after 10 years of management. These values, together with those of N were far higher than reported by other researchers in agricultural areas (Hernanz et al., 2002; Hernández et al., 2005) but similar to those found by Jarecki & Lal (2005) for forest soils in Ohio.

Texture plays a major role in SOC accumulation in the soil. In many works, the losses of SOC from agricultural soils were lower in clayey soils, since these tended to accumulate this fraction more rapidly and retain it longer (Percival et al., 2000; Arrouays et al., 2006). In our work, a correlation between the SOC and the clay was found only in under the canopy (Table 6), since the input of plant debris changed this relationship. Similar results have been reported by Castro et al. (2008).

According to Hevia et al. (2003) the NFOM presented high correlations with the clay content only in the soils UC (Figure 4). This showed that the addition of plant debris and other residues with high carbon content alter the relation between SOC and the fine-size particle. The C:N relationship indicates the rate of the mineralization of the soil organic matter. According to Giménez and Bratos (1985), C:N values higher than 15 indicate a very low N release. Our results show high values at the uppermost soil levels of PD+CR, and thus humification processes predominated. At greater depths and in UC, mineralization processes predominating (values lower than 10). It bears mentioning that a high C:N relationship does not necessarily signify N deficiencies in soil, as pointed out by Rhoton et al. (1993), as the progressive increase in organic matter of the soil augments the availability of many nutrients, including N. In this case, the high superficial values cause SOC to act as a protector and store of N, releasing this nutrient little by little into the soil.

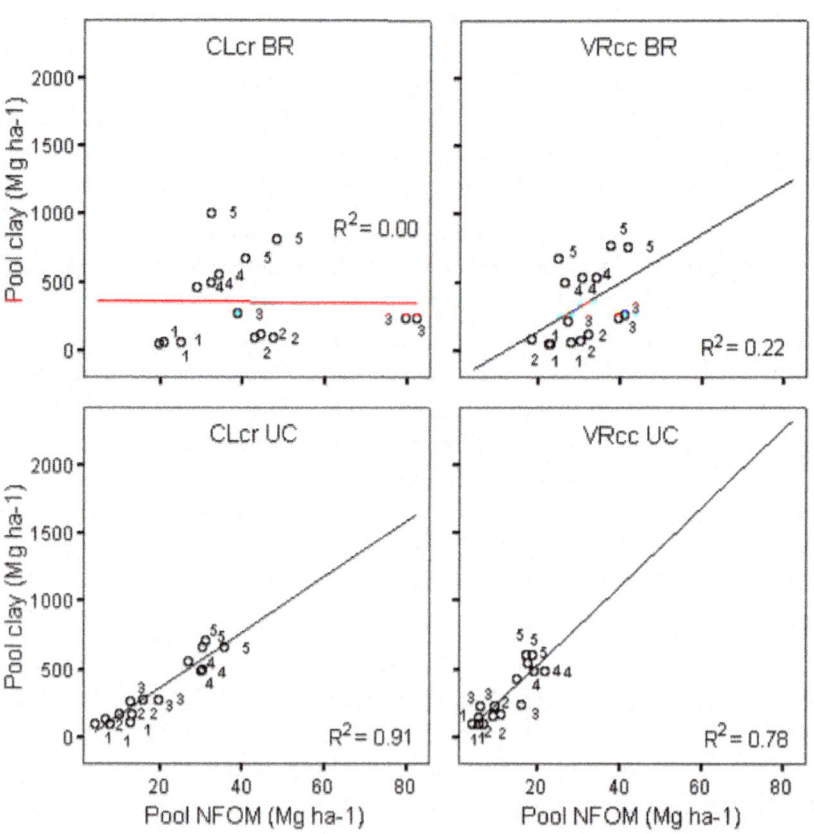

Figure 4. Linear regression between clay and NFOM (non-floatable organic matter) for each soil type and location (1: 0-2 cm, 2: 2-5 cm, 3: 5-10 cm, 4: 10-15 cm, 5: 15-30 cm).

One of the most important characteristics of this management is the continued non-tillage, which affects the distribution of certain properties such as SOC throughout the soil profile. Franzluebbers (2002) showed that SOC stratification ratios higher than 2 are infrequent under degradation conditions. The data compiled in our study, lower than 2 in conventional tillage and higher in PD+CR cover, confirm the increase in soil quality as well as a control of erosion and degradation processes.

The high values of ρb UC indicated a compaction process related to cultivation (Soria et al., 2005). Between rows, the spreading of olive-pruning debris and the maintenance of the material on the surface (non-tillage) generated a mulching effect (Bescansa, 2006; Lal, 2008) that protected the soil effectively with respect to nutrient loss and sediment production (Rodríguez-Lizana et al., 2008). Thus, ρb diminished in the uppermost 2 cm of PD+CR by 44% in CLcr and by 31% in VRcc, with respect to UC. This difference between the two soil types is related to the SOC content, determined by the application time of the plant residues. Similar results have been reported by Ordóñez et al. (2001) after 6 years of applying olivepruning debris.

The K^+ content raised with the SOC (Rhoton et al., 1993; Thomas et al., 2007) by 3.4-fold in the VRcc and by 1.5-fold in the CLcr over the entire depth studied with respect to UC. This higher increase observed in the VRcc was linked to the greater content in smectite clays (Bhonsle et al., 1992; Ghosh & Singh, 2001).

According to Oorts et al. (2003) organic matter can be responsible for as much as 85% of the CEC of the soil. The correlation between CEC and SOC found in the present work (Table 6) was similar to those documented by Caravaca et al. (1999) for calcareous soils. The higher values of CEC were detected in VRcc, with lower duration of the experiment and greater fine-fraction content than CLcr (Table 3). These results coincide with those of Leinweber et al. (1993), who indicated that the order of factors that affect the CEC is: particle size of the fraction, management, and duration of the experiment.

The water-storage capacity of the soil changed too with the organic matter content since the pore size and distribution changes (Rawls et al., 2003; Bescansa et al., 2006). However, the effect of SOC on water retention was high in sandy soils and marginal in fine-textured soils (Bauer & Black, 1981). Our results match those of Rawls et al. (2003) in the regression tree for the SWC at -33 and -1500 kPa and in the variation of the moisture with the change in organic-carbon content, which, for high SOC contents, increased for all the textures but in a higher proportion with greater sand contents, as occurred in our case in the CLcr.

Soil Carbon Sequestration

The effectiveness of changing the management from traditional tillage to cover crop between trees is manifest in this study in the resulting high SOC and N values. The SOC rose from 34.0 and 46.2 Mg ha^{-1} to 158.0 and 113.6 Mg ha^{-1} for CLcr and VRcc, respectively, indicating greater in soil fertility.

The RothC model estimated the carbon input into the soil of the tilled olive grove in equilibrium as being 1.0 Mg C ha^{-1} yr^{1}, which falls within the range of 1-2 Mg C ha^{-1} yr^{1} estimated by Jenkinson & Rayner (1977). Romanyà et al. (2000) registered similar results for a vineyard in the Mediterranean area, with an annual carbon input of 1.4 Mg C ha^{-1} yr^{1}. The only estimates of carbon input for olive groves were reported by Sofo et al. (2005), who registered an annual input as senescent leaves of 0.4 Mg C ha^{-1} yr^{1} but did not account for other inputs such as root turnover and rhizodeposition.

In our work, the SOC content modelled for the VRcc soil (Figure 3) was higher than that predicted for the CLcr after eight years of mulching. This might be explained by the higher percentage of clay in the VRcc soil, which is masked by the large quantities of organic matter added in the form of PD+CR. The relationships between fine soil fractions and organic-carbon sequestration have been addressed by other authors such as Paustian et al. (2000). It has been noted that carbon is physically protected against biodegradation when it is contained in clay- or silt-sized micro-aggregates (Balesdent et al., 2000).

The RothC model fitted carbon turnover satisfactorily for the change from tillage to PD+CR cover due to the absence of erosion (Gottschalk et al., 2010). During the 10 and 6 years that the experiment lasted, we registered a much higher annual input of residues into the soil between trees than that reported by other authors, due to the fact that our management system included the addition of cleaning residues as well as pruning debris. For this scenario, RothC predicts an annual input of 25.3 and 23.6 Mg C ha^{-1} yr^{-1} for CLcr and VRcc soils, respectively (Table 8), thus confirming that the model closely fits this kind of management.

Turnover time is defined as the migration of organic carbon through a given volume of soil (Jenkinson & Rayner, 1977). The high values found in tillage (Table 8) indicate carbon stabilization in the soil, signifying that carbon migrates slowly from one pool to another. Our results are higher than those modelled for Kenyan savanna (~16 yr) or dry forest in Zambia (~8 yr) by Jenkinson et al. (1999). For PD+CR the turnover times were lower than those found by these latter authors. This showed rapid migration of carbon and thus a soil that is not close to equilibrium.

During the first years of olive-tree establishment, CO_2 is distributed preferentially in the permanent structures and the root system, but in mature olives trees, fixed CO_2 is located to a greater extent in the leaves and fruit, and consequently also in the pruning debris (Sofo et al., 2005). Thus pruning debris is an important carbon reservoir that can be returned to the soil rather than the atmosphere in the form of CO_2 (Figure 3). When the pruning debris is shredded and spread on the ground, the RothC model predicts a decrease in CO_2 emission of ~2 Mg CO_2 ha^{-1} yr^{-1} for each soil (Table 10). This value is within the range of the data collected by Smith et al. (2008) for warm-dry zones, who registered an emission reduction of 3.45 Mg CO_2 ha^{-1} yr^{-1} when degraded lands are restored and 0.33 Mg CO_2 ha^{-1} yr^{-1} for croplands under tillage and residue management.

Abandoning tillage in favour of using organic waste to cover the ground is considered to be an efficient way of increasing carbon sequestration in agricultural soils (cf. Smith et al., 2008; Lal, 2008). In our experiment, we measured a carbon sequestration of 0.5 and 0.6 Mg C ha^{-1} yr^{-1} with pruning debris cover in CLcr and VRcc soils, respectively (Table 10). Fairly wide ranges have been estimated for carbon sequestration in agricultural soils. Álvaro-Fuentes et al. (2009) measured a carbon sequestration of 0.46 and 0.15 Mg C ha^{-1} yr^{-1} for continuous barley and barley-fallow rotation in Mediterranean area, respectively. For the European area, Smith et al. (2000) estimated a carbon sequestration of 0.7 Mg C ha^{-1} yr^{-1} with crop residues and 0.4 Mg C ha^{-1} yr^{-1} with no tillage. Using organic residues Hutchinson et al. (2007) registered average rates of potential carbon gain from 0.1 to 0.5 Mg C ha^{-1} yr^{-1}. After changes in cropland use, and introduction of the best series of management techniques for every land use and climate zone, IPCC (2000) suggested a carbon sequestration potential of 0.3 Mg C ha^{-1} yr^{-1}. Our results coincide with these values, supporting the idea that the recycling of pruning debris in olive groves is an effective way of storing carbon in the soil. As with SOC, carbon sequestration was greater in VRcc soil, with its higher clay content.

Apart from the carbon sequestration from the recycling of pruning debris, the addition of the fruit-cleaning residues and the absence of tilling resulted in an increase in SOC (Figure 3). Although it may be difficult to introduce the reuse of the large quantities of CR + PD described in this work as a commonplace practice in olive-grove management, merely recycling the pruning, which is produced in the grove itself, together with a policy of zero tillage, will increase the SOC content considerably compared to T soils.

CONCLUSIONS

This work shows the improvement in soil quality and fertility after the soil management in olive orchards was changed from conventional tillage to non-tillage with plant residues cover. The application of shredded olive-pruning debris and the plant residues and soil from olive-fruit cleaning increased the organic fraction in both soils, CLcr and VRcc, with respect the tillage soils. The changes affected the uppermost 10 cm of the soils although the SOC content was greater in the CLcr, where the management spanned a longer time period.

With the change in SOC, some soil properties were affected. The spreading of plant residues lowered the pH and the ρb in the uppermost soil depths between trees, where there was an effect of mulching with respect to under canopy soils, and the N content increased. The slow release of this N ensured plant nutrition and the soil fertility despite the high values of the C:N relationship.

The non-tillage generated a highly fertile surface layer in PD+CR, which also protected the soil physically, as demonstrated by the high stratification observed, similar to that described in forest systems. In addition, the NFOM percentage, although higher under the tree canopy, was high in both types of soils, indicating the stability of the material applied. The addition of plant debris also increased the K^+ content, CEC and water-storage capacity, improving the soil quality.

When the soil management was changed from conventional tillage to cover crop with PD+CR, carbon storage in the soil improved considerably together with its general quality. Carbon turnover in Mediterranean olive-grove with PD+CR cover was quite accurately predicted by the RothC model. Over the long term, the carbon sequestration was higher in soils with greater quantities of clay. A soil-management system that abandons tillage in favour of reusing both pruning debris and the residue from cleaning the olives to cover the ground, constitutes the most effective way of increasing soil quality and diminishing CO_2 emissions in one of the most extensive agricultural enterprises in the entire Mediterranean area.

ACKNOWLEDGMENTS

We would like to thank F. J. Garcia de Zuñiga, the farm owner, for his collaboration. The work has been financed with investigation projects INIA RTA2007-010-C3-01.

REFERENCES

1. Angers, D.A., Bolinder, M.A., Carter, M.R., Gregorich, E.G., Drury, C.F., Liang, B.C., Voroney, R.P., Simard, R.R., Donald, R.G., Beyaert, R.P. & Martel, J. (1997). Impact of tillage practices on organic carbon and nitrogen storage in cool, humid soils of eastern Canada. Soil and Tillage Research, Vol. 41, pp. 191-201.

2. Álvaro-Fuentes, J., López, M.V., Arrúe, J.L., Moret, D. & Paustian, K. (2009). Tillage and cropping effects on soil organic carbon in Mediterranean semiarid agroecosystems: Testing the Century model. Agriculture, Ecosystems and Environment, Vol. 134, pp. 211-217.

3. Arrouays, D., Saby, N., Walter, C., Lemercier, B. & Schvartz, C. (2006). Relationships between particle-size distribution and organic carbon in French arable topsoils. Soil Use and Management, Vol. 22, pp. 48-51.

4. Balesdent, J., Chenu, C. & Balabane, M. (2000). Relationship of soil organic matter dynamics to physical protection and tillage. Soil and Tillage Research, Vol. 53, pp. 215-230.

5. Bauer, A. & Black, A.L. 1981. Soil carbon, nitrogen, and bulk density comparisons in two cropland tillage systems after 25 years and in virgin grassland. Soil Science Society of America Journal, Vol. 45, pp. 1166-1170.

6. Bescansa, P., Imaz, M.J., Virto, I., Enrique, A. & Hoogmoed, W.B. (2006). Soil water retention as affected by tillage and residue management in semiarid Spain. Soil and Tillage Research, Vol. 87, pp. 19-27.

7. Bhonsle, N.S., Pal, S.K. & Sekhon, G.S. (1992). Relationship of K forms and release characteristics with clay mineralogy. Geoderma, 54, pp. 285-293.

8. Blake, G.R. & Hartge K.H. (1986). Bulk density. In: Klute, A. (Ed.), Methods of Soil Analysis. Part 1. Physical and Mineralogical Methods. 2nd ed. ASA, SSSA Monograph No. 9, Madison, pp. 363-375.

9. Bremner, J.M. (1965). Nitrogen availability indexes. In: Black, C.A., Evans, D.D., Esminger, T.E., Clark, F.E. (Eds.), Methods of Soil Analysis. Part. 2. Chemical and Microbiological Properties. American Society Agronomy, Madison, pp. 1324-1345.

10. Buyanovsky, G.A., Aslam, M. & Wagner, G.H. (1994). Carbon turnover in soil physical fractions. Soil Science Society of America Journal, Vol. 58, pp. 1167-1174.

11. Caravaca, F., Lax, A. & Albaladejo, J. (1999). Organic matter, nutrient contents and cation exchange capacity in fine fractions from semiarid calcareous soils. Geoderma, 93, pp. 161-176.

12. Cassel, D.K. & Nielsen, D.R. (1986). Fields capacity and available water capacity. In: Klute, A. (Ed.), Methods of Soil Analysis. Part 1. Physical and Mineralogical Methods. 2nd ed. ASA, SSSA Monograph No. 9, Madison, pp. 901-926.

13. Castro, J., Fernández-Ondoño, E., Rodríguez, C., Lallena, A.M., Sierra, M. & Aguilar, J. (2008). Effects of different olive-grove management systems on the organic carbon and nitrogen content of the soil in Jaén (Spain). Soil and Tillage Research, Vol. 98, pp. 56-67.

14. Coleman, K. & Jenkinson, D.S. (1996). RothC-26.3 – a model for the turnover of carbon in soil. En D.S. Powlson, P. Smith, J.U. Smith, eds. Evaluation of soil organic matter models using existing, long-term datasets. Springer-Verlag, Berlin, pp. 237-246.

15. Falloon, P., Smith, P., Coleman, K. & Marshall, S. (1998). Estimating the size of inert organic matter pool from total soil organic carbon content for use the Rothamsted Carbon Model. Soil Biology and Biochemistry, Vol. 30, pp. 1207-1211.

16. Falloon, P. & Smith, P. (2002). Simulating SOC changes in long-term experiments with RothC and CENTURY: model evaluation for a regional scale application. Soil Use and Management, Vol. 18, pp. 101-111.

17. FAO (2006). World Reference Base for Soil Resources. IUSS – ISRIC – FAO, Roma, 117 p.

18. Fontaine, S., Bardoux, G., Abbadie, L. & Mariotti, A. (2004). Carbon input to soil may decrease soil carbon content. Ecology Letters, Vol. 7, pp. 314-320.

19. Franzluebbers, A.J. (2002). Water infiltration and soil structure related to organic matter and its stratification with depth. Soil and Tillage Research, Vol. 66, pp. 197-205.

20. Ghosh B.N. & Singh, R.D. (2001). Potassium release characteristics of some soils of Uttar Pradesh hills varying in altitude and their relationship with forms of soil K and clay mineralogy. Geoderma, Vol. 104, pp. 135-144.

21. Giménez, M. & Bratos, J. (1985). Análisis de suelos. Ed. ERT. Spain.

22. Gómez, J.A, Sobrinho, T.A., Giráldez, J.V. & Fereres, E. (2009) Soil-management effects on runoff, erosion and soil properties in an olive grove of Southern Spain. Soil and Tillage Research, Vol. 102, pp. 5-13.

23. Gottschalk, P., Bellarby, J., Chenu, C., Foereid, B., Smith, P., Wattenbach, M., Zingore, S. & Smith, J. (2010). Simulation of soil organic carbon response at forest cultivation sequences using 13C measurements. Organic Geochemistry, Vol. 41, pp. 41-54.

24. Hernández, A.J., Lacasta, C. & Pastor, J. (2005). Effects of different management practices on soil conservation and soil water in a rainfed olive orchard. Agricultural Water Management, Vol. 77, pp. 232-248.

25. Hernanz, J.L., López, R., Navarrete, L. & Sánchez-Girón, V. (2002). Long-term effects of tillage systems and rotations on soil structural stability and organic carbon stratification in semiarid central Spain. Soil and Tillage Research, Vol. 66, pp. 129-141.

26. Hevia, G.G., Buschiazzo, D.E., Hepper, E.N., Urioste, A.M. & Antón, W.L. (2003). Organic matter in size fractions of soils of the semiarid Argentina. Effects of climate soil texture and management. Geoderma, Vol. 116, pp. 265-277.

27. Hutchinson, J.J., Campbell, C.A. & Desjardins, R.L. (2007). Some perspectives on carbon sequestration in agriculture. Agricultural and Forest Meteorololgy, Vol. 142, pp. 288- 302.

28. Ingram, J.S. & Fernandes, E.C.M. (2001). Managing carbon sequestration in soils: concepts and terminology. Agriculture, Ecosystems and Environment, Vol. 87, pp. 111–117.

29. IPCC (2000). Special report on land use, land-use change and forestry. Cambridge University Press, Cambridge.

30. Jarecki, M.K. & Lal, R. (2003). Crop management for soil carbon sequestration. Critical Reviews in Plant Sciences, Vol. 22, No. 5, pp. 471-502.

31. Jarecki, M.K. & Lal, R. (2005). Soil organic carbon sequestration rates in two long-term no-till experiments in Ohio. Soil Science, Vol. 170 No. 4, pp. 280-291.

32. Jenkinson, D.S. & Rayner, H. (1977). The turnover of soil organic matter in some of the Rothamsted classical experiments. Soil Science, Vol. 123, No. 5, pp. 298-305.

33. Jenkinson, D.S., Meredith, J., Kinyamario, J.I., Warren, G.P., Wong, M.T.F., Harkness, D.D., Bol, R. & Coleman, K. (1999). Estimating net primary production from measurements made on soil organic matter. Ecology, Vol. 80 No. 8, pp. 2762-2773.

34. Lal, R. (2008). Soils and sustainable agriculture. A review. Agronomy for Sustainable Development, Vol. 28, pp. 57-64.

35. Leinweber, P., Reuter, G. & Brozio, K. (1993). Cation exchange capacities of organo-mineral particle-size fractions in soils from long-term experiments. Journal of Soil Science, Vol. 44, pp. 111-119.

36. Ministerio de Agricultura, Pesca y Alimentación - MAPA (1989). Caracterización agroclimática de la provincia de Jaén. Dirección General de Producción Agraria, Madrid.

37. Nieto, O.M., Castro, J., Fernández, E. & Smith, P. (2010). Simulation of soil organic carbon stocks in a Mediterranean olive grove under different soil-management system using the RothC model. Soil Use and Management, Vol. 26, pp. 118-125.

38. Oorts, K., Vanlauwe, B. & Merckx, R. (2003). Cation exchange capacities of soil organic matter fractions in a Ferric Lixisol with different organic matter inputs. Agriculture, Ecosystems and Environment, Vol. 100, pp. 161-171.

39. Ordóñez, R., Ramos, F.J., González, P., Pastor, M. & Giráldez, J.V. (2001). Influencia de la aplicación continuada de restos de poda de olivo sobre las propiedades físicoquímicas de un suelo de olivar. In: López, J.J., Quemada, M. (Eds.), Temas de Investigación en la Zona no Saturada, Vol. 5. Univ. Pública Navarra, Spain.

40. Pastor, M. (2004). Sistemas de manejo del suelo. In: Barranco, D., Fernández-Escobar, R., Rallo, T. (Eds.), El cultivo del olivo. Mundi-Prensa, Spain, pp. 231-285.

41. Paustian, K., Six, J., Elliott, E.T. & Hunt, H.W. (2000). Management options for reducing CO_2 emissions from agricultural soils. Biogeochemistry, Vol. 48, pp. 147-163.

42. Percival, H.J., Parfitt, R.T. & Scott, A.N. (2000). Factors controlling soil carbon levels in New Zealand grasslands: is clay content important? Soil Science Society of America Journal, Vol. 64, pp. 1623-1630.

43. Rawls, W.J., Pachepsky, Y.A., Ritchie, J.C., Sobecki, T.M. & Bloodworth, H. (2003). Effect of soil organic carbon on soil water retention. Geoderma, Vol. 116, pp. 61-76.

44. Rhoton, F.E., Bruce, R.R., Buehring, N.W., Elkins, G.B., Langdale, C.W. & Tyler, D.D. (1993). Chemical and physical characteristics of four soil types under conventional and notillage systems. Soil and Tillage Research, Vol. 28, pp. 51-61.

45. Rodríguez-Lizana, A., Espejo-Pérez, A.J., González-Fernández, P. & Ordóñez- Fernández, R. (2008). Pruning residues as an alternative to traditional tillage to reduce erosion and pollutant dispersion in olive groves. Water, Air and Soil Pollution, Vol. 193, pp. 165-173.

46. Romanyà, J., Cortina, J., Falloon, P., Coleman, K. & Smith, P. (2000). Modelling changes in soil organic matter after planting fast-growing Pinus radiata on Mediterranean agricultural soils. European Journal of Soil Science, Vol. 51, pp. 627-641.

47. Schomberg, H.H. & Steiner, J.L. (1999). Nutrient dynamics of crop residues decomposing on a fallow no-till soil surface. Soil Science Society of American Journal, Vol. 63, pp. 607- 613.

48. Six, J., Elliott, E.T. & Paustian, K. (1999). Aggregate and SOM dynamics under conventional and no-tillage system. Soil Science Society of American Journal, Vol. 63, pp. 1350-1358.

49. Smith, P., Nabuurs, G.J., Janssens, I.A., Reis, S., Marland, G., Soussana, J.F., Christensen, T.R., Heath, L., Apps, M., Alexeyev, V., Fang, J., Gattuso, J.P., Guerschman, J.P., Huang, Y., Jobbagy, E., Murdiyarso, D., Ni, J., Nobre, A., Peng, C., Walcroft, A., Wang, S.Q., Pan, Y. & Zhou, G.S. (2008). Sectoral approaches to improve regional carbon budgets. Climatic Change, Vol. 88, pp. 209-249.

50. Sofo, A., Nuzzo, V., Palese, A. M., Xiloyannis, C., Celano, G., Zukowskyj, P. & Dichio, B. (2005). Net CO_2 storage in Mediterranean olive and peach orchards. Scientia Horticulturae, Vol. 107, pp. 17-24.

51. Soil Conservation Service (SCS). 1972. Soil Survey Laboratory Methods and Procedures for Collecting Soil Samples, Soil Survey Report 1. US Department of Agriculture: Washington, DC.

52. Soria, L. (2002). Fertilización y riego en el olivar de la provincia de Jaén: comarcas de La Loma y Sierra Morena. Ph.D. thesis, University of Jaén, Spain.

53. Soria, L., Fernández, E., Pastor, M., Aguilar, J. & Muñoz, J.A. (2005). Impact of olive-orchard cropping systems on some soil physical and chemicals properties in southern Spain. In: Faz, A., Ortiz, R, Mermut, A.R. (Eds.), Advances in GeoEcology, Catena Verlag, Vol. 36, pp. 428-435.

54. Thomas, G.A., Dalal, R.C. & Standley, J. (2007). No-till effects on organic matter, pH, cation exchange capacity and nutrient distribution in a Luvisol in the semi-arid subtropics. Soil and Tillage Research, Vol. 94, pp. 295-304.

55. Tyurin, I.V. (1951). Analytical procedure for a comparative study of soil humus. Trudy. Pochr. Inst. Dokuchaeva 38.

56. Williams, D.E. (1948). A rapid manometric method for the determination of carbonate in soils. Soil Science Society American Proceeding, Vol. 13, pp. 27-129.

Chapter 8

SOIL CARBON SEQUESTRATION UNDER BIOENERGY CROPS IN POLAND

Magdalena Borzecka-Walker, Antoni Faber, Katarzyna Mizak, Rafal Pudelko, and Alina Syp
Institute of Soil Science and Plant Cultivation-State Research Institute Poland

INTRODUCTION

Agriculture practices have an important role to play in mitigating climate change due to atmospheric enrichment of carbon dioxide, and other greenhouse gases (GHG). Land management can strongly influence soil carbon stocks and careful management can be used to sequestered soil carbon. It is important to propose contemporary management practises to farming, like the conversion from a tillage system to no-tillage, incorporation of cover crops and forages in the crop rotation, use of crop residues and biosolids e.g. mulch, implementation of biocrops, as well as integrated nutrient management which including compost/manures as well as the precision use of fertilizers and integrated pest management. Sustainable management in agriculture should reduce and avoid the introduction of carbon dioxide (CO_2) to the atmosphere, which is one of three most prevalent GHGs directly emitted by human activity. CO_2 is the most important anthropogenic GHG, and according to IPCC Fourth Assessment Report (2007), anthropogenic CO_2 emissions grew by about 80% between 1970 and 2004. Carbon sequestration is a process through which agricultural and forestry practices remove carbon dioxide (CO_2) from the atmosphere into a form that does not affect atmospheric chemistry (Lal, 2004a). A natural way to trap atmospheric CO_2 is by photosynthesis, where carbon dioxide is absorbed by plants and turned into carbon compounds, stored or fixed C as soil organic carbon (SOC). The SOC pool consist litter, humads and humus, which it is comprised of mixtures of plant and animal residues at various stages of decomposition along with microbial by-products (Lal, 2004a). Agriculture is responsible for 13.5% of global anthropogenic GHG emissions (IPCC, 2007), but if sustainable land management practices are implemented, agricultural soils could become a carbon sink (Dumanski et al., 1998). There are five principal global carbon pools. The oceanic pool (38 Gt) is the largest, followed by the geologic (5 Gt), pedologic (2.5 Gt), biotic (0.56 Gt), and the atmospheric pool (0.76 Gt). The soils beneath the oceans are the most important reservoir of carbon in the terrestrial biosphere and contain three times the amount as compared with those that are found in vegetation (Lal, 2004b; SEC, 2009). Soils contain more than twice the carbon that can be found in the atmosphere and the loss of carbon from soils can have a significant effect on atmospheric CO_2 concentrations, which can influence the climate (Smith, 2008). Many studies have examined the sequestration potential in agriculture and forestry in Europe (Smith et al., 1997; Smith et al., 2000; Vleeshouwers & Verhagen, 2002; Freibauer et al., 2004; Smith, 2004;) and globally (Smith 2004; IPCC 2007; Lal, 2004a), as well as in other regions of the world such as North America (Dumanski et al., 1998; Franzluebbers & Follett, 2005) or Africa (Ringius, 2002). The potential for carbon sequestration in the European Union (EU) is approximately 90–120 Mt C/y, in the US cropland is 75–208 Mt C/y, in Canada is approximately 24 Mt C/y, to obtain this potential, optimal land management practices have to be implemented (Hutchinson et al., 2007). It is estimated that the global potential scale of carbon sequestration in soils used for agricultural purposes is around 0.3 t C/ha/y on arable lands, and around 0.5 – 0.7 t C/ha/y on grasslands (IPCC, 2000). The conducted researches indicate existence of a high potential for carbon sequestration in soils under agricultural crops. Depending on the used method for its evaluation and it range between from 0.15- 0.22 t C/ha/y for willow (Bradley & King, 2004) up to 0.93 t C/ha/y for Miscanthus (Matthews & Grogan 2001). The net soil carbon sequestration simulated for biocrops in Poland was around 0.38 – 0.95 t C/ha/y Miscanthus crops and 0.22 – 0.39 t C/ha/y for willow coppice (Borzecka-Walker et al., 2008).

There are many policies, directives, standards, as well as norms in the EU designed to stimulate and support the reduction of GHG emission and to improve the carbon mitigation potential. The publication of a Green paper "Towards a European strategy for the security of energy supply" (2000) started a debate on energy security, which is considered a key element of politico-economic independence of the EU. It stressed the need to improve the organisation's strategic stocks of raw materials and coordinate its use. Additionally, the European Commission presented a White Paper that sets out the actions necessary to strengthen the Union's ability to adapt to a changing climate. To support the biofuels industry, the Energy Taxation Directive allows exemptions or reductions from energy taxation for biofuels (Directive,

2003/96/EC). The aims of the recently released European Parliament and the Council directive on the promotion of the use of energy from renewable sources amending and subsequently repealed Directives 2001/77/EC and 2003/30/EC (Directive 2009/28/EC); are to achieve by 2020 a 20% share of energy from renewable sources in the EU's final consumption of energy and a 10% share of energy from renewable sources in each member state's transport energy consumption. Moreover the GHG emission saving from the use of biofules and bioliguids shall be at least 35%, 50% in 2017 and 60% in 2018 yrs.

The aim of this review is an evaluation into the current knowledge of carbon sequestration and to present potential bioenergy crops for carbon sequestration in Poland.

THE SOIL'S ORGANIC MATTER BALANCE

There are several methods and simulation models useful for defining the content of soil organic matter. The following work will present two methods which were applied for the Polish territory. It should be stressed that the results obtained by using both methods are comparable. Based on these results, it can be concluded that the coefficient method used in the assessment of carbon balance has a small error and gives equally reliable results as while using the soil profiles method.

The Soil's Organic Matter Balance Based on the Determination of Soil Profiles

The content of organic matter in soils of agricultural land is highly variable. The results of determinations carried out in Poland show that it varies in the arable layer within the limits of 0.5-10% with an average of 2.2%. According to the division used in Polish soil with low humus content (2% of approximately 33%. The global balance of organic matter in Polish soils is negative in all regions (-0.06 to - 1.05 t C/ha/y), with the average for the country of -0.47 t C/ ha/y. This means that in large areas of Poland we note the CO_2 emissions from soil to the atmosphere (Terelak et al., 2001; Stuczynski et al., 2007). In the years 2000-2004, a preliminary analysis of soil humus content trends were carried out under repeated testing of standard profiles. Studies have shown a significant decline in humus, mainly in soils initially rich in organic matter. A decline in soil organic matter is associated with the change of soil water relations, i.e. more intensive use and drainage. In contrast, a large part of the light soils of the last 30 years recorded an increase of humus content associated with an increased level of fertilization and increase in quantity of crop residues (Stuczynski et al., 2007). Based on measurements taken in the years 1968-1983 and in 2003, the changes in soil organic matter and humus loss risk were able to be calculated. The results presented in figure 1 show both accumulation and humus loss in soil as well as soil organic matter balance. The highest losses of soil organic matter were calculated for the Kujawsko-pomorskie voivodeship, whilst the lowest was in the Malopolskie voivodeship. Voivodeships of the North Western part of the country have the lowest soil organic balance, and this indicates a greater share of soils with a higher risk of loss of function due to mineralisation of soil humus.

The Soils Organic Matter Balance Based on Coefficients

The amount of organic matter in soils is a key indicator for the quality, and is significant for their physicochemical properties such as sorption. Maintaining high humus content in soil is important because of its impact on soil carbon sequestration. Increasingly popular intensive use of soils, combined with a simplified crop rotation, increased predominance of cereal plants with reduced amounts of livestock, leads to a reduction in the amount of organic residues entering the soil, which in turn leads to reduced carbon sequestration in soil.

The basic principle of good farm management is to maintain a positive, or at least a sustainable balance of soil organic matter. This balance can be obtained by the selection of species of cultivation plants, their participation in the crop structure, and the quantity of manure and organic. The various species of crops leave different amounts of crop residue. The soil carbon in cropland can be increased by planting more forages, and increasing residue inputs from plants with high biomass potential. Approximately, it can be concluded that the weight of cereal crop residues is about 3-fold greater than the root, and legumes with grasses, by up to 6-fold. In addition, a different duration and degree of shading the soil surface and the number of tillage performed and care, which affects the mineralisation of humus.

The cultivated plants can be separated in three groups depending of the impact on the balance of humus in the soil.

The first group includes plants with a potential in enriching the soil with organic matter. Among them are primarily long-term forage legumes and their mixtures with grasses and grasses grown in the field as well as crops grown for energy sources like tall grasses, fast growing trees. In addition, legumes and intercrops ploughed as green fertilizers have little positive effect on the balance of humus. The reproduction rate of soil for this group of plants ranges from 0.21 to 2.10, depending on the type of soil they are grown on (Fotyma & Mercik, 1992).

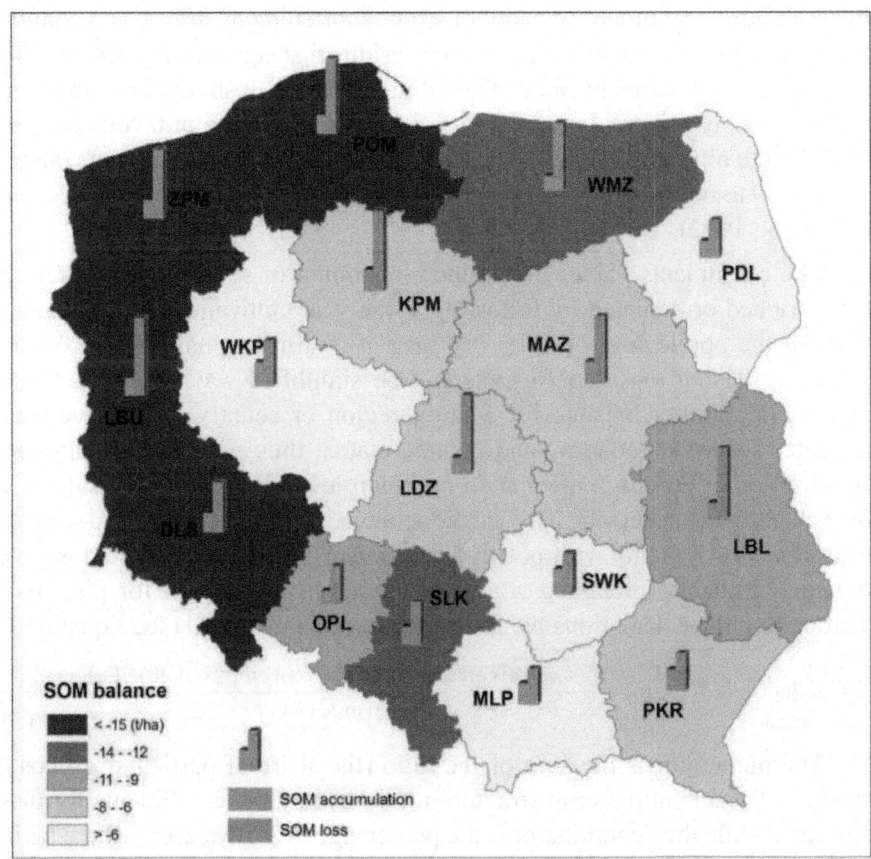

Figure 1. Forecast loss of soil organic matter (SOM) from agricultural land1. Source: own work based on Stuczynski et al., (2007).

The second group includes plants with a potential in degrading the soil organic matter. This group includes mainly root crops, root vegetables and corn. The characteristics for this group of crops is very little crop residue, seeded in wide rows, intercrops, heavy maintenance, and the short canopy (cover spacing) increases the distribution of humus resulting in increased erosion. The soil degradation rate for this group of plants ranges from -0.12 to -1.54, and is dependent on the type of soil they are grown on (Fotyma & Mercik, 1992). The mineralisation for these types of plants per year is about 1.0-1.5 t/ha of humus. To compensate for this loss about 15-16 t/ha manure should be used.

The third group include plants with a small negative or neutral impact on soil organic matter. This group of plants includes cereals and oilseeds. Cereals previously were treated as plants degrading the soil organic matter, but changes in agricultural techniques (density of straw, shortening of straw), and combine harvester collection, leaves a lot of crop residue that significantly reduces their negative impact on the balance of soil organic matter. It should be emphasized that the quality of cereal crop residues is worse due to the unfavourable ratio of carbon to nitrogen. The soil degradation rate for this group of plants ranging from -0.49 to -0.56, is dependent on the type of soil they are grown on (Fotyma & Mercik, 1992).

The coefficients values determine the amount of soil organic matter t/ha can enriched or depleted by following a one-year cultivation of the plants or through the application of 1t/ha dry matter of different natural and organic fertilizers. Using these coefficients can be simplified way to determine the soil organic matter balance for a farm, region or country. A positive result indicates a normal economy and organic matter, thus ensuring the long-term stabilisation of humus content at an optimum level. If the balance is negative then changes are necessary. This can be achieved by changing the crop structure (introduction of plants with positive coefficient), or increasing the dosage of organic fertilizers (ploughed straw) or intercrops cultivation for ploughing. Throughout the calculations the following formula was used (see Equation 1):

$$\text{Degradation coefficient} = \frac{\sum (\% \text{ cereals area } x - 0.53) + (\% \text{ root crops} \times -1.40) + (...)}{\text{sown area (\%)}} \quad (1)$$

The numerator is the sum of the ratio (the share of particular groups or species of plants in the crop structure multiplied by the coefficients for these species), while the denominator is the percentage of a sown area (where taking into account all the sown land as 100%). Based on those coefficients following the agricultural use of arable land, allows us to calculate the decreases in the soil organic matter amount by 0.39 to 0.66 t/ha for particular voivodeships and approximately 0.53 tonnes per one ha per year in Poland (Kus et al., 2006). Figure 2 presents carto-diagram which shows spatial differentiation, presented by standard deviation methods.

The best situation of soil organic matter was calculated for the voivodeships of Warminskomazurskie, Podlaskie followed by the Malopolskie, and their positive situation is associated with a high share of legumes or their mixtures with grasses. An adverse situation appears in the Dolnoslaskie and Opolskie voivodeships, where there is a large share of root crops and maize. To offset this loss, approximately six tons of manure should be applied on every hectare of arable land used. The calculations (Kus et al., 2006) show that the national average production of natural fertilizers (manure) was approximately 7.3 t/ha of sowing area (Fig. 3).

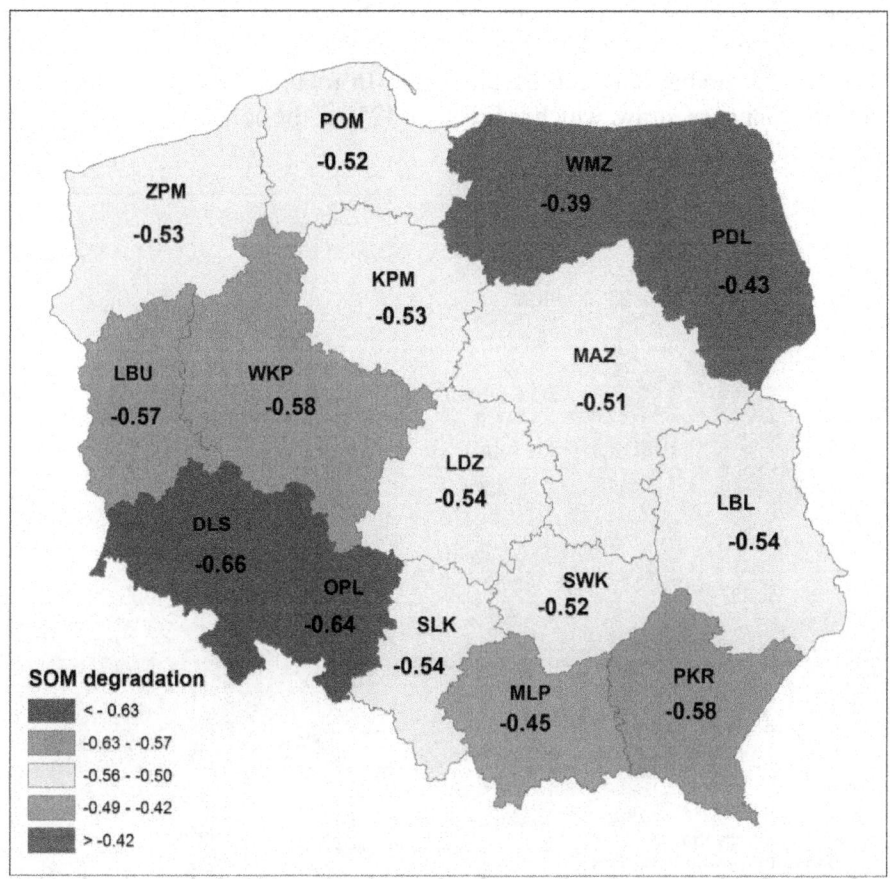

Figure 2. The coefficient values of soil organic matter degradation for the individual regions calculated for the cropping system average for 2002-2005[1]. Source: own work based on Kus et al., (2006).

The highest amount of manure was produced in the Podlaskie voivodeship due to high livestock and low sowing area. This led to a significant surplus of soil organic matter in the Podlaskie voivodeship. The lowest production of manure was in the Dolnosląskie and Zachodniopomorskie voivodeships where the sowing area is high and livestock low.

In some voivodeships with a negative value of SOM, a balance can be best achieved through ploughing the straw. Particularly large quantities of straw can be ploughed in four voivodeships, (0.9 - 1.0 t/ha in Opolskie and Lubuskie, about 1.2 t/ha in Zachodniopomorskie and to 1.9 t/ha Dolnoslaskie (Fig. 4). However, in some voivodeships such as Lubelskie, only 0.2 tons of the straw per 1 ha of arable land can be ploughed. In total, across the country three million tonnes of straw, which is less than 12% of the collected straw, could be allocated for ploughing.

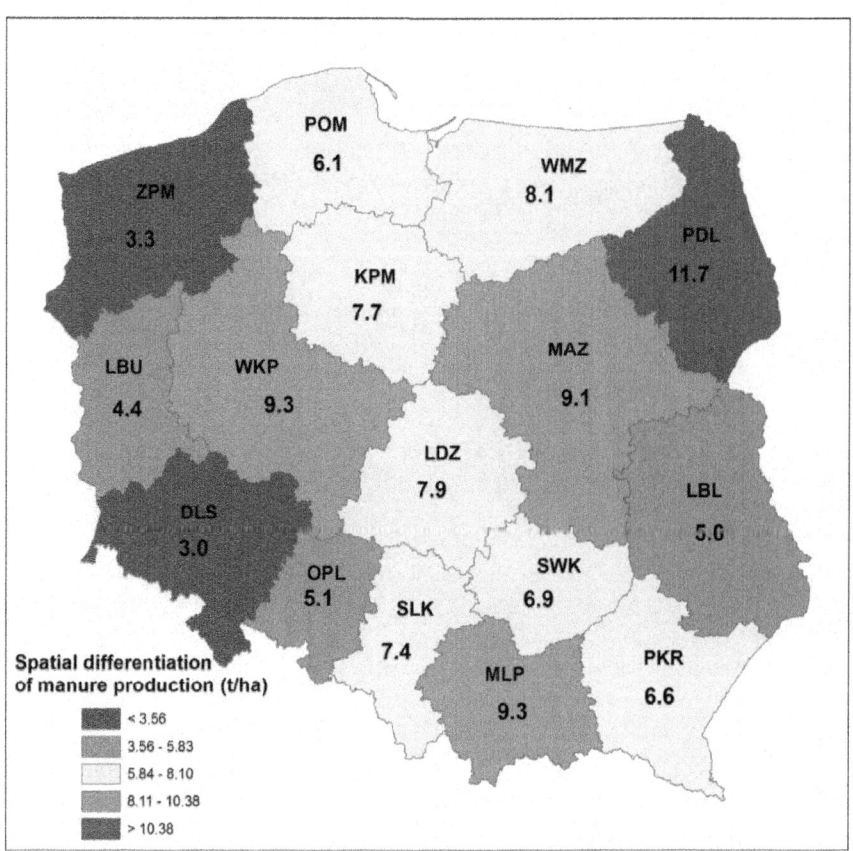

Figure 3. The production of manure per 1 ha of sowing area (average 2002-2005)1. Source: own work based on Kus et al., (2006).

Assessment of C Organic Balances

As indicators for determining a positive, neutral or negative balance of carbon in the soil the scale developed by Korschens et al., (2004) can be used - presented in the table 1. The presented scale has a range between below -200 kg/ha/y and values above 300 kg/ha/y. Comparing the results with this scale allows judging the impact on soil functions and potential yield performance of plants. It is important to emphasize that the carbon balance of more than

300 increases high emissions of nitrogen. Therefore, soil carbon sequestration cannot be considered in isolation from the nitrogen emissions.

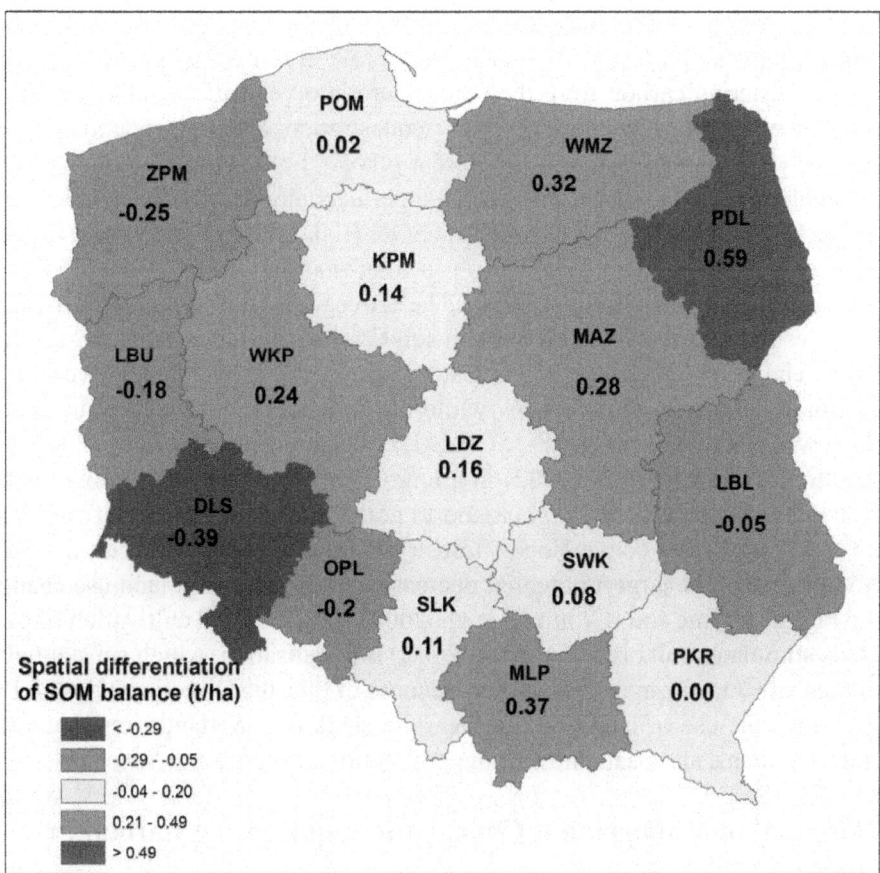

Figure 4. Balance of soil organic matter (t/ha) in voivodeship1. Source: own work based on Kus et al., (2006).

Table 1. Evaluation of the carbon organic balance (VDLUFA, 2004)

Balance		Impact
kg C/ha/y	group	
<-200	very low	unfavourable influence of soil functions and yield performance
-200- -75	low	medium-term tolerable, especially on soils enriched with humus
-75- +100	optimal	optimal in terms of yield loss at low risk of long-term site-adapted setting humus content
100-+300	high	medium-term tolerable, especially on impoverished soils with humus
>300	Very high	increased risk of nitrogen loss, low efficiency

The Potential Land Use Change for Sequestering Carbon in Soils

Land use change significantly affects soil carbon stock (Guo & Gifford, 2002). Most longterm research shows significant changes in SOC (Smith, 2008). Land use change can play a positive or negative role in mitigating global warming by sequestering carbon from the atmosphere into vegetation and soils. Many land use activities generate carbon sequestration and thus counteract the impact of emissions made in a different place. There are two components of estimated emissions from land use change: decomposition of vegetation and mineralisation/oxidation of humus or SOC (Lal, 2004b). The conversion of arable land to woodland may result in a substantial increase in soil carbon sequestration from 0.3 to 0.6 t C/ha/y. The conversion of arable land to grassland may result in a substantial increase in soil C sequestration from 1.2 to 1.7 t C/ha/y. The potential carbon sequestration rate in the conversion of woodland to arable land was -0.6 t C/ha/y, while the conversion of grassland to arable land was at a rate of between -1.0 and -1.7 t C/ha/y (Freibauer et al., 2004). In addition, Guo & Gifford (2002) in a long-term experiment had shown that a conversion of forestland or grassland to arable land caused a significant loss of SOC, whereas a conversion of forestry to grassland did not result in a loss for all cases. The largest potential decrease of SOC loss is in land use change on highly organic soil (Gronlund et al., 2008). Drainage and cultivation of peat soils stimulates soil organic matter (SOM) mineralisation, which substantially increases CO_2 emissions from soils. Because of this, the Directive 2009/28/EC prohibits the use of land with high carbon stock (i.e. wetlands, continuously forested areas, and peat land) for the production of biofuels.

Potential of Management Change for Sequestering Carbon in Soils

Correct agricultural practices of the soil can have a significant influence for carbon sequestration. A change in conventional cultivation practises has an important role in improving the soil's structure. Implementing modern practises like reduced tillage, no tillage or conservation tillage can significantly improve the soil's organic matter. Conventional tillage is defined as the mechanical manipulation (ploughing, disking and harrowing) of the top soil that leaves no more than 15% of the ground cover with crop residues. Such tillaging tends to disrupt the soil structure, accelerating the decomposition of soil organic matter, and making the bared topsoil vulnerable to erosion by rain and wind (Hillel & Rosenzweig, 2009). The alternative for conventional tillage is conservation tillage. The European Conservation Agriculture Federation (ECAF) defines conservation tillage as soil management practices, which minimise the disruption of the soil's structure, composition and natural biodiversity, thereby also minimising erosion and degradation, and water contamination (ECAF, 2002). By avoiding deep ploughing, this can increase the sequestration rate by 1.4 to 4.1 t C/ha/y. There is a growing interest in the impact of conservation tillage practices on carbon sequestration in recent years. According to Holland (2004), agriculture can act both as a sink and a source of CO_2 emission and the use of conservation practices by agriculture could decrease this emission. The coverage of the soil surface with straw and cover crops, increases biomass productivity and turns the soil into a tremendous carbon sink. Reducing the intensity of soil cultivation lowers energy consumption and the emission of carbon dioxide, while carbon sequestration is raised through the increase of soil organic matter (Holland, 2002). On the basis of long-term experiments, West & Post (2002) concluded that conversion of conventional tillage to no-till sequesters an average of 0.57 ± 0.14 t C/ha/y. Long-term field experiments are the most reliable source information about GHG emissions from different agricultural systems. However, they are difficult to manage and limited by time and costs (Li et al., 2009). Reduced tillage, enhanced crop residue incorporation, and farmyard manure application each increased soil C-sequestration, increased N2O emissions, and had little effect on CH4 uptake. Over 20 years, increases in N2O emissions, which were converted into CO_2- equivalent emissions with 100-year global warming potential multipliers, offset 75–310% of the carbon sequestered, depending on the scenario (Li et al., 2005). Simulation models provide an alternative method of assessment of agricultural practices effects (Farge et al., 2007). Many models have been developed to describe the responses of crop growth, soil water dynamics and soil biogeochemistry such as Roth C (Groenigen et al., 2010) for organic carbon turnover, CENTURY (Grant et al., 2004) or DNDC (Li et al., 1992; Giltrap et al., 2010) for carbon and nitrogen cycles. In order to calculate the management influence of carbon changes, we used 'Tool for Estimation of Changes in Soil Carbon Stocks associated with management Changes in Croplands and Grazing Lands based on IPCC Default Data'. At the moment Poland is assumed as having a cold temperate climate with both maritime and continental elements impact. The conversion from full tillage to reduced tillage under cold a temperate climate with maritime influence can cause the annual carbon stock to change by approximately 0.56 t C/ha, while the conversion from full tillage to no tillage 0.76 t C/ha depending on type of climate, soil, and the amount of fertiliser applied while under a cold temperate climate with a continental influence is 0.20 and 0.28 t C/ha respectively.

POTENTIAL OF CULTIVATION ENERGY PLANTS FOR SEQUESTERING CARBON IN SOILS

The soil carbon in croplands can be increased by planting more forages, and increasing residue inputs from plants with a high biomass potential. Hopes for increased soil carbon sequestration are associated with an increase in large-scale energetic crops cultivation. Energy crops are characterised by rapid growth and large amount of biomass produced and consequently a very large amount of crop residue left at the field. Annually about 30% of the senescent leaves and post harvest remnants are entering the soil (Matthews & Grogan 2001). One factor that is highly important for the amount of carbon sequestration in soil is that these perennial plants are grown for about 20 years on one field. The cultivation of energy crops is associated with greenhouse gas emissions (burning fuel, the production of fertilizers, crop protection). The assumption is that carbon sequestration of around 0.25 t C/ha/y resulting from energetic crops cultivation, allows biomass combustion to be neutral in terms of greenhouse gas emissions (Volk et al., 2004). Nevertheless, different authors found that the carbon sequestration rates for these cultivars are different. Bradley and King (2004), determined the carbon sequestration in forests and willows cultivations at 0.15-0.22 t C/ha/y, whereas in Miscanthus cultivation it was at 0.13-0.20 t C/ha/y. According to Matthews and Grogan (2001), carbon sequestration in the surface layer of the soil (0-23 cm) was at 0.31 for forests, and 0.41 for the cultivation of willow, whereas for Miscanthus it was measured at 0.93 t C/ha/y. Freibauer et al., (2004) and Smith (2004), determined the carbon sequestration in cultivations of energy crops at appropriately 0.60 and 0.62 t C/ha/y. Unfortunately for the carbon sequestration the liquidation of biocrops plantations causes large losses of accumulated carbon. This topic is not yet completely understood and requires further work from researchers.

Materials and Methods

As mentioned before Poland has a temperate climate on the Western side of the Vistula River with larger influence of maritime climate than on the Eastern side of the river that has a larger influence of continental climate. For research on soil carbon sequestration dependant on the climate, we have divided Poland along the Vistula River into two regions of climatic influence. We have selected randomly ten experimental positions with available climate data from the MARS Database elaborated by the Joint Research Centre European Commission – 5 grids cells (50 km x 50 km) in each region.

For a simulation of total inflows of carbon into soil under willow and Miscanthus crops in a 19 year period, the DNDC model was used. The model was calibrated based on the data from experimental fields established in 2003 at two Experimental Stations of the Institute of Soil Science and Plant Cultivation, Puławy. The experimental fields are located in the Experimental Station Pulawy-Osiny on heavy black earth, and at the Experimental Station in Grabow on medium-heavy soil where five genotypes of Miscanthus and four clones of willow were planted.

Results and Discussion

The results have shown that the significant difference in yield of Miscanthus grown under two different climate influences, as well as organic matter input, while soil organic matter did not denote a statistically significant difference. In comparing the case of willow, the yield was not significantly different, but the input of organic matter and soil organic matter was significantly different. This might be explained as the temperature had an impact in the case of Miscanthus but it did not influence willow plants. The opposite situation was found in the effect of precipitation. Borzęcka-Walker et al., (2008), did not find any significant differences between the yields of willow clones. The yield of willow grown in two different localisations, ranged from 11.1 - 13.7 t/ha/y. There were lower simulated yields for willow cultivations (13.8-18.1 t/ha/y) located on very good soils of Eastern Europe (Fischer et al., 2005). It can be assumed that the limited water influenced the experimental yield of willow. The dry matter yield of Miscanthus genotypes was significantly different within an average of 10.2 - 20.7 t/ha/y at both localisations. The yield for the first year of the experiment was low; this could be because it was the second year of cultivation when the plants are still not mature enough to obtain an economic yield (Clifton-Brown & Lewandowski 2000). In the second year of the experiment, the yield witnessed a high increase. In the third year of experiment this was characterised with very bad weather conditions, including a late spring ground-frost and long summer draught. The yields were approaching the presupposed simulated yields for Miscanthus cultivations (17.7-21.8 t/ha/y) located on very good soils of Eastern Europe (Fischer et al., 2005). It can be assumed with a high probability that the limited water in 2005 did not influence the experimental Miscanthus yield, but there was an influence from weather condition in 2006 (Borzecka-Walker et al., 2008).

Table 2. Soil carbon (C) balance (t C ha y) under Miscanthus and Willow cultivation

	Miscanthus		Willow	
	maritime	continental	Maritime	continental
yield t C/ha/y	4.8	5.1	4.8	5.0
organic matter input t C/ha/y	5.4	6.1	2.9	3.5
soil organic carbon t C/ha/y	1.5	1.7	0.71	0.85

The aboveground biomass has a high influence on the amount of carbon sequestration, which enters the soil usually in the form of senescent leaf mass and postharvest remnants. Kahle (2001) measured in Germany that about 3.0-7.5 t/ha/y aboveground biomass full to soil. Matthews and Grogan (2001) in Great Britain estimated the inflow of the organic matter at a level of 7.5 t/ha/y, while for Poland it was calculated at from 2.63 to 6.58 t/ha/y (Borzecka-Walker et al., 2008). The organic matter input table 2 shows a greater potential for carbon mitigation for Miscanthus than for willow. Despite the very different movement of C into the soil plantations of willow and Miscanthus, most of this element accumulates in the litter. This is labile to C fraction, which in total will be mineralised and released as CO_2 in a short time after the restoration of conventional land use. Moderately stable fractions of carbon in the form of living organisms (humads) is the second largest fraction of sequestrated carbon. In almost all cases it will be transformed into humus after a change of use in the plantation. The stable fraction C (humus) in the lifetime of the plantation rose to a negligible extent. So it can be concluded that an effective carbon sequestration expressed the sum of humus fractions and humads. For willow and Miscanthus (tab. 3), during the period of cultivation it was respectively at 0.20, 0.21 for the maritime climate and the 0.23 and 0.25 in the continental climate. The obtained values from the model are close to the obtained values by Bradley & King (2004), who have determined the carbon sequestration in forests and willows cultivations at 0.15-0.22 t C/ha/y, whereas in Miscanthus cultivations they were at 0.13-0.20 t C/ha/y. According to Matthews & Grogan (2001), carbon sequestration in the surface layer of the soil (0-23 cm) was at 0.31 for forests, and 0.41 for the cultivation of willow, whereas for Miscanthus it was measured at 0.93 t C/ha/y. The net soil carbon sequestration in Miscanthus crops was around 0.38-0.95 t C/ha/y and 0.22-0.39 t C/ha/y for coppice willow (Borzecka-Walker et al., 2008). Freibauer et al., (2004) and Smith (2004) have determined the carbon sequestration in cultivations of energy crops appropriately at 0.60 and 0.62 t C/ha/y. Much of the carbon mitigation potential associated with the use of SRC willow and Miscanthus as bioenergy crops arises from their indefinite capacities as 'carbon neutral' alternatives to fossil fuel combustion (Grogan & Matthews, 2001). The assumption is that carbon sequestration of around 0.25 t C/ha/y resulting from energy crops cultivation makes biomass combustion neutral in terms of greenhouse gas emissions (Volk et al., 2004). The new cultivations will result in changes in fossil-fuel use, agricultural inputs, and carbon emissions with fossil fuels and other inputs. Management practices that alter crop yields and land productivity can affect the amount of land use crop production with further significant implications for both emissions and sequestration potential (West & Marlnd, 2003; Schneider & Mccarl, 2003).

Table 3. Soil organic matter (SOM) pools under Miscanthus and willow cultivation

SOM	Litter	Humads	Humus
Maritime			
Willow	0.28	0.19	0.01
Miscanthus	0.86	0.20	0.01
Continental			
Willow	0.32	0.21	0.02
Miscanthus	0.91	0.22	0.03

The organic matter input (tab. 2) is basically consisting of litter and dead roots fractions. Miscanthus harvested in autumn delivers approximately 20% of leaf mass and some of underground biomass while 30 % of leaf mass and some underground biomass from spring harvest is entering the soil. In compare 100 % of willow leaf mass and some of underground biomass is entering the soil. Soil organic matter includes tree pools: very labile fraction of litter, labile humads, and passive humus (tab.3).

CONCLUSION

Agriculture practices have an important role to play in mitigating climate change due to atmospheric enrichment of CO_2 and other greenhouse gases. To improve the negative balance of soil carbon sequestration in Poland, corrective action should be taken. Land management can strongly influence soil carbon stocks and careful management can be used to increase soil carbon sequestration. It is important to propose contemporary management practises to farming like the conversion from tillage to no tillage systems, incorporation of cover crops, forages in crop rotation, as well as a liberal use of crop residues and biosolids like mulch. Special care should be taken of integrated nutrient management including compost/manures and precision use of fertilizers and integrated pest management. A very important role can be played by the implementation of biocrops which are characterised with very high potential of carbon sequestration and much lower GHG emission during the cultivation.

When considering carbon sequestration it should be mandatory to combine these analyses with nitrogen (N) as carbon and nitrogen move through terrestrial ecosystems coupled with biogeochemical cycles, and increasing C stocks in soils and vegetation which have an impact on the N cycle.

ACKNOWLEDGMENT

The work was done based on the results obtained within the research projects N N315 759240 and N N313 436839, funded by the Ministry of Science and Higher Education.

REFERENCES

1.   Borzecka-Walker, M.; Faber, A. & Borek, R. (2008). Evaluation of carbon sequestration in energetic crops (Miscanthus and coppice willow). Int. Agrophysics, No 22, 185-190.

2.   Bradley, R.I. & King, J.A. (2004). A review of farm management techniques that have implications for carbon sequestration – validating an indicator. OECD Expert Meet. Farm Management Indicators and the Environment, 8-12 March, Palmerston North, Avaialble from http://webdomino1.oecd.org/comnet/agr/farmind.nsf/22afaebba539ba74c1256a3 b004d5175/b3b 8d25f219f4ae3c1256bd5004874f1/$FILE/Bradley1.pdf

3.   Clifton-Brown, J.C.; & Lewandowski, I.; (2000). Overwintering problems of newly established Miscanthus plantations can be overcome by identifying genotypes with improved rhizome cold tolerance. New Phytol. No 148, 287-294.

4.   COM, (2000) 769: Green Paper: Towards a European strategy for the security of energy supply

5.   Directive, 2003/96/EC. (2003). The European Parliament and of The Council on restructuring the Community framework for the taxation of energy products and electricity.

6.   Directive, 2009/28/EC. (2009). The European Parliament and of The Council on the promotion of the use of energy from renewable sources amending and subsequently repealing Directives 2001/77/EC and 2003/30/EC.

7.   Dumanski, J.; Desjardins, R.L.; Tarnocal, C.; Monreal, EG.; Gregorich, E.G.; Kirkwood, V. &Campbell, C.A. (1998). Possibilities for future carbon sequestration in Canadian agriculture in relation to land use changes. Climatic Change, Vol. 40, 81-103.

8.   ECAF (European Conservation Agriculture Federation). Observations on

the Communication form the European Commission: Towards a Thematic Strategy for Soil Protection (COM (2002) 179 final) www.ecraf.org

9. Farge, P.K.; Ardő, J.; Olsson, L.; Rienzi, E.A. & Pretty, J.N. (2007). The potential for soil carbon sequestration in the three tropical dryland farming systems of Africa and Latin America: a modeling approach. Soil Tillage Res, Vol. 94, 457-472.

10. Fischer, G.; Prieler, S.; & van Velthuizen, H.; (2005). Biomass potentials of Miscanthus, willow and poplar: results and policy implications for Eastern Europe. Northern and Central Asia. Biomass and Bioenergy. No 28,119-132.

11. Fotyma, M. & Mercik, S. (1992). Współczynniki reprodukcji glebowej materii organicznej. In: Chemiarolna, PWN Warszawa.

12. Franzluebbers, A.J. & Follett, R.F. (2005). Greenhouse gas contributions and mitigation potential in agricultural regions of North America: Introduction. Soil & Tillage Research, Vol. 83, 1–8.

13. Freibauer, A.; Rounsevell, M.; Smith, P. & Verhagen A. (2004). Carbon sequestration in European agricultural soils. Geoderma, Vol. 122, 1–23.

14. Giltrap, D.L.; Li, Ch. & Saggar, S. (2010). DNDC: A process-based model of greenhouse gas fluxes from agricultural soils. Agriculture Ecosystems and Environment, Vol. 136, 292- 300.

15. Grant, B.; Smith, W.N.; Desjardins, R.; Lemke, R. & Li, C. (2004). Estimated N20 and CO_2 emissions as influenced by agricultural practices in Canada. Climatic Change, Vol. 65, 315-332.

16. Groenigen, K.J.; Hastings, A.; Forristal, D.; Roth, B.; Jones, M. & Smith, P. (2010). Soil C storage as affected by tillage and straw management: An assessment using field measurements and model predictions. Agriculture Ecosystems and Environment, Vol. 140, 218-225.

17. Grogan, P. & Matthews, R. (2001). Review of the potential for soil carbon sequestration under bioenergy crops in the U.K. Report for U.K. Dept. of Environment, Food, and Rural Affairs. Contract NF 0418.

18. Gronlund, A.; Hauge, A.; Hovde, A. & Rasse, D.P. (2008). Carbon loss estimates from cultivated peat soils in Norway: a comparison of three methods Nutr Cycl Agroecosyst, Vol. 81, 157–167.

19. Guo, L.B. & Gifford, R. M. (2002). Soil carbon stocks and land use change: a meta analysis. Global Change Biology, Vol. 8, 345-360.

20. Hillel, D. & Rosenzweig, S. (2009). Soil and carbon climate change. CSA News, Vol. 54, No 6, 4-11.

21. Holland, J.M. (2004). The environmental consequence of adopting

conservation tillage in Europe: reviewing the evidence. Agriculture Ecosystems and Environment, Vol. 103, 1- 25.

22. Hutchinson, J.J.; Campbell, C.A. & Desjardins, R.L. (2007). Some perspectives on carbon sequestration in agriculture. Agricultural and Forest Meteorology, Vol. 142, 288–302.

23. IPCC, (2000). IPCC special report on Impacts. Adaptation and Vulnerability. http://www.grida.no/climate /ipcc_t ar/wg2/index.htm

24. IPCC, (2007). Fourth Assessment Report. Synthesis Report. http://www. ipcc.ch/pdf/assessment-report/ar4/syr/ar4_syr.pdf

25. Kahle, P.; Beuch, S.; Boelcke, B.; Leinweber, P.; & Schulten, H.R.; 2001: Cropping of Miscanthus in Central Europe: biomass production and influence on nutrients and soil organic matter. Eur. J. Agron. No 15: 171-184.

26. Knowler, D. & Bradshaw, B. (2007). Farmers' adoption of conservation agriculture: A review and synthesis of recent research. Food Policy, Vol. 32, 25-48.

27. Kus, J.; Madej, A. & Kopinski, J. (2006). Bilans słomy w ujęciu regionalnym. Studia i Raporty IUNG-PIB, Wyd. IUNG, Pulawy, No 3, 211-225.

28. Lal, R. (2004a). Agricultural activities and the global carbon cycle. Nutrient Cycling in Agroecosystems, Vol. 70, 103–116.

29. Lal, R. (2004b). Soil carbon sequestration to mitigate climate change. Geoderma, Vol. 123, 1 – 22.

30. Li, C.; Frolking, S. & Butterbach-Bahl, K. (2005). Carbon sequestration in arable soils is likely to increase nitrous oxide emissions, offsetting reductions in climate radiative forcing. Climatic Change, Vol. 72, 321-338.

31. Li, C.; Frolking, S. & Frolking, T.A. (1992). A model of nitrous-oxide evolution from soil driven by rainfall events: model structure and sensitivity. Journal of Geophysical Research–Atmospheres, Vol. 97, 9759–9776.

32. Li, T.; Feng, Y. & Li, X. (2009). Predicting crop growth under different cropping and fertilizing management practices. Agricultural and Forest Meteorology, Vol. 149, 985- 998.

33. Matthews, R.B. & Grogan, P. (2001). Potential C-sequestration rates under short-rotation coppiced willow and Miscanthus biomass crops: a modelling study. Aspects Appl. Biol., Vol. 65, 303-312.

34. Ringius, L. (2002). Soil carbon sequestration and the CDM: opportunities and challenges for Africa. Climatic Change, Vol. 54, 471–495.

35. Schneider, U.A. & Mccarl, B.A. (2003). Economic potential of biomass based fuels for greenhouse gas emission mitigation. Environmental and Resource Economics, Vol. 24, 291–312.

36. SEC. Commission Staff Working Document. (2009). The role of European agriculture in climate change mitigation. Brussels, 23.7.2009, 1093 final.

37. Smith, P. (2004). Carbon sequestration in croplands: the potential in Europe and the global context. Europ. J. Agronomy, Vol. 20, 229–236.

38. Smith, P. (2008). Land use change and soil organic carbon dynamics. Nutr Cycl Agroecosyst, Vol. 81, 169–178.

39. Smith, P.; Powlson, D.S.; Glendining, M.J. & Smith, J.U. (1997). Potential for carbon sequestration in European soils: preliminary estimates for five scenarios using results from long-term experiments. Global Change Biol, Vol. 3, 67– 79.

40. Smith, P.; Powlson, D.S.; Smith, J.U.; Falloon, P.D. & Coleman, K. (2000). Meeting Europe's climate change commitments: quantitative estimates of the potential for carbon mitigation by agriculture. Global Change Biol, Vol. 6, 525–539.

41. Stuczynski, T.; Kozyra, J.; Lopatka, A.; Siebielec, G.; Jadczyszyn, J.; Koza, P.; Doroszewski, A.; Wawer, R. & Nowocien, E. (2007). Przyrodnicze uwarunkowania produkcji rolniczej w Polsce Studia i Raporty IUNG-PIB, Wyd. IUNG, Pulawy, No 7, 77-115.

42. Terelak, H.; Motowicka-Terelak, T.; Wroblewska, E.; Gawrysiak, L. & Pietruch, C. (2001). Mapa zawartości substancji organicznej w glebach użytków rolniczych Polski. Wyd. IUNG, Puławy.

43. VDLUFA (2004): VDLUFA Standpunkt Humusbilanzierung, Methode zur Beurteilung und Bemessung derHumusversorgung von Ackerland, Bonn 2004, http://www.vdlufa.de, (accessed April 2004).

44. Vleeshouwers, L.M. & Verhagen, A. (2002). Carbon emission and sequestration by agricultural land use: a model study for Europe. Global Change Biol, Vol. 8, 519–530.

45. Volk, T.A.; Verwijst, T.; Tharakan, P.J.; Abrahamson, L.P. & White, E.H. (2004). Growing fuel: a sustainability assessment of willow biomass crops. Front Ecol. Environ, Vol. 2, No 8, 411- 418.

46. West, T.O. & Marlnd, G. (2003). Net carbon flux from agriculture: Carbon emissions, carbon sequestration, crop yield, and land use change. Biogeochemistry, Vol. 63, 73-83.

47. West, T.O. & Post, W.M. (2002). Soil Organic carbon sequestration rates by tillage and crop rotation: a global analysis. Soil Science Society of America Journal, Vol. 66, 1930-1946.

Chapter 9

SOIL-LANDSCAPE MODELLING – REFERENCE SOIL GROUP PROBABILITY PREDICTION IN SOUTHERN ECUADOR

Mareike Ließ[1], Bruno Glaser[2], and Bernd Huwe[1]

[1]University of Bayreuth, Department of Geosciences/ Soil Physics, Germany

[2]Martin-Luther University Halle Wittenberg, Soil Biogeochemistry, Germany

INTRODUCTION

Since long, soils are understood as a function of their genetic factors: parent material, relief, climate, organisms and time, a concept first described by Dokutschajew (1883) and better known from Jenny (1941). The complex interaction of these factors activates particular soil forming processes, which in dependence of their intensity and duration, lead to characteristic soil properties. The resulting profile reflects the balance of these processes in its properties (Grunwald, 2006). Soil-landscape modelling uses the knowledge about soil genesis to predict soil distribution in a landscape based on continuously available environmental parameters by statistical models. The early conceptual models have resulted into quantitative soil-landscape models, which do not only make the spatial prediction of continuous soil properties possible, but include model uncertainty. Being at first an unwelcome nuisance that reduced map reliability, gradually soil variation and its unpredictability was seen as a key soil attribute by itself (Burrough et al., 1994).

Soil research within the area of the scientific research station San Francisco in the southern Ecuadorian Andes has been carried out for many years. A first preliminary soil map was provided in 2009 (Liess et al., 2009). This soil map, based on Reference Soil Groups (RSGs) from the World Reference Base for Soil Resources (WRB) (FAO, IUSS Working Group WRB, 2007) does not include prediction uncertainty and neglects Cambisols. However, Cambisols are part of the dataset and have also been described by Yasin (2001) and Wilcke et al. (2002, 2003). Other soils that occur under natural vegetation within the area are Histosols, Stagnosols, Umbrisols and Regosols (FAO, IUSS Working

Group WRB, 2007). Histosols were described by Yasin (2001) and Schrumpf et al. (2001) as Haplosaprists according to Soil Taxonomy classification (Soil Survey Staff, 2006). Yasin (2001) investigated forest soils only between 1900 – 2240 m a.s.l., whereas Schrumpf et al. (2001) explored soils along an altitudinal gradient from 1850 – 3050 m a.s.l. Thus, Histosols were found on slope angles varying from 10 – 50° at 1850 – 2700 m a.s.l.; Stagnosols were described between 2080 – 2850 m a.s.l. (Yasin, 2001; Schrumpf et al., 2001; Liess et al., 2009). Umbrisols were assigned by Schrumpf et al. (2001) and Liess et al. (2009).

Prediction of soil types from terrain factors by statistical models is a standard approach within the field of soil-landscape modelling. Lagacherie & Holmes (1997) as well as Moran & Bui (2002) assigned soil classes by CTs based on parameters calculated from a digital elevation model (DEM). Skidmore et al. (1996), Thomas et al. (1999) and Dobos et al. (2000) spatially predicted soil types from terrain analysis. Furthermore, Gessler et al. (1995), Moore et al. (1993) and Odeh et al. (1994) predicted soil attributes from terrain parameters. Bourennane et al. (2000) and Hengl et al. (2004) regionalised soil horizon and topsoil thickness from a DEM. Several statistical models are available to relate soils to environmental predictors. Bishop & Minasny (2006) compared some of them: Linear, generalized linear (GLM) and generalized additive (GAM) models, classification and regression trees (CART) and artificial neural networks (ANN). Among the considered model types, only ANN were assigned a better predictive power than CART, but lack the ease of use, parsimony, interpretability and computational efficiency that applies for CART.

By extending the dataset of Liess et al. (2009) and constructing various classification trees (CT), we expect to develop a more precise RSG map and include prediction uncertainty by displaying the RSG probability. The investigated soils will be related to terrain parameters by a CT (Breimann et al., 1984) that organises the dataset according to the respective RSG. The tree model can then be used to assign the RSG probabilities to the whole area covered by a DEM.

MATERIAL AND METHODS

Research Area

The research area is situated between the provincial capitals Loja and Zamora (Figure 1) in the southern Ecuadorian Andes from 1670 to 3160 m a.s.l. It extends in UTM-Zone 17M from west to east between 710500 and 716000, and from north to south between 9561500 and 9557000 (Figure 1). The San

Francisco River divides the area into two parts: The north-west facing slopes south of the river are covered by montane rain forest and subpáramo vegetation above the tree line. Within this area, Homeier et al. (2002) differentiated various forest types according to their altitude and position on the ridge or in the valley. The southeastern facing slopes north of the river are mainly covered by pastures and succession vegetation after fire clearance when sites were left unused. For soil model development, only sites under natural vegetation were considered.

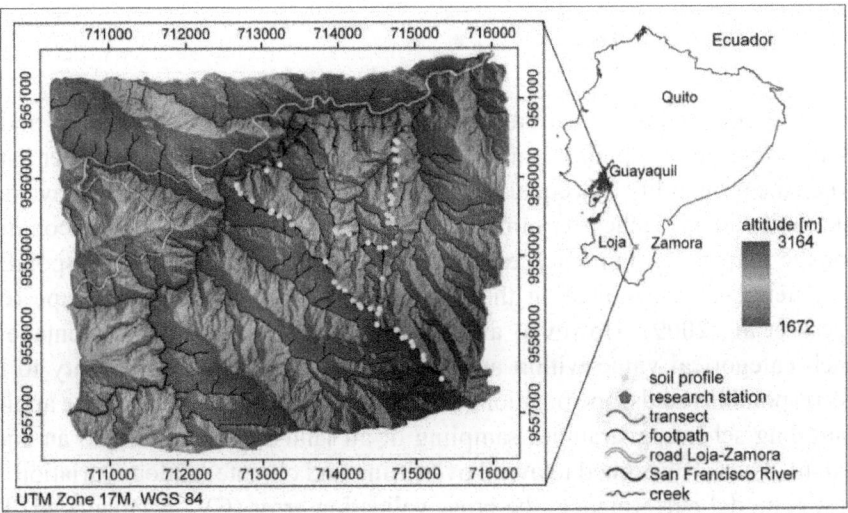

Figure 1. Research area. Overlaid hill shading with light source from north-east (adapted from Liess et al., 2009).

As part of the Chiguinda unit, the research area is lithologically covered by metasiltstones, siltstones and quartzites which are intermixed with layers of phyllite and clay schists (Litherland et al., 1994). Furthermore, it is influenced by the regular occurrence of landslides. Average total annual rainfall increases from 2050 mm at an altitude of 1960 m a.s.l. to approximately 4400 mm at 3100 m a.s.l. (Rollenbeck, 2006). Average air temperature decreases with increasing altitude from 19.4 to 9.4 °C (Fries et al., 2009).

Classification Trees

Classification trees (CTs), a method first described by Breimann et al. (1984), were used to relate the RSGs to terrain parameters. It was conducted with the rpart library of the RProject for Statistical Computing (Therneau & Atkinson, 2003).

In CTs subdivision is based on a categorical response variable, i. e. RSG. The final subsets, also called end nodes, should be as pure as possible. This is done by trying to assign them to only one category in the response variable, e.g. to Histosol. The Gini criterion (Equation 1) is applied as a measure of purity (Breiman et al., 1984). It serves as a decision criterion, to determine which terrain parameter best separates the dataset continuously into always two subsets to create the purest end nodes.

$$gini(t) = 1 - \sum_{i=1}^{k} P_i^2$$

(1)

The Gini-Index (Equation 1) reaches its maximum in a particular node t if all categories k within this node are equally represented. On the other hand, when the probability P_i is equal to zero for all but one category within any node, the Gini-Index reaches its minimal value. The categorical value accounting for the majority within each end node is then assigned to the corresponding parameter values, indicating the typical position within the landscape (e.g. Liess et al., 2009). However, another option is to assign the percentage of each categorical value within an end node as occurrence probability to the corresponding landscape position. This is of course only justified if the applied sampling scheme guarantees sampling of all landscape positions to an equal extent. The CT is pruned to avoid overfitting and obviate random variation. To assess model performance, the cross validation error (CV) is calculated. The dataset is subdivided into 10 subsets, and the process is repeated 10 times with 9 parts for model training and the 10th part as the evaluation dataset. Eventually, among all trees considered for the final model, the tree with the lowest cross validated error rate is chosen. CV and model pseudo R^2 are calculated. Pseudo stability indices are constructed to satisfy the different interpretations, e.g. explained variance or square of correlation. They are similar to R^2 in that they also range between 0 and 1 and a higher value represents a better adaptation to the data.

Dataset and GIS Methodology to Gain Terrain Data

Topographic data for the research area is available on a continuous landscape level. The DEM used to obtain terrain parameters for the establishment of a prediction model of RSG occurrence has 2 m cell size (Liess et al., 2009). For model application, this accuracy was reduced to 10 m to decrease calculation time. The used terrain parameters include altitude a.s.l., aspect, slope angle, terrain curvature, upslope contributing catchment area and overland flow distance to the channel network (OFD).

Slope angle, aspect and curvature were computed with a 2nd degree polynomial fit from Zevenbergen and Thorne (Zevenbergen & Thorne, 1987; Cimmery, 2007). The contributing area was calculated with two methods; (1) based on the Kinematic Routing Algorithm (KRA CA) (Lea, 1992) and (2) based on the Braunschweiger Digital Relief Model (BS CA) (Bauer et al., 1985). In addition to the OFD, the horizontal (HOFD) and vertical (VOFD) overland flow distances were also calculated. The channel network itself was assessed applying the Strahler stream order ≥ 5 as initiation threshold (Strahler, 1957). Terrain curvature was computed using directly adjacent cells. Finally, the terrain parameters were calculated and the RSGs were predicted for each individual raster grid cell. The free and open source GIS software, SAGA, was used (Böhner et al., 2006).

The research area was sampled at 367 sites, including 311 auger points and 56 soil profiles. Soil sampling covered 24 sampling classes produced by an overlay of four altitudinal, three slope angle and two aspect classes to guarantee representative area coverage. Transects for auger sampling (Figure 1) were laid according to the catena concept (Milne, 1935) from hilltop to valley bottom. For more detailed information on the applied sampling design, see Liess et al. (2009).

Two methods were used to assign terrain parameters to the soil dataset. On the one hand, the nearest neighbour (n. n.) value was allocated to each soil profile or auger point. On the other hand, a buffer representing the radius of GPS accuracy was placed around the sampled location, and the calculated mean value of the corresponding area was assigned. This assignment was completed for each of the described parameters apart from the slope angle and aspect. These were directly measured in the field. The slope angle and aspect which were computed from the DEM were solely used for model application.

Probability Calculation

The probability of each RSG was predicted via a CT which grouped the soil sampling points regarding the existence or absence of that RSG. Thus, the percentage of sampling points assigned to the corresponding RSG in each end node of the tree was used to predict the probability of that RSG. Thereby, the diagnostic properties necessary for assigning the particular RSG were used, whereas the necessary absence of other properties was neglected. This was done in particular to establish a good prediction scheme for Stagnosols. It was decided that the occurrence and thickness of a sufficient stagnic colour pattern and/ or albic horizon is more important than the limitation in organic layer thickness. As a consequence, soils with a 40 cm organic layer displaying also a thick stagnic horizon were classified as Histosols and Stagnosols. Any other proceeding would have made the development of a Stagnosol prediction scheme incomplete and complex.

To sum the individual probabilities and standardize them by relating each RSG to the total probability sum, is one option. This option neglects WRB (FAO, IUSS Working Group WRB, 2007) hierarchy, because all RSGs are competing on an equal level and no soil process is given dominance over another. As a consequence, the probabilities refer to the probability of the diagnostic property necessary for RSG assignation. Later we will refer to these as WRB independent probabilities.

Figure 2 shows the probability calculation scheme based on WRB (FAO, IUSS Working Group WRB, 2007) hierarchy.

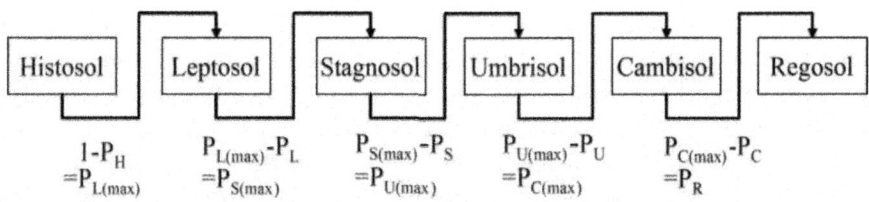

Figure 2. Hierarchical calculation scheme for the maximum possible probability of each RSG according to WRB hierarchy. P_x is the actual probability of the respective RSG: H Histosol, L Leptosol, S Stagnosol, U Umbrisol, C Cambisol, R Regosol. $P_{X(max)}$ is the maximum possible probability of the RSG.

It is used to calculate the maximal possible probability for each RSG from the probability predicted by the CTs. Maximal Leptosol probability is left after subtracting Histosol probability from 1. Maximal Stagnosol probability is left after also subtracting the actual Leptosol probability and so on. Equation 2 shows the calculation of the actual probability, P_x, according to the CT probability, $P_{X(tree)}$, and the maximal possible probability, $P_{X(max)}$.

$$P_{x(max)} \cdot P_{x(tree)} = P_x \qquad (2)$$

RESULTS AND DISCUSSION

Classification Tree Models and Digital Soil Maps

Figure 3 presents the CT models to predict Histosol, Leptosol and Stagnosol occurrence probability from nearest neighbour (n. n.) and mean terrain values.

The RSG Histosol is assigned to soils with an organic layer ≥ 40 cm (FAO, IUSS Working Group WRB, 2007). Its probability within the research area was found to depend on two hydrological parameters (Figures 3a and 3d): KRA CA and VOFD. Probability is predicted with at least 0.2 (Figure 3a, d) throughout the research area. The highest probability (0.87) as predicted by n. n. relief values (Figure 3a) was obtained for small catchments (KRA CA < 258 m^2) within a distance of 14 – 23 m from the channel network. Though, probabilities are also high, 0.65, for small catchments (KRA CA < 258 m^2) within a VOFD of 54 – 176 m. The latter is a more conservative prediction, since it is based on 206 sampled sites and not only 15 as for the first differentiation criteria (Figure 3a). Sites seem to coincide in some parts with upper slope areas and ridges (Figure 4a).

Prediction by mean terrain values (Figure 3d) again shows high probabilities in similar landscape positions, i.e. for small catchments < 254 m^2 from 54 – 175 m VOFD (0.65) and < 26 m VOFD (0.70). The former is the safest prediction similar to the Histosol prediction from n. n. terrain values (188 sampled sites). Areas likely to be covered by Histosols with this 0.65 probability are again found along ridges. In contrast to the CT from n. n. terrain values, the highest probabilities, 0.85, by mean relief values (Figure 3d) are assigned to large catchments (≥ 254 m^2) with a VOFD from 103 – 145 m, dominating in dark colours as broad belts at 103 m distance around the creeks (Figure 4b). This also accounts for the major difference between the two models (Figure 4c). But since the corresponding end node in the tree model (Figure 3d) is only supported by 13 sampled sites, this finding is not representative for the research area.

Leptosols refer to soils limited to 25 cm depth by continuous rock (FAO, IUSS Working Group WRB, 2007). During soil sampling continuous rock was rarely attained, and refusal typically occured at the C horizon. This made the establishment of a model predicting soil depth to continuous rock impossible. Therefore, to calculate Leptosol occurrence probability expert knowledge was applied in addition to the CT methodology. From field work and data review it was known that Leptosols are found on steep slopes ≥ 50° and close to the creeks at approximately < 20 m HOFD. Other soils, which occurred at the same landscape positons with even higher probability, were excluded for model development. Afterwards, they were included again to calculate the probabilities of the tree end nodes. This explains the rather untypical appearance of the Leptosol CTs (Figures 3b and 3e). Usually, for any final subdivision into two end nodes, one of them would always display a probability > 0.5 and the other < 0.5. However, for the reason of adding more datasets after tree development this is not the case. This procedure was necessary in order to develop a reasonable model and account for true probabilities. Leptosol CTs established with n. n. and mean terrain values are very similar. In the already mentioned positions, Leptosol probability was assumed 0.30 – 0.36 (Figure 3b and 3e).

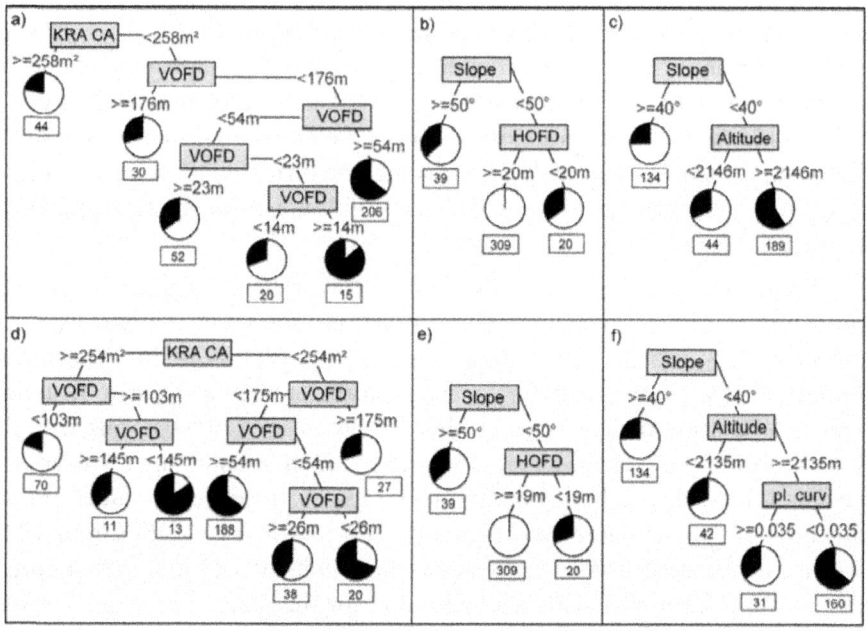

Figure 3. Classification trees predicting RSG probability. The pie charts' black parts represent the occurrence probability in the corresponding landscape positions. The numbers in the boxes underneath the charts refer to the number of sampling sites used for the probability prediction in each end node. Prediction by n. n. terrain values: a) Histosol probability, b) Leptosol probability and c) Stagnosol probability. Prediction by mean terrain values: d) Histosol probability, e) Leptosol probability and f) Stagnosol probability. (KRA CA = upslope contributing catchment area according to the Kinematic Routing Algorithm, VOFD = vertical overland flow distance, HOFD = horizontal overland flow distance, pl. curv = plan curvature).

The lighter colours in Figure 4b compared to Figure 4a are due to the fact that a probability of 0.20 (< 103 m VOFD, Figure 3d) falls into a smaller mapping class than 0.23 (Figure 3a) in the map layout. The similarity between the two models for the mentioned sites is indicated by yellow colours in Figure 4c. The sites mapped in red colours refer to a 0.1 – 0.3 higher probability as predicted by mean relief values. Comparison of the two tree models (Figures 3a and 3d) shows that differences are not higher than 0.13. The models differ only by a probability of 0.03 – 0.13, neglecting the mentioned 13 sites.

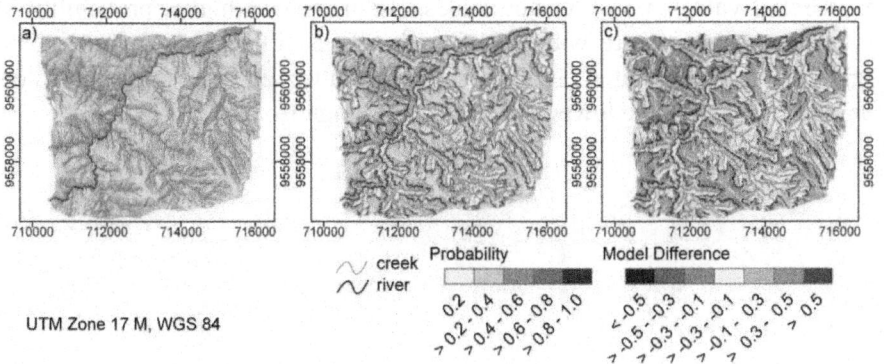

Figure 4. Maps of Histosol occurrence probability (Overlaid hill shading with light source from north-east): a) Prediction by n. n. terrain values, b) prediction by mean terrain values and c) model difference.

Figures 5a and 5b show the Leptosol probability distribution within the research area after model application. With the inclusion of WRB (FAO, IUSS Working Group WRB, 2007) hierarchy, Leptosol probability also depends on Histosol probability. But since Histosol probability close to the creeks (< 103 m VOFD) is predicted with only 0.2, model from mean relief values (Figure 3d), and 0.3, model from n. n. relief values (Figure 3a), it does not influence Leptosol probability much for those sites. Model difference regarding prediction by n. n. and mean terrain values (Figure 5c) is always $\leq \pm 0.1$ (0.05); including WRB hierarchy (FAO, IUSS Working Group WRB, 2007), model difference (Figure 5f) is increasing (hardly recognisable in the map). Model difference regarding probability predicted directly by the CTs and probability being calculated based on WRB hierarchy (Figures 5g and 5h) shows a similar picture. The difference between the WRB independent and dependent prediction by n. n. values (Figure 5g) is $\leq \pm 0.1$, but higher regarding the prediction difference by mean terrain values (Figure 5h).

Stagnosols are "soils exhibiting hydromorphic features for some time during the year in some part within 50 cm of the mineral soil surface and show a stagnic colour pattern and/ or an albic horizon in half or more of the soil volume" (FAO, IUSS Working Group WRB, 2007). Planosols are classified by similar diagnostic properties, but in addition display an abrupt textural change, which could not be confirmed for the investigated soils. Stagnosol probability is predicted throughout the research area with at least 0.25 (Figures 3c and 3f). The probability in both models depends on slope angle and altitude. It is higher on slopes < 40°. Above 2146 m a.s.l. for the prediction by n. n. and above 2135 m a.s.l. by mean terrain values, the probability increases even further. While curvature is of no importance for Stagnosol probability prediction by

n. n. relief values, mean terrain values assign an even higher probability for concave plan curvature with 0.64. Landscape positions < 2146 m a.s.l. for prediction by n. n. and < 2135 m a.s.l. by mean terrain values, and high slope angles account for the lowest probability of Stagnosols.

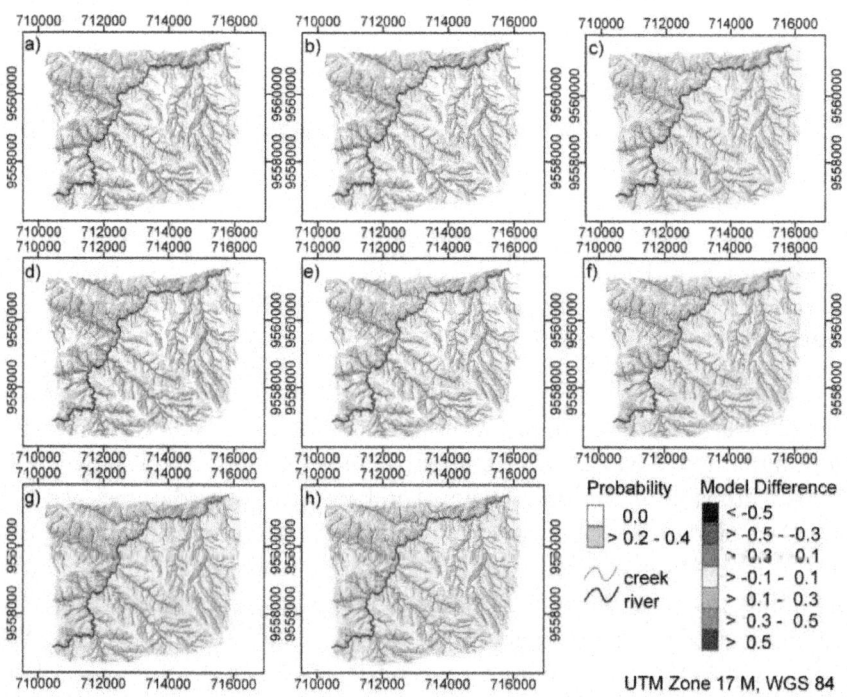

Figure 5. Maps of Leptosol occurrence probability (Overlaid hill shading with light source from north-east). Independent on WRB hierarchy: a) prediction by n. n. terrain values, b) prediction by mean terrain values and c) model difference. Dependent on WRB hierarchy: d) prediction by n. n. terrain values, e) prediction by mean terrain values and f) model difference. Difference between independent and WRB hierarchy dependent prediction: g) n. n. terrain values and h) mean terrain values.

Model application to the research area is shown in Figure 6. Stagnosols reach higher probabilities by the mean terrain values model (Figure 6b) compared to the prediction from n. n. terrain values (Figure 6a). Figures 3c and 3f show that the difference between the probability prediction by n. n. and mean relief values (Figure 6c), + 0.1 – 0.3, is not due to this higher Stagnosol probability on high altitudes as predicted by mean terrain values on concave sites. This difference accounts for only 0.05. However, it is due to the reduced probability assigned to convex sites ≥ 2135 m a.s.l. (0.24 difference). As a conclusion to this, the two models are quite similar, mainly differing by the dependence on curvature, which is not included in the model from n. n. relief values. Including WRB (FAO, IUSS Working Group WRB, 2007) hierarchy in the probability prediction, a site classified as Histosol or Leptosol cannot be classified as Stagnosol. Accordingly, Histosol probability reduces Stagnosol probability to a perceptible extent (Figures 6d and 6e). Figures 6g and 6h show that these differences account for 0.1 to 0.3 for most of the research area with the prediction by n. n. terrain values (Figure 6g) still yielding less differences in the lower altitudes compared to the prediction by mean terrain values (Figure 6h). Differences between the two models are extended while including WRB hierarchy (Figure 6f), compared to that being independent of WRB hierarchy (Figure 6c).

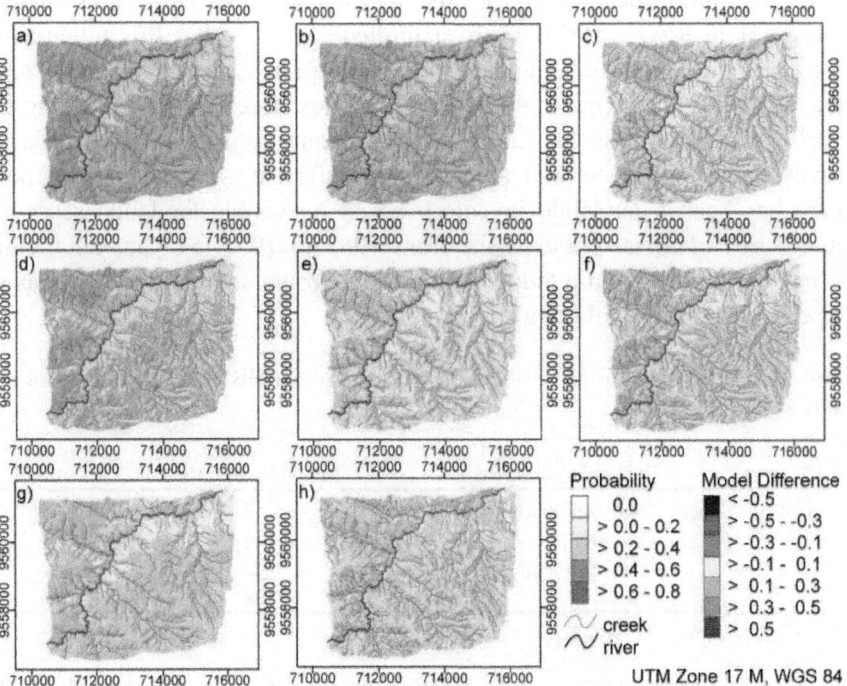

Figure 6. Maps of Stagnosol occurrence probability (Overlaid hill shading with light source from north-east). Independent on WRB hierarchy: a) prediction by n. n. terrain values, b) prediction by mean terrain values and c) model difference. Dependent on WRB hierarchy: d) prediction by n. n. terrain values, e) prediction by mean terrain values and f) model difference. Difference between independent and WRB hierarchy dependent prediction: g) n. n. terrain values and h) mean terrain values.

CTs for Umbrisols, Cambisols and Regosols cannot be provided. Umbrisol prediction was impossible, since the used dataset contains only 7 Umbrisols among 367 sampled sites and is not enough to gain a clear prediction scheme. Furthermore, not all but some of the determined Umbrisols are situated within the accumulation zone of former landslides so that an additional variable to predict their occurrence would be necessary. Cambisols and Regosols, on the other hand, are rather unspecific RSGs which makes their prediction difficult. Cambisols need a cambic horizon, but apart from that they are rather determined by the absence of diagnostic criteria that would classify the soil for another RSG. Regosols are even worse, since they do not have any characteristic on their own, but refer to all soils that do not classify as another RSG.

Model Performance and Uncertainty

Overall CT model performance is limited (Table 1). Terrain attributes can likely only explain RSG distribution to a limited extent within this mountainous tropical landscape. Unfortunately, no information is available about parent material distribution, but rapid bedrock changes were discovered during field work. The profound influence of landslides causes shifts in soil material and mixes it with rock material, leading to quite different soil properties. Although there has been a landslide inventory based on visible landslide scars on a time series of aerial photographs from 1962 to 1998 (Stoyan, 2000), most former landslides remain hidden under the regrown dense forest cover as was experienced during field work.

Table 1. Model quality of classification trees to predict Histosol and Stagnosol probability

RSG	Terrain Parameters	Model Pseudo R^2	CV Pseudo R^2
Histosol	nearest neighbour	0.34	0.22
	mean	0.35	0.21
Stagnosol	nearest neighbour	0.22	0.19
	mean	0.28	0.13

RSG= Reference Soil Group, CV = cross validation

CTs in general have certain disadvantages: (1) They are very dependent on the dataset used, i.e. some sample points more or less may lead to rather different models and (2) they predict abrupt values due to the grouping into end nodes. A continuous probability distribution of the RSGs in reality therefore is replaced by some probability classes according to Figure 3. What makes WRB RSG prediction in general problematic is the character of the WRB itself. Assignment of some RSGs requires exceeding an absolute (Histosols) and for others a relative (Stagnosols) thickness value of a diagnostic horizon. If a soil has an organic layer $\geq$ 40 cm, it is classified as Histosol independent of its mineral properties. If the organic layer is 1 cm less, these mineral properties abruptly become important. Relating the extent of the stagnic horizon to soil depth obviously is not characteristic enough to allow for a good model relating the Stagnosol occurrence pattern to terrain parameters. This is probably the reason why model accuracy is limited. As a consequence the low R^2 are not considered as a problem, but as a natural phenomenon in predicting complex entities such as RSGs.

Furthermore, the calculated CT R^2 refers to a one value prediction. As was described earlier, a CT model usually assigns the category which forms the majority within each end node to the respective landscape position. It does not consider other categories assigned to that end node as classification possibility, but neglects them. Any soil map has a certain degree of uncertainty. Usually boundaries between soil units are drawn according to expert knowledge or GIS interpolations. However, the degree of uncertainty which is a logical phenomenon in any below ground investigation usually is not included within the soil map. The new generation of digital soil maps provides a new development in this area. Accordingly, our digital soil maps include this model uncertainty through assigning RSG occurrence probabilities instead of unique values. Other authors mainly used fuzzy-logic to include this uncertainty, e.g. McBratney & De Gruiter (1992), Hannemann (2010).

Another aspect to be considered, is that generally soil maps are gained on a much larger scale. Lagacherie & Holmes (1997) use a spatial resolution of 50 m, Moran & Bui (2002) use 250 m. Therefore, the small scale, 10 m resolution, in our soil maps might be another reason for the low R^2. The soils within the research area change within a few meters radius as typical for tropical soils. Accordingly, the highest possible resolution was used. This way low scale soil variability is included within the models, which would be neglected while working on a larger scale. To conclude, the size of the applied dataset is not enough to represent the investigated soil-landscape at this high precision.

Comparison with Earlier Soil Map

A RSG probability prediction is also possible from a single CT which predicts all RSGs at once. Liess et al. (2009) established such a CT for the research area (Figure 7), but did not predict probabilities from it. The percentage of the RSGs within each end node of this tree was interpreted as occurrence probability for the RSGs according to the related landscape position and compared it to the findings from the various CTs of this study. The difference between RSG probability by the tree model from Liess et al. (2009) and our predictions is displayed in Figure 8. The first column maps the RSG probabilities according to Liess et al. (2009), the second column presents the differences between the latter and our prediction from n. n. relief values (WRB dependent), and the third column shows the differences regarding the prediction from mean relief values.

The model from Liess et al. (2009) (Figure 7) assigned a very high Histosol probability with 0.6 – 0.8 to about half of the research area. For some sites the predicted probability was even higher. In our new model, Histosol probability was less, 0.2 – 0.4 for most of the area (Figure 4b), but continuous on all sites with at least 0.2 (Figure 3a and 3d). It was shown that Histosol probability is high within some landscape positions and for a VOFD from 54 – 175 m this is supported by a high number of sampled sites. In contrast to this, the end nodes in the tree model from Liess et al. (2009) mostly contain only a very limited number of sampling sites, e.g. the end nodes that predict particularly high Histosol probabilities (≥ 0.8) only contain 12 – 15 sampled sites. The end node with the most sites predicting Histosol probability with 0.78, refers to landscape positions in small catchments < 214 m HOFD, similar to our findings. The importance of the catchment size as first subdividing variable for model development was confirmed. For smaller catchment sizes, i.e. sites through which a smaller area discharges, Histosol occurrence is more likely. Leptosols were predicted with low probability on steep slopes and close to the creeks (< 20, < 19 m HOFD). The latter is confirmed by Liess et al. (2009) who predicted Leptosols < 21 m HOFD, but with a high probability of 0.71 (Figure 8d). 0.71 of 7 sampled sites that are contained within the respective end node no. 7 (Figure 7) are 5 sampled sites. To use only five Leptosol sites to predict such a high probability seems unreasonable. On steep slopes, especially in an area influenced by landslides, soils have less chance to develop. Hence, it is no surprise to find Leptosols in these landscape positions. Close to the creeks soil material is probably removed downslope within the channel system during times of high rainfall; through these sites a high amount of water discharges due to a high contributing catchment area. On many sites the organic layer directly overlies continuous rock.

Stagnosols were predicted with a higher probability by n. n. relief values compared to the model from Liess et al. (2009) (Figure 8h). This is due to the fact that Stagnosols were predicted as all soils that display sufficient stagnic properties, but it was neglected that some of them carry a sufficiently thick organic layer to qualify as Histosols. Stagnic properties and thick organic layers occur at the same landscape position: The WRB (FAO, IUSS Working Group WRB, 2007) describes Histosols as soils in "poorly drained basins and depressions" and "highland areas with a high precipitation–evapotranspiration ratio".

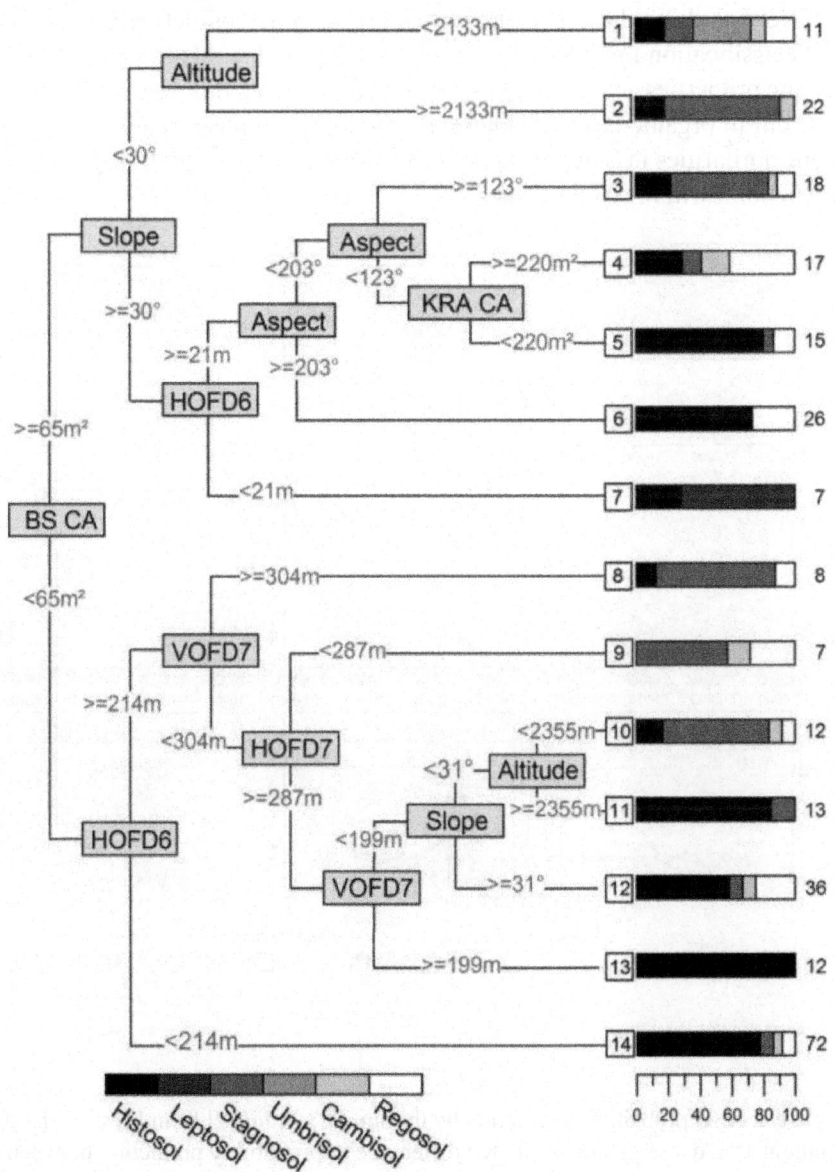

Figure 7. Classification tree model to predict RSG probability within the research area. Numbers before the boxplots indicate the node number, numbers behind the boxplots indicate the number of sample sites per end node. BS CA and KRA CA upslope contributing catchment area according to the Braunschweiger relief model and kinematic routing algorithm, HOFD horizontal and VOFD vertical overland flow distance, 6, 7 refer to different precision in channel network (adapted from Liess et al., 2009).

Nevertheless, these two properties are seen as competing if it comes to soil classification by WRB. Two soils showing both a thick organic layer and stagnic properties are assigned to different RSGs even if they are different only by 1 cm in organic layer thickness. Prediction from mean relief values shows more similarities in Stagnosol probability to Liess et al. (2009) (Figure 8i) than prediction by n. n. terrain values.

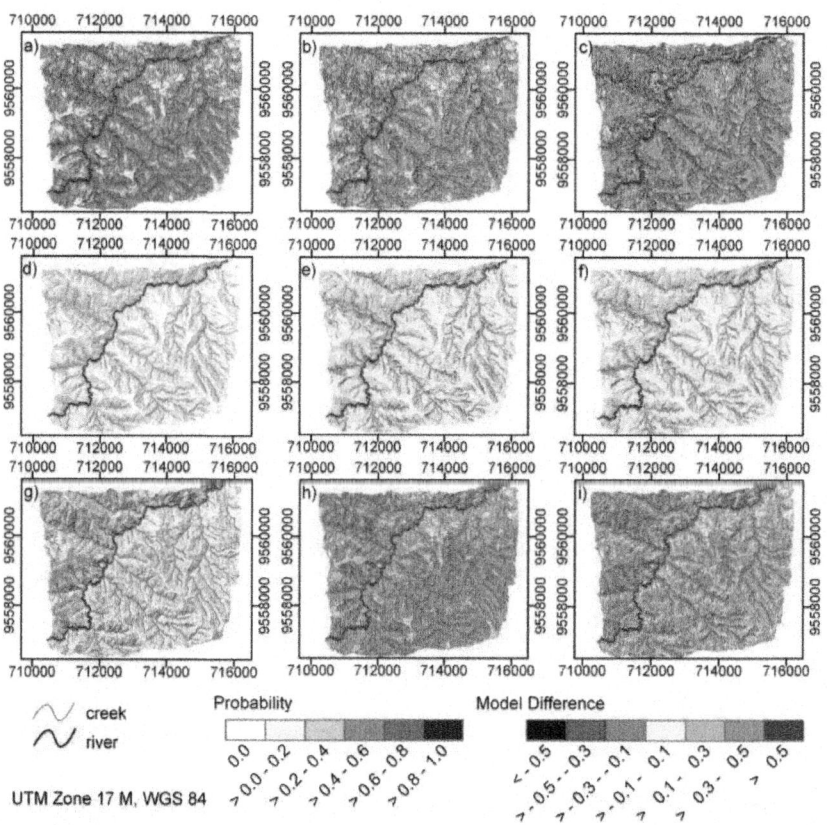

Figure 8. RSG probability prediction by the simple tree model from Liess et al. (2009) (column 1: a, d and g) and calculated difference in probability prediction between that model and the WRB dependent model from n. n. (column 2: b, e and h) and mean terrain values (column 3: c, f and i). Histosol (1st row), Leptosol (2nd row) and Stagnosol (3rd row). Overlaid hill shading with light source from north-east.

This is because Liess et al. (2009), who used a subset of our dataset, predicted the RSGs by mean relief values, too. Stagnosol probability increases above an altitude of 2146 m a.s.l. on slope angles < 40°. An increase in Stagnosol abundance with increasing altitude and decreasing slope angle was also described by Liess et al. (2009). Schrumpf et al. (2001) stated an increase in hydromorphic properties with increasing altitude and designated soils as Humaquepts (Soil Survey Staff, 2006). The increase with altitude can be attributed to the increasing rainfall (Rollenbeck, 2006). Lesser steep slope angles account for a slower discharge.

We assume the RSG probability predicted by various CTs, to better represent soil reality within the research area, since the dataset does not consist of all RSGs to an equal extent so that some are preferred over others during the tree subdivision process. Furthermore, the multiple CTs rather predict probabilities of soil diagnostic properties, which can occur simultaneously at one site within the soil profile. Accordingly, the model from Liess et al. (2009) overestimated Histosol probability for most sites as can be seen by the mainly green colours in Figure 8b and c. However, at the same time it underestimated Stagnosols in most of the area as can be deduced from the prevailing red colours in Figures 8h and i. In a similar way, Leptosols are overestimated by the model from Liess et al. (2009).

CONCLUSIONS

Models adapted for n. n. compared to those adapted for mean terrain values showed only minor differences. We conclude that predicting all RSGs at once is not as good as predicting each RSG on its own by a CT. The dataset does not consist of all RSGs to an equal extent, so some RSGs are preferred over others during the tree subdivision process.

Model performance might be improved by choosing a lower resolution to exclude small scale diversity, reducing model dependence on the dataset, applying a different statistical model or predicting soil properties instead of the complex RSG entities. However, further research is needed to prove these assumptions.

Model uncertainty in the digital soil maps is represented by the occurrence probabilities of the RSGs. Probabilities of various RSGs at the same landscape position can be understood as competing RSGs. But the probabilities of the various RSGs can also be interpreted as a soil composed of the various RSGs, i.e. various diagnostic horizons or various soil processes running simultaneously or successively as has been part of soil genesis theory for a long time (Simonson, 1959; Schelling, 1970). Thereby, this provides a good means to acknowledge inter-relations between the RSGs. An even better chance to acknowledge this would be the prediction of the diagnostic properties necessary for WRB classification by themselves. In accordance with McBratney & De Gruiter (1992), who thought to improve the existing soil classification systems via fuzzy sets, we would like to contribute the above-mentioned ideas to the development of a continuous soil systematisation system.

ACKNOWLEDGEMENTS

The authors are indebted to the German Research Foundation (DFG) for funding the study in the framework of the Research Unit FOR816. Logistic support of the foundation Nature and Culture International (NCI, San Diego – Loja) is gratefully acknowledged. Furthermore, we would like to thank Christopher L. Shope for English language revision.

REFERENCES

1.   Bauer, J., Rohdenburg, H., & Bork, H.-R. (1985). Ein digitales Reliefmodell als Vorraussetzung für ein deterministisches Modell der Wasser- und Stoff-Flüsse. In: Landschaftsgenese und Landschaftsökologie H. 10, Parameteraufbereitung für deterministische Gebiets-Wassermodelle, Grundlagenarbeiten zur Analyse von Agrar- Ökosystemen, Bork, H.-R., Rohdenburg, H., pp. (1 – 15). Selbstverlag Abteilung für Physische Geographie und Landschaftsökologie der Technischen Universität Braunschweig, ISSN 0170-7299, Braunschweig

2.   Bishop, T. F. A., Minasny, B., 2006. Digital Soil-Terrain Modeling: The Predictive Potential and Uncertainty. In: Grunwald, S. (Ed.) Environmental Soil-Landscape Modeling. ISBN 0-8247-2389-9, CRC Press, Boca Raton.

3.   Böhner, J., McCloy; K.R., & Strobl, J. (2006). Göttinger Geographische Abhandlungen 115: SAGA – Analysis and Modelling Application, Geographisches Institut der Universität Göttingen, Verlag Erich Goltzt GmbH & Co. KG, Göttingen

4.   Bourennane, H., King, D., Couturier, A., 2000. Comparison of kriging with external drift and simple linear regression for predicting soil horizon thickness with different sample densities. Geoderma 97, pp.(255– 271).

5.   Breimann, L., Friedmann, J.H., Olshen, R.A., & Stone, C.J. (1984). Classification and regression trees, CRC press, Wadsworth.

6.   Burrough, P. A., Bouma, J., Yates, S. R., 1994. The state of the art in pedometrics. Geoderma, Vol 62, pp.(311 – 326), ISSN 0016-7061

7.   Cimmery, V. (2007). User guide for SAGA, version 2.0, Available from: http://sourceforge.net/ projects/saga-gis/ files/(access: 25.11.2009)

8.   Dobos, E., Micheli, E., Baumgardner, M.F., Biehl, L., & Helt, T. (2000). Use of combined digital elevation model and satellite radiometric data for regional soil mapping. Geoderma, Vol. 97, pp. (367–391), ISSN 0016-7061

9. Dokutschajew, W. W., 1883. Russkij Cernozem, St. Petersburg.

10. FAO, IUSS Working Group WRB (2007). World Reference Base for Soil Resources, ISRIC, Rome

11. Fries, A., Rollenbeck, R., Göttlicher, D., Nauss, T., Homeier, J., Peters, T., & Bendix, J. (2009). Thermal structure of a megadiverse Andean mountain ecosystem in southern Ecuador, and its regionalization. Erdkunde, Vol. 63, pp. (321–335), ISSN 0014-0015

12. Gessler, P., Moore, I., McKenzie, N., & Ryan, P. (1995). Soil-landscape modelling and spatial prediction of soil attributes. International Journal of Geographical Information Systems, Vol. 9, 4, pp. (421– 432), ISSN 1365-8816

13. Grunwald, S., 2006. What Do We Really Know about the Space-Time Continuum of Soil Landacapes? In: Grunwald, S. (Ed.), Environmental Soil-Landscape Modeling, pp. (3 – 36), ISBN 0-8247-2389-9, CRC Press, Boca Raton.

14. Hannemann, J. (2010). Die Berücksichtigung inhaltlicher und räumlicher Unschärfe bei der GISgestützten Erstellung von Bodenkarten. Dissertation Universität Bayreuth, Geowissenschaft. Shaker Verlag, Aachen

15. Hengl, T., Heuvelink, G.B.M., & Stein, A. (2004). A generic framework for spatial prediction of soil variables based on regression-kriging. Geoderma, Vol. 120, No. 1-2, pp. (75– 93), ISSN 0016-7061

16. Homeier, J., Dalitz, H., & Breckle, S.-W. (2002). Waldstruktur und Baumartendiversität im montanen Regenwald der Estacón Cientíca San Franscisco. Südecuador. Ber. d. Reinh. Tüxen-Ges, Vol. 14, pp. (109–118), ISSN 0940-418X

17. Jenny, H. (1941). Factors of soil formation. A system of quantitative pedology, McGraw-Hill, New York.

18. Lagacherie, P. & Holmes, S. (1997). Addressing geographical data errors in a classification tree soil unit predicton. International Journal of Geographic Information Science, Vol. 11, pp. (183–198), ISSN 1365-8816

19. Lea, N. L. (1992). An aspect driven kinematic routing algorithm. In: Overland Flow Hydraulics and Erosion Mechanics. Parsons. A. J. and Abrahams, A. D., pp. (393 – 407), London

20. Liess, M., Glaser, B., & Huwe, B. (2009). Digital Soil Mapping in Southern Ecuador. Erdkunde, Vol. 63, 4, pp. (309–319), ISSN 0014-0015

21. Litherland, M., Aspen, J.A., & Jemielita, R.A., 1994. The metamorphic belts of Ecuador. Overseas Memoirs - British Geological Survey, Vol. 11,

pp. (1–147), ISSN 0951-6646

22. McBratney, A.B. & DeGruiter, J.J. (1992). A continuum approach to soil classification by modified fuzzy k-means with extragrades. Journal of Soil Science, Vol. 43, pp. (159– 175), ISSN 0038-075X

23. Milne, G. (1935). Some Suggested Units of Classification and Mapping, Particularly for East African Soils. Soil Research, Vol. 4, pp. (183–198), ISSN 0309-133

24. Moran, C.J. & Bui, E.N. (2002). Spatial data mining for enhanced soil map modelling. International Journal of Geographic Information Science, Vol. 16, pp. (533–549), ISSN 1365-8816

25. Moore, I., Gessler, P., Nielsen, G., Peterson, G., 1993. Soil attribute prediction using terrain analysis. In: Soil Science Society of America Journal, 57/ 2, pp.(443– 452).

26. Odeh, I., McBratney, A., Chittleborough, D., 1994. Spatial prediction of soil properties from landform attributes derived from a digital elevation model. Geoderma, 63/3-4, pp.(197– 214).

27. Rollenbeck, R. (2006). Variability of precipitation in the Reserva Biólogica San Francisco / Southern Ecuador. Lyonia, A Journal of Ecology and Application, Vol. 9, No. 1, pp. (43– 51), ISSN 0888-9619

28. Schelling, J. (1970). Soil genesis, soil classification and soil survey. Geoderma, 4, 3, pp. (165– 193), ISSN 0016-7061

29. Schrumpf, M., Guggenberger, G., Valarezo, C., & Zech, W., 2001. Tropical montane rainforest soils. Development and nutrient status along an altitudinal gradient in the South Ecuadorian Andes. Die Erde, Vol. 132, pp. (43–59), ISSN 0013-9998

30. Simonson, W.R. (1959). Outline of a Generalized Theory of Soil Genesis. Soil Science Society of America Journal, Vol. 23, pp. (152–156), ISSN 0361-5995

31. Skidmore, A.K., Watford, F., Luckananurug, P., & Ryan, P.J. (1996). An operational GIS expert system for mapping forest soils from a geographical information system. International Journal of Geographic Information Science, Vol. 5, pp. (431–445), ISSN 1365-8816

32. Soil Survey Staff (2006). Keys to Soil Taxonomy. 10th ed. United States Department of Agriculture, Natural Resources Conservation Service, Available from: http://soils.usda.gov/technical/classification/ taxonomy/ (access: 22/10/2007)

33. Stoyan, R., 2000. Aktivität, Ursachen und Klassifikation der Rutschungen

in San Francisco/ Südecuador. Diplomarbeit Universität Erlangen.

34. Strahler, A.N. (1957). Quantitative analysis of watershed geomorphology. Transactions of the American Geophysical Union, Vol. 38, No. 6, pp. (913–920), ISSN 0002-8606

35. Therneau, T.M. & Atkinson, B. (2003). The rpart Package, Available from: http://cran.rproject.org/web/ packages/ rpart/ rpart.pdf (access: 28/02/2008)

36. Thomas, A.L., King, D., Dambrine, E., Couturies, A., & Roque, A. (1999): Predicting soil classes with parameters derived from relief geologic materials in a sandstone region of the Vosges mountains (northeastern France). Geoderma, Vol. 90, pp. (291– 205), ISSN 0016-7061

37. Wilcke, W. ,Yasin, S., Abramowski, U., Valarezo, C., & Zech, W. (2002). Nutrient storage and turnover in organic layers under tropical montane rain forest in Ecuador. European Journal of Soil Science, Vol. 53, pp. (15–27), ISSN 1351-0754

38. Wilcke, W., Valladarez, H., Stoyan, R., Yasin, S., Valarez, C., & Zech, W. (2003). Soil properties on a chronosequence of landslides in montane rain forest, Ecuador. Catena, Vol. 53, pp. (79–95), ISSN 0341-8162

39. Yasin, S. (2001). Water and Nutrient Dynamics in Microcatchments under Montane Forest in the South Ecuadorian Andes. Bayreuther Bodenkundliche Berichte, Band 73.

40. Zevenbergen L.W. & Thorne C.R. (1987). Quantitative Analysis of Land Surface Topography. Earth Surface Processes Landforms, Vol. 12, pp. (47 – 56), ISSN 0197-9337

Chapter 10

CLASSIFICATION AND MANAGEMENT OF HIGHLY WEATHERED SOILS IN MALAYSIA FOR PRODUCTION OF PLANTATION CROPS

J. Shamshuddin[1] and Noordin Wan Daud[2]

[1]Department of Land Management, Faculty of Agriculture, Universiti Putra Malaysia, 43400 Serdang, Selangor, Malaysia

[2]Department of Crop Science, Faculty of Agriculture, Universiti Putra Malaysia, 43400 Serdang, Selangor, Malaysia

INTRODUCTION

In Malaysia, Ultisols and Oxisols containing kaolinite, gibbsite, goethite and hematite in the clay fraction are very common especially in the upland areas, occupying about 72 % of the country's land area. The soils are highly weathered as they exist under tropical environment with high rainfall and temperature throughout the year, resulting in leaching of plant nutrients and accumulation of sesquioxides (Anda et al., 2008a). They are by nature devoid of basic cations (Ca and Mg) and available P (due to fixation by the oxides) and hence, their productivity is generally considered as low. The soils are mainly utilized for oil palm and rubber cultivation with great success due to excellent soil management practices. With the expertise available in the country, palm oil and rubber are produced in large amounts for the world market. However, cocoa growing on these soils produces low yield which are attributed to low pH and aluminum and/or manganese toxicity.

The problems of low productivity can be overcome by liming using ground magnesium limestone (Shamshuddin et al., 1991; Ismail et al., 1993; Shamshuddin and Ismail, 1995; Shamshuddin et al., 1998; Shamshuddin et al., 2009; Shamshuddin et al., 2010) or by applying basalt (Gillman et al., 2001; Anda et al., 2009; Shamshudin & Kapok, 2010). Basalt releases Ca, Mg, K, P and S on its dissolution into the soils (Gillman et al., 2002). Out of the six macronutrients needed by the growing crops in the field only N is not present in basalt.

Most of the Ultisols and Oxisols in the tropics is lacking in organic matter which can supply essential plant nutrients as well as improve soil structures. Normal organic matter applied for alleviating the infertility of Ultisols and Oxisols in Malaysia is compost (Anda et al., 2008b; Anda et al., 2010) and palm oil mill effluents.

This paper intends to classify the highly weathered soils in Malaysia and discusses the management of the soils for sustainable production of oil palm, rubber and cocoa. The information given in this paper is useful to students and researchers alike.

HIGHLY WEATHERED SOILS OF MALYSIA

Classification

Due to high temperature in Malaysia silicate minerals become unstable due to the changes in the chemistry of their environment. The new chemical conditions are dominated by aqueous state. The consequence of such an environment is a tendency to hydrate the hightemperature silicate minerals. Major effect of hydration is a large portion of the minerals is dissolved integrally into the altering aqueous solution and it is transported as such into lakes and ocean. This process of chemical weathering is very intense in Malaysia due to the prevailing high rainfall and temperature. Residual products of rock and mineral weathering under this condition are quartz, secondary phyllosilicates and sesquioxides with or without muscovite, depending on the degree of weathering.

A chemical reaction denoting dissolution of feldspar (orthoclase) in water is as follows (Duff, 1993):

$$6H_2O + CO_2 + 2KAlSi_3O_8 \rightarrow Al_2Si_2O_8(OH)_4 + 4SiO(OH)_2 + K_2CO_3$$

$$\text{(feldspar)} \qquad \text{(kaolinite)}$$

In this reaction, K is lost via leaching into the groundwater. Further weathering of the clay mineral results in the formation of colloidal materials according to the following reaction (Duff, 1993):

$$H_2O + Al_2Si_2O_8(OH)_4 \rightarrow Al_2O_3.nH2O + SiO(OH)_2$$

Exposure of granite, shale, schist, sandstone, basalt, andesite and serpentinite in Malaysia over a long period of time to the forces of weathering results in the formation of highly weathered materials, dominated by kaolinite, halloysite, gibbsite, goethite and hematite in the clay fraction (Tessens & Shamshuddin, 1983). The overall products are collectively termed as soil materials, which are usually devoid of plant nutrients.

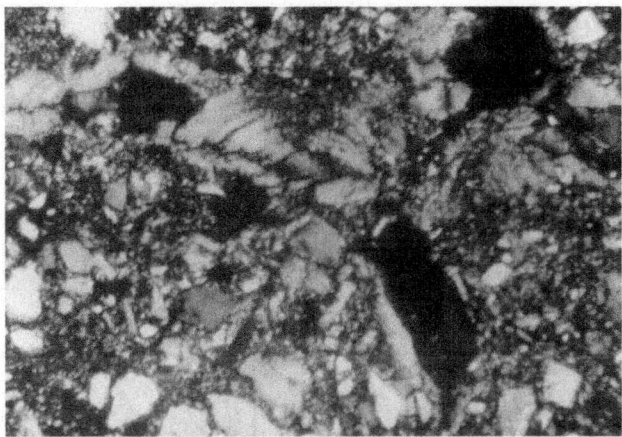

Figure 1. Argillic horizon present in the B horizon of Ultisols (Courtesy of S Paramananthan).

The soils are classified as either Ultisols or Oxisols according to Soil Taxonomy (Soil Survey Staff, 2010). Ultisol is defined by the presence of argillic horizon in the subsoil (Figure 1). Clays from the topsoil are moved and accumulated in the B horizon under intense leaching environment in the tropics. By nature, Oxisols are considered as more weathered than Ultisols. They are dominated by kaolinite and sesquioxides (Anda et al., 2008a). The soils are defined by the presence of oxic horizon in the subsoil (Figure 2). The CEC of the soils is extremely low, with value < 16 cmolc/kg clay. As such, plant nutrients are mostly lost via leaching, further lowering the productivity of the soils.

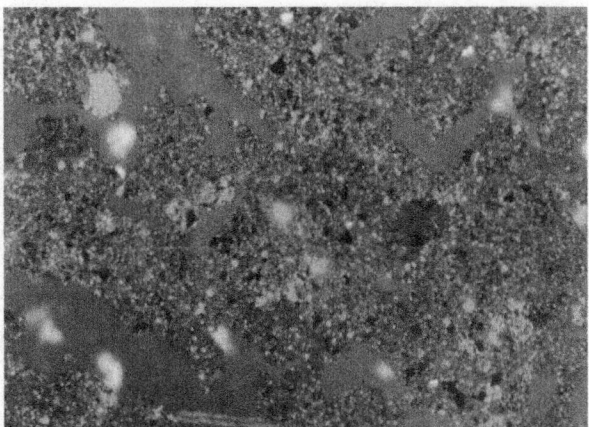

Figure 2. Oxic horizon present in the B horizon of Oxisols (Courtesy of S Paramananthan).

Charge Properties of Ultisols and Oxisols

It is known that soil materials are negatively- and/or positively-charged. A certain amount of negative charge in the soils is derived from within the phyllosilicates themselves via a process called isomorphic substitution. For instance, replacement of Si by Al in kaolinite present in the soils would result in excess of negative charges. Such charges are termed as negative permanent charges. Positive permanent charges are also existed in soils, being produced by isomorphic substitution of Fe by Ti in soils containing high amount of oxides of Fe (Tessens & Shamsuddin, 1983). This kind of isomorphic substitution is common in Oxisols having high amount of hematite and goethite, the active minerals in the soils.

Oxides of Fe and Al and the broken edges of phyllosilicates have another kind of charge known as variable charge. As the pH of the ambient solution changes the charge on the surfaces of these minerals also changes. When the pH is low, protons are chemisorbed onto the minerals to become net positively-charged. On the other hand, the minerals are net negatively-charged at high pH. The pH at which the net charge of the variable-charge mineral is zero is termed as pHo.

Each mineral has its own pHo value. The value for silicate is low, but for oxides it is high. Generally, soil is composed of many variable-charged minerals. The value reported for a particular soil is actually the resultant pHo value of the whole minerals in the soils. Silicate is abundant in Ultisols, while oxides are abundant in Oxisols. Hence, the pHo value of Ultisols is lower than that of Oxisols (Tessens & Shamsuddin, 1983). This means that the mineralogy of the soil affects its pHo value.

We do not have soils completely composed of variable-charged minerals. Thus, overall charges in the soils have to be considered for the meaningful interpretation of the soil properties. The pH at which the net charge is zero, taken into account the whole soil materials, is termed as point of zero net charge (PZNC). Studies in Malaysia showed that PZNC is lower than pHo in the Ultisols and is higher than pHo in the Oxisols (Tessens & Shamsuddin, 1983). It is also observed that both pHo and PZNC values increase with increasing stage of soil weathering. As the soils weather more oxides are formed, leading to an increase in pHo and PZNC. The soils then become less productive and need special management practices for sustainable crop production.

Total charge in the soil can be subdivided into permanent and variable charge components, and can be represented by the following equation:

$$Qt = Qp + Qv$$

where

Q = charge

t = total

p = permanent v = variable

The amount of total charge in the soils can be manipulated by changing the Qv.

Qv is related to (pHo - pH)

Taking soil having net negative charge as an example, Qt can be increased by increasing the difference between pHo and pH. This can readily be done either by lowering pHo or increasing pH. The former can be implemented by incorporating basalt into the soils (Anda et al., 2009), while the latter is easily accomplished by liming (Shamshuddin et al., 1991) or applying basalt (Shamshuddin & Kapok, 2010). pHo can also be lowered, to a certain extent, by incorporating organic matter into the soil (Shamshuddin et al., 1987; Anda et al., 2008b). A soil system is stable when the charge in it is low or at the minimal, suggesting that soil potential decreases as the charge decreases. Therefore, the potential in the soil is very low when pH is near its pHo. Under natural condition, soil pH tries to move to its pHo in order to achieve maximal stability.

As soil pH increases with weathering, its pHo increases. It has been shown clearly that soil pH increases with increasing pHo (Tessens & Shamshuddin, 1983). Primary minerals break up during the course of weathering and new minerals are formed, meaning that the silicates in the soils changes to oxides of Fe and Al, and consequently the pHo of the soils increases.

Soil pH

We have analyzed pH of hundreds of soil samples from all over Peninsular Malaysia and found that the values are mostly between 4 and 5 (Tessens & Shamshudin, 1983). The values are generally lower for the Ultisols than that of the Oxisols. The former ranges from 4.0 to 4.5, while the latter ranges from 4.5 to 5.0. The explanations for this phenomenon are as follows:

a. Soils will be at their greatest stability when the potential in them is zero, that is at the pH = pHo. The pHo of majority of the highly weathered soils in Malaysia is 4-5. Therefore, pH of the soils tries to approach 4-5 in order to remain stable; and

b. On weathering, Al^{3+} and Fe^{3+} in the phyllosilicates are released into the soil solution. Al^{3+} in the soil solution undergoes hydrolysis.

This allows us to determine the pK_a of Al which is 5. Likewise, Fe^{3+} hydrolysis releases proton into the soil solution. The pK_a value of Fe is 3.0. If these free reactions are allowed to take place without interruption, then the pH of the soil solution will go near their pK_a value in order to achieve equilibrium

Effects of Low pH on Al and Mn Availability

As the soil pH of highly weathered Malaysian soils is low (< 5), Al on the exchange complex of the soils readily dissolves into the soil solution. In many cases, the Al in the soil solution is present at toxic level. The Al concentration in the soil solution increases as the pH lowers. Likewise, Mn may exist at toxic level at low pH. Exchangeable Al is lower in the Oxisols than in the Ultisols. It follows that there is less soil solution Al in Oxisols than in the Ultisols. This is consistent with the higher pH of the Oxisols as compared to that of the Ultisols. Low pH and high exchangeable Al have little effect on the growth of either oil palm or rubber, but they are expected to reduce the yield of cocoa significantly.

OIL PALM CULTIVATION

The most important agricultural crop in Malaysia right now is oil palm (Figure 3). The area covered by the so-called golden crop is estimated to be about 4.6 million ha, sporadically distributed throughout the length and breadth of the country. Currently, most of the oil palm is grown on upland areas where Ultisols and Oxisols occur. With good soil management practices oil palm grows very well, contributing to the wealth of the nation. At the current of production rate and high price in the marketplace some USD 16.7 billion is added into the economy annually, and so oil palm is indeed helping the economy of Malaysia going for a long run. In the case of Malaysia, the industry has to be protected at any cost in order to become a developed country comes the year 2020. By then Malaysia would have become a country with high income. If we keep doing what we do now and stick to the principle of sustainable crop production, it certainly will.

For oil palm cultivation, fertilizer input is very necessary where NPK fertilizers need to be applied regularly at the appropriate rates. This is because the Ultisols and Oxisols in the country are devoid of macronutrients, resulting from extreme leaching and weathering. As the soils are acidic in nature P-fertilizer recommended for application is phosphate rock, which is a slow release fertilizer of sort. As it dissolves slowly under low pH condition P is released and made available to the growing oil palm in the field. Phosphate is known to react with oxides/hydroxides of Fe in the soils, resulting in a slight increase in negative charge and pH (Tessens & Zaharah, 1983). This phenomenon has, to some extent, improved the productivity of the soils. However, we should not forget the fact that some of the applied P is fixed by the oxides of Fe, especially in Oxisols and consequently lost into the soils indefinitely or until some other reactions that dissolve the so-formed $FePO_4$. As such, phosphate fertilizer efficiency study is a popular topic of research in the tropical region, such as Malaysia and Brazil.

Oil palm is found to be acid tolerant, and hence, it can still grow even at the soil pH of 4.3 and at high Al concentration (Auxtero & Shamshuddin, 1991). Due to that Ultisols and Oxisols in Malaysia are considered suitable for oil palm cultivation. There is no need to apply lime onto the soils for oil palm cultivation as the area under oil palm is very large and therefore not economical. For sure, soil moisture has to be maintained at the optimal level for oil palm growth. We know for sure that oil palm can only be grown if the annual rainfall exceeds 1800 mm and it must be evenly distributed throughout the year. There should not be a dry period exceeding a few months. That it is so because oil palm is originated from swampy areas in African countries. The best growing areas in Peninsular Malaysia for cultivating oil palm is southern part (Johor) where rainfall is evenly distributed at 3000 mm/year. Kedah and Perlis (northern part) are probably a bit too dry for oil palm cultivation.

Under estate management, the yield of oil palm grown on Ultisols and Oxisols ranges from 20 to 30 tones fresh fruit bunches (FFB) per hectare per year. It will be higher on well managed soils planted with high yielding oil palm clone. According to a reliable source, this special clone can produce yield up to 40 t ffb/ha/year under special soil management practices. In the near future, we can expect more of this clone to be planted in Malaysia. On the average, the rate of oil extracted by the oil palm factory in Malaysia right now is 20 % of the ffb. So, we can get at least 4 tones of oil/ha/year. Malaysia is now aiming for a rate of 25 % extraction in the near future considering the advent of new extracting technology coming from R&D by the industry.

Figure 3. A well-managed oil palm estate in Malaysia.

As seen in Figure 3 oil palm fronds are placed in between the planting rows. This practice helps maintain soil moisture in the oil palm estates. When the organic matter is decomposed (mineralized) plant nutrients (N, P, K) are released into the soils and can be taken up by the growing oil palm in the field.

After 25 years of production it is time to replant the oil palm. At this age the palm is no longer productive. Furthermore, it is too tall for harvesting the fresh fruit bunches using standard practice (knife stuck to a long wooden pole). Under the current practice burning of oil palm fronds and trunks are not allowed, not eco-friendly. Oil palm estates are now forced to practice zero-burning technology. In this technique of estate management, the fronds and trunks are cut and chopped off into small pieces and are buried into the ground. In this practice, we return the oil palm biomass to the soils for good. Again on mineralization, essential nutrients are added into the soils. It has been found that zeroburning technology is able to cut down the cost of production considerably where less fertilizer is applied to keep the oil palm growing. In the beginning a teething problem has arisen. The burying of plant biomass has encouraged the outbreaks of pests and diseases. However, these problems have since been taken care of by the estate management in the country.

Many areas under oil palm cultivation in Malaysia are undulating to steep land. These areas need to be protected lest soil erosion removes the fertile topsoil comes rainy season. In these areas contour terraces are constructed perpendicular to the slope of the land. The bare areas in between the terraces are planted with leguminous cover crops before oil palm canopy closes up. The cover crops fix some nitrogen from the air and added to the soils when they die. This is an excellent management practice for sustaining the productivity of Ultisols and Oxisols in Malaysia.

RUBBER CULTIVATION

Rubber has been grown commercially in Malaysia since 1903 when the first rubber estate was established in Melaka. Malaysia is the biggest consumer of pure latex and fifth in the consumption of natural rubber in the world. Asia is the biggest producer of natural rubber, led by Thailand, Indonesia and Vietnam. Surgical latex gloves and condoms are known for their high quality and accepted by the consumers. Currently, Malaysia is the biggest exporter of these products. The export earning from rubber in 2009 was worth USD 8.3 billion. Demand for natural rubber is expected to increase and it is projected that there will be a shortage of natural rubber in the near future. The price of SMR 20 was USD 0.77/kg in 2003 and USD 3.52/kg in 2010 (Malaysian Rubber Board, 2010). Rubber prices reached their highest in decades, fuelled by strong demand from China's auto industry.

Clones, in the case of rubber, refer to rubber plants produced from selection breeding process, which have been field-tested. There are several selection criteria in rubber breeding. These include vigor, resistance to diseases, resistance to wind damage, fewer and higher branches, bark thickness and yield. The clones are produced to improve the yield of rubber trees. In recent time, yield is not only latex but also timber from the trees, which has high demand in the marketplace. To satisfy for demand in latex and timber, the Malaysian Rubber Board has conducted research to produce rubber trees that can give high yield of latex and timber, resulting in the introduction of RRIM 2000 and RRIM 3000 series of clones. Clone RRIM 3000 is superior to that of RRIM 2000. Clone RRIM 3001, because of its vigorous growth and high yield, has been assigned a new name, i.e. Klon 1 Malaysia. Rubber plants exhibit a variety of responses to different water regimes. The ability of rubber plants to cope with water stress varies across and within clones (Shafar Jefri Mokhatar & Noordin Wan Daud, 2011).

For many years Malaysia was the major exporter of rubber. Now the country is number three in terms of latex production; Thailand and Indonesia have overtaken us of late. However, we still export rubber (scrap or solid rubber) to the world market, although we import some rubber (latex) in order to feed our own growing rubber industry. Due to the acute shortage of labor in Malaysia some rubber trees are left not tapped. The current area under rubber in Malaysia stands at 1.2 million ha, much less than what it used to be some 40 years ago. Most of our rubber are grown on Ultisols and Oxisols with little problem (Figure 4). Rubber seems to grow quite well on soils under well drained condition. Like oil palm it is acid tolerant and pH of 4-5 does not affect its growth. Under normal fertilizer input, rubber grows very well, giving high yield and contributing to the country's economic growth.

Like oil palm, the P-fertilizer for rubber cultivation is phosphate rock. This fertilizer has been applied in rubber estates for as far as we could remember. The benefit of phosphate rock has been clearly explained before. It works very well under acidic condition prevailing in the highly weathered soils of Malaysia. The preferred N-fertilizer in rubber estates is $(NH_4)_2SO_4$. The ammonium from this fertilizer undergoes nitrification which releases H^+ and hence, its long-term application would increase soil acidity slightly. Beside nitrogen, this fertilizer supplies S to the soils, which is another macronutrient. Rubber plant is said to be acid tolerant and so the acidity so produced by the application of the fertilizer would not affect its growth much. Unlike oil palm, Kedah and Perlis are quite suitable for rubber cultivation. If the rainfall is too high it will affect latex production because farmers are reluctant to tap rubber trees.

Figure 4. A well managed rubber estate in Malaysia.

Many rubber estates in the country are located on undulating or steep land, which are prone to soil erosion during rainy season. As such, contour terraces are needed to be constructed to prevent or reduce soil erosion, very much the case of oil palm estate. If this practice is carried out coupled with modern tapping technique there is no reason why Malaysia should not be a great rubber producer again, like it once was. Right now the price of rubber in the marketplace is very good and many farmers have since returned to the industry, for the good of the country.

Like oil palm estates, rubber estates are practicing zero-burning technology without fail. Malaysia is now in the forefront in promoting this eco-friendly technology for the good of us all. As has been said, practicing zero-burning in rubber estate management has reduced the cost of production and improved or protected the environment. This technology has won worldwide recognition and as such other countries are learning from us. Zero-burning technology is Malaysia's contribution to the world in reduction of global warming.

COCOA CULTIVATION

The fortune of cocoa industry in Malaysia is in the balance or in limbo. Way back in the 1970s, Malaysia was the 6th biggest cocoa bean producer in the world. The area under cocoa then was > 400,000 ha. Since then it is in the downward trend. Many cocoa plantations are either abandoned or replaced with oil palm. The cocoa area now is about 50,000 ha with production not enough to feed our own factories producing cocoa products. The government is now encouraging farmers to go back to planting cocoa by offering lucrative subsidy.

Some of the cocoa trees are planted on highly weathered soils with low fertility (Figure 5). Beside low price in the marketplace, outbreak of diseases has created havoc in cocoa industry. Furthermore, the Europeans are so used to the good taste of cocoa imported from Ivory Coast, Africa, which is grown on fertile soils with high pH and low Al saturation. Our soils, Ultisols and Oxisols, are acidic with pH ranging from 4 to 5, lower pH for the Ultisols. These soils contain toxic amount of Al and/or Mn, which also affect the taste of cocoa although it has been improved somewhat via innovative fermentation technology.

Current national average rate of cocoa production in Malaysia is low, only 0.8 t/ha/year. This is far too low compared to that of our counterpart in African countries. The target rate of production for country is 1.5 t/ha, which is still way below the potential cocoa yield of > 10 t/ha/year. In Sabah, a yield of > 2 t/ha/year can be obtained for cocoa planted on weathered soils with good soil management practice. At this rate of cocoa production, farmers can make money because the price of cocoa in the marketplace now is quite high.

For growing cocoa on the Ultisols and Oxisols, soil pH needs to be raised to above 5. At this pH, Al in the soil solution starts to precipitate as inert Al-hydroxides, rendering it unavailable to the growing cocoa in the field. Simultaneously, Mn in the soil solution is eliminated. Soil pH can be increased by liming (Shamshuddin et al., 1991; Shamshuddin et al., 2010). Likewise, soil pH can be increased by ground basalt application (Gillman et al., 2002; Shamshuddin & Kapok, 2010). Besides increasing pH, basalt supplies Ca, Mg, K, P and S into the soils for the crop's requirement. This agronomic practice helps reduce the cost of production as less fertilizer needs to be applied. The best technique is to apply basalt in combination with organic fertilizer. The organic fertilizer supplies nitrogen needed by cocoa for its healthy growth. Basalt is, however, taking a long time to disintegrate and dissolve completely. But it gives a long-term ameliorative benefit to the cocoa plant. As soil pH increases negative charges on the exchange complex of the variable charge minerals increases (Shamshuddin & Ismail, 1995). This helps retain basic cations in the topsoil and hence, soil productivity is further improved.

If the rainfall at an area is too much, the area is less suitable for cocoa cultivation. Under this condition, cocoa can be infested by diseases, resulting in the abortion of cherelles. This in the end would reduce the yield of cocoa.

This being the case, it is not wise to grow cocoa in areas with heavy rainfall like Johor where the annual rainfall can exceed 3,000 mm. As it is Sabah is the best growing area in Malaysia for cocoa production in terms of soils and climatic conditions.

Figure 5. A well managed cocoa estate in Malaysia (Courtesy of Malaysian Cocoa Board).

CONCLUSION

Oil palm and rubber are suitable to be grown on Ultisols and Oxisols in Malaysia because the crops are acid and Al tolerant, but cocoa is not unless the soils are amended with lime or basalt. However, climate can somewhat affect their production except oil palm. With proper soil management practices, the yield of oil palm and rubber are high, contributing money that sustains the Malaysian economy. As the crops are planted either on undulating and/ or sloping land, contour terraces are needed to be constructed to reduce or to prevent soil erosion that removes the fertile topsoil. Planting leguminous cover crops in between the rows of oil palm or rubber trees further helps improve soil conservation. On the other hand, cocoa requires lime or basalt application at the appropriate rate and time to increase soil pH, lower Al and increase basic cations. In so doing, cocoa can be grown sustainably on Ultisols and Oxisols in Malaysia.

ACKNOWLEDGEMENTS

The authors would like to acknowledge Universiti Putra Malaysia for technical and financial support.

REFERENCES

1.   Anda, M., Shamshuddin, J., Fauziah, C.I & Syed Omar, S.R. (2008a). Mineralogy and factors controlling charge development of three Oxisols developed from different parent materials. Geoderma, 143, 153-167.

2.   Anda, M., Shamshuddin, J., Fauziah, C.I. & Syed Omar, S.R. (2009). Dissolution of ground basalt and its effects on Oxisol chemical properties and cocoa growth. Soil Science, 174, 264-271.

3.   Anda, M., Shamshuddin, J., Fauziah, C.I. & Syed Omar, S.R. (2010). Increasing the organic matter content of an Oxisol using rice husk compost: changes in the composition and its chemistry. Soil Science Society of America Journal, 74, 1167-1180.

4.   Anda, M., Syed Omar, S.R., Shamshuddin, J. & Fauziah, C.I. (2008b). Changes in properties of composting rice husk and their effects on soil and cocoa growth. Communication in Soil Science and Plant Analysis, 39, 2221-2249.

5.   Auxtero, E.A. & Shamshuddin, J. (1991). Growth of oil palm (Elaies guineensis) seedlings on acid sulfate soils as affected by water regime and aluminum. Plant and Soil, 137, 243-257.

6.   Duff, D. (1993). Holmes' Principles of Physical Geology. Chapman and Hall, London.

7.   Gillman, G.P., Burkett, D.C. & Coventry, R.J. (2001). A laboratory study of basalt dust to highly weathered soils: effect on soil chemistry. Australian Journal of Soil Science, 39, 799-811.

8.   Gillman, G.P., Burkett, D.C. & Coventry, R.J. (2002). Amending highly weathered soils with finely ground basalt rock. Applied Geochemistry, 17, 987-1001.

9.   Ismail, H., Shamshuddin, J. & Syed Omar, S.R. (1993). Alleviation of soil acidity in a Malaysian Ultisol and Oxisol for corn growth. Plant and Soil, 151, 55-65.

10. Malaysian Rubber Board. (2010). Prices of SMR 20 by Malaysian Rubber Board. Http://www3.lgm.gov.my/mre/YearlyChart.aspx. Accecced on November 15, 2010.

11. Shafar Jefri Mokhatar & Noordin Wan Daud. (2011). Performance of Hevea brasiliensis on Haplic Ferralsols as affected by different water regimes. American Journal of Applied Sciences, 8(3), 206-211.

12. Shamshuddin, J. & Ismail, H. (1995). Reactions of ground magnesium limestone and gypsum in soils with variable-charge minerals. Soil Science Society of America Journal, 59,106-112

13. Shamshuddin, J. & Kapok, J.R. (2010). Effect of ground basalt application on the chemical properties of an Ultisol and Oxisols in Malaysia. Pertanika Journal of Tropical Agricultural Science, 33, 7-14.

14. Shamshuddin, J., Che Fauziah, I. & Sharifuddin, H.A.H. (1991). Effects of limestone and gypsum applications to a Malaysian Ultisol on soils solution composition and yields of maize and groundnut. Plant and Soil, 134, 45-52.

15. Shamshuddin, J., Fauziah, C.I. & Bell, L.C. (2009). Soil solution properties and yield of corn and groundnut grown on Ultisols as affected by dolomitic limestone and gypsum applications. Malaysian Journal of Soil Science, 13, 1-12.

16. Shamshuddin, J., Jamilah, I. & Mokhtaruddin, A.M. (1987). Chemical changes in soils affected by the application of palm oil mill effluent. Pertanika, 10, 41-47.

17. Shamshuddin, J., Sharifuddin, H.A.H. & Bell, L.C. (1998). Longevity of magnesium limestone applied to an Ultisol. Communication in Soil Science and Plant Analysis, 29, 1299-1313.

18. Soil Survey Staff. (2010). Keys to Soil Taxonomy. United States Department of Agriculture, Washington DC.

19. Tessens, E. & Shamshuddin, J. (1983). Quantitative Relationship between Mineralogy and Properties of Tropical Soils. UPM Press, Serdang, Malaysia.

20. Tessens, E. & Zaharah, A.B. (1983). The residual influence of P-fertilizer application on soil pH values. Pedologie, 32, 367-368.

Chapter 11

POULTRY LITTER FERTILIZATION IMPACTS ON SOIL, PLANT, AND WATER CHARACTERISTICS IN LOBLOLLY PINE (PINUS TAEDA L.) PLANTATIONS AND SILVOPASTURES IN THE MID-SOUTH USA

Michael A. Blazier[1], Hal O. Liechty[2], Lewis A. Gaston[1], and Keith Ellum[2]

[1]Louisiana State University Agricultural Center, USA
[2]University of Arkansas Monticello, USA

INTRODUCTION

Increasing global human populations and wealth have resulted in increased demands for animal protein and widespread use of confined animal feeding operations to meet added animal protein consumptive demands. Disposal of animal wastes from these operations can be ecologically and environmentally problematic (Kellogg et al., 2000; Roberts et al., 2004; Shober & Sims, 2003). Poultry production is an important source of this protein and is a major agricultural industry in the United States. The United States is the world's largest producer and second largest exporter of poultry meat (UDSA Economic Research Service, 2009). Four-fifths of the United States poultry industry is comprised of broiler meat production. Broiler meat production is largely concentrated in Southeastern states (Alabama, Arkansas, Florida, Georgia, Kentucky, Louisiana, Mississippi, North Carolina, Oklahoma, South Carolina, Tennessee, Texas and Virginia), with 82% of U.S. broiler production occurring in these states (National Agricultural Statistic Service, 2008).

Broiler production results in the generation of massive amounts of litter, a mixture of feces, feed, feathers and bedding materials such as straw, peanut or rice hulls, and wood shavings (Gupta et al., 1997; Weaver, 1998). The U.S.A. poultry industry produces more than 11 million Mg of litter per year (Cabrera & Sims, 2000). Broiler poultry litter contains several plant macro- and micronutrients (Table 1), which makes it desirable as an agricultural fertilizer (Sistani et al., 2008). Following removal from poultry production facilities, litter is commonly applied to nearby pastures, hay meadows, and agricultural crops such as corn and cotton to increase crop production and quality (Harmel et al., 2004; Sims & Wolf, 1994). Applications of poultry litter ranging from 4.5 to 11.2 Mg ha^{-1} yr^{-1} are common to supplement or replace inorganic annual fertilizer additions to pastures (Adams et al., 1994). Thus, poultry litter application is an efficient and potentially cost-effective method for improving forage production within the vicinity of production facilities, which helps to sustain non-poultry related agriculture economies in poultry producing regions. Substitution of broiler litter for inorganic fertilizers continues to increase in the southeastern U.S.A. as prices of inorganic fertilizers escalate (Funderberg, 2009).

Table 1. Ranges of reported nutrient concentrations in broiler poultry litter on an oven-dry basis. Adapted from Eichhorn, 2001; Ekinci et al., 2000; Kingery et al., 1994; Mitchell & Donald, 1995; Pote et al., 2003; Sauer et al., 2000; Sims, 1986; Williams et al., 1999

Element	Nutrient Concentration	
	(g kg^{-1})	(mg kg^{-1})
C	280-320	
N	31-49	
P	4-13	
K	2-28	
Ca	2-28	
Mg	0.4-6	
Fe		1950-2395
Mn		277-424
Cu		263-332
Zn		252-404
B		45-55

Poorly planned, excessive, or long-term applications of broiler litter to pastures and other agroecosystems can result in excessive nutrient losses, reductions in surface water quality, and potential risks to human health. Poultry litter is typically applied as a nitrogen fertilizer, but N availability from litter is relatively difficult to predict because only one-third of the N in litter is in exchangeable forms such as NH_4-N and NO_3-N. Two-thirds of N in litter is in organic form, which must be mineralized before it is plant available. Mineralization of N in litter varies from 40 to 90% with edaphic and environmental conditions, particularly conditions at the time of litter application (Mitchell & Donald, 1995). Gaseous losses of N from litter via volatilization can vary from 5 to 20% of total N, which reduces the amount of N available for plant use (Mitchell & Donald, 1995). While the amounts and forms of N in litter can vary considerably, those of other nutrients, particularly P, are relatively stable. As a result, if litter is applied at rates that supply sufficient N to meet crop demand soils can become saturated with P as well as K, Ca, Mg, Cd, Cu, Mn, and Zn (Edmeades, 2003; Kingery et al., 1994). Surface water runoff or soil water leaching associated with these nutrient-saturated soils can reduce water quality in watersheds (Friend et al., 2006; Gallimore et al., 1999; Gaston et al., 2003; Kellogg et al., 2000; Sauer et al., 1999; Sims & Wolf, 1994). Excess nutrients are transported to surface waters via runoff either in particulate forms or sorbed to soil particles which are suspended in surface runoff. Soluble P and C, NO_3-N, NH_4N, and some organic N species have been demonstrated to be transported by runoff as a solution. NH_4N and P are often sorbed to soil particles and conveyed by runoff through erosion, and organic C, P, and N have been shown to be moved by runoff in particulate form (Edwards & Daniel, 1992). Repeated applications of poultry litter can lead to accumulations of N and P in soil as well as elevated levels of one or both of these nutrients in surface runoff and subsurface water (McLeod and Hegg, 1984; Sharpley and Menzel, 1987; Kingery et al., 1994). The potential for P saturation and leaching may be particularly high for highly-fertilized and sandy soils (Breeuwsma and Silva, 1992; Nair and Graetz, 2004). Large or chronic accumulations of N and P can contribute to accelerated eutrophication of water bodies, impairing their use and potentially leading to fish mortality and growth of algae (Schindler, 1978; Lemunyon and Gilbert, 1993). Elevated concentrations of N and P in surface water and eutrophication of water bodies have been found in areas with high levels of confined poultry and other animal production (Daniel et al., 1998; Sharpley, 1999; Fisher et al., 2000). State and federal environmental protection agencies have responded to these environmental concerns by implementing regulations requiring poultry operations to develop nutrient management plans, which will frequently reduce the allowable amounts of litter that can be applied (Friend et al., 2006). It is estimated that 50% of the litter produced from areas with high concentrations of poultry production facilities cannot be applied to grasslands and croplands in these same areas due to environmental or economic constraints, which has led to surpluses of manure N and P production in some parts of the southeastern U.S. (Kellogg et al., 2000).

The environmental impacts of poultry litter fertilization could be reduced by applying surplus litter to terrestrial ecosystems other than pasture and cropland where nutrient levels in soils are low and which have a low risk of nutrient transport. Disposal of poultry litter to these ecosystems could increase the area of litter dispersal, reduce the spatial concentration of poultry litter application, and decrease risks to water quality in watersheds. Forests may be a viable alternative to pastures and croplands for broiler litter application.

Similar to agroecoystems, forests are often limited by soil N and P supplies (Elser et al., 2007). Forests also have a high potential for nutrient uptake (O'Neill & Gordon, 1994) and have been successfully used to mitigate environmental impacts of municipal waste, municipal effluent, and mill waste disposal (Henry et al., 1993; Polglase et al., 1995; Falkiner & Polglase, 1997; Jackson et al., 2000). In addition, infiltration rates are much higher in forested landscapes than many agricultural landscapes, which could increase potential retention of nutrients and reduce losses through surface runoff in comparison to agricultural crops.

Loblolly pine (Pinus taeda L.) has been identified as a practical species to receive poultry litter application (Beem et al., 1998; Friend et al., 2006; Samuelson et al., 1999). Much of the poultry-producing regions of the southeastern U.S.A. are within the natural range of loblolly pine (Pinus taeda L.), so transportation of litter to land with loblolly pine is likely minimal (Friend et al., 2006). Loblolly pine is a prevalent and economically important species in the southeastern U.S.A.; it is used within the region to produce 18% of the world's supply of industrial timber (Allen et al., 2005; Prestemon and Abt, 2002). Loblolly pine growth is often limited by soil supplies of N and P (Binkley et al., 1999), and tree- and forest-level growth responses to N and P fertilization have been well demonstrated (Blazier et al. 2006; Colbert et al., 1990; Haywood et al., 1997; Murthy et al., 1997; Vose & Allen, 1988). Single applications of poultry litter, ranging between 2 to 23 Mg ha^{-1}, have been shown to increase loblolly pine growth rates (Dickens et al., 2004; Friend et al., 2006; Lynch and Tjaden, 2004; Roberts et al., 2006; Samuelson et al., 1999) and economic value (Dickens et al., 2004) of Nand P-deficient loblolly pine. Loblolly pine forests have a high capacity for fertilizer retention due to high plant biomass and soil organic matter. Will et al. (2006) reported that 90 to 100% of annually applied N and P was sequestered in aboveground biomass, soil organic matter, or the uppermost 10 cm of soil of a loblolly pine plantation. Due to this high fertilizer retention capacity, minimal offsite movement of nutrients associated with poultry litter application is expected. Furthermore, water runoff potential of forests is lower than that of grasslands and croplands due to their relatively higher infiltration (Zimmermann et al., 2006) and evapotranspiration rates (Farley et al., 2005). Friend et al. (2006) found that nutrients from an application of 4.6 Mg ha^{-1} of poultry litter (on a dry matter basis) was substantially contained within a loblolly pine forest and did not impair water quality.

The principal limitation to fertilizing loblolly pine with poultry litter is the relatively poor accessibility and maneuverability within dense plantations by ground-based fertilizer application equipment. Conventional manure-spreading equipment (manure trucks, tractordrawn spreaders) cannot be driven through typical pine plantations due to close tree spacing, dense understory vegetation, and/or stumps high enough to cause equipment damage. These issues likely reduce the number of spreader contractors willing to operate within forests (Dickens et al., 2003).

Silvopasture is an alternative land management system that would allow the application of poultry litter to loblolly pine trees by largely circumventing maneuverability limitations of conventional loblolly pine plantations. Silvopasture management systems consist of forage grasses established and cultivated beneath trees in order to simultaneously produce timber and livestock (Clason & Robinson, 2000; Clason & Sharrow, 2000). Silvopasture regimes are currently the most popular form of agroforestry in the southeastern U.S. (Clason & Sharrow, 2000; Zinkhan & Mercer, 1996). Silvopastures are created by either planting trees in pastures (Robinson & Clason, 2002) or by establishing forage crops in forests (Clason & Robinson, 2000). Forage management in silvopastures is conducted similarly to conventional grasslands in the southeastern U.S.A.; herbicides and/or prescribed burning are used to reduce herbaceous and woody competition and fertilization is carried out to optimize forage yields. Due to land ownership and use patterns, there is high potential for conversion between agriculture and forestry in the southeastern U.S.A. (USDA SCS, 1989). Clason (1995) determined that loblolly pine was compatible with several forage crops in silvopasture systems. The relatively wide spacing of trees and forage understory in silvopastures make navigation of manure-spreading equipment possible (Figure 1).

Figure 1. Applying broiler poultry litter to a silvopasture at the Louisiana State University Agricultural Center Hill Farm Research Station in northwest Louisiana, U.S.A. Picture by Terry Clason, USDA Natural Resource Conservation Service.

A pine plantation in which straw is harvested is another management system where applications of poultry litter could increase commodity production. Pine straw mulch has emerged as a substantial commercial product for horticultural crops and landscaping in urban and suburban areas (Duryea and Edwards, 1989). Adding straw harvesting to conventional timber management regimes has been shown to markedly increase profits, with straw revenue potentially exceeding that of traditional forest products (Haywood et al., 1998; Lopez-Zamora et al. 2001; Roise et al,. 1991). These plantations are typically designed to allow access by conventional agronomic equipment to harvest the straw and are thus well suited for application of the poultry litter by small to mid-size manure or litter spreaders. Harvests of straw on large plantations are usually performed using a hay or pine straw rake, tractor, and mechanical baler (Mills and Robertson, 1991). Understory biomass is typically suppressed in straw harvesting management regimes to improve straw quality by eliminating woody and herbaceous debris (Mills and Robertson, 1991). Coarse and fine woody debris is also removed from the forest floor prior to baling to improve the economic value of baled pine straw (Minogue et al., 2007). This suppression of vegetation and woody debris removal between rows of trees fosters navigation of the plantations with tractor-drawn straw raking and baling equipment (Figure 2) as well as poultry litter application equipment.

Figure 2. Mechanically baled straw in a 19-year-old loblolly pine plantation at the Louisiana State University Agricultural Center Calhoun Research Station in northeast

Louisiana. Inset: Tractor-drawn straw baler used for mechanically baling straw in the plantation. Pictures by Keith Ellem, University of Arkansas Monticello.

Poultry litter can be highly beneficial when applied to plantations in which straw is harvested because it can replenish nutrients lost in straw harvesting. The nutrient content in pine needles is substantial, and repetitive harvesting of pine straw removes significant amounts of nutrients from the soil. One metric ton of harvested straw contains approximately 21.3 kg nitrogen (N), 1.8 kg phosphorus (P), 4.5 kg potassium (K), 9.0 kg calcium (Ca), and 1.8 kg magnesium (Mg) (Pote & Daniel, 2008). Since fallen leaves are major sources of nutrient inputs to soils, repeated raking can reduce soil nutrient availability, particularly N, unless nutrients are replenished through management activities (Jorgenson & Wells, 1986; Lopez-Zamora et al., 2001). As such, periodic fertilization has been recommended to remedy nutrient removals that can occur with straw harvesting (Haywood et al., 1998; Lopez-Zamora et al., 2001).

Since poultry litter contains organic matter, it could potentially replenish some of the organic matter removed by pine straw raking. Fallen pine straw is a prominent source of organic matter in the soil organic horizon of pine forests, and it is the major reservoir of labile carbon used by soil microbes in the synthesis of new cells, a process that also mineralizes N (Pritchett & Fisher, 1987; Sanchez et al., 2006; Wagner & Wolf, 1999). Soil microbial biomass and activity are highly sensitive to changes in soil organic matter and are thus used as indicators of soil quality and sustainability (Fauci & Dick, 1994; Harris, 2003; Powlson & Brookes, 1987). Removal of the soil organic horizon decreased soil microbial biomass carbon (C_{mic}) due to reduced substrate availability in a study simulating organic matter removals associated with tree harvesting and site preparation in a boreal forest (Tan et al., 2005). Activities other than soil organic matter removal associated with straw harvesting may also impact soil biological properties. The suppression of understory vegetation prior to straw raking can reduce microbial biomass and activity because understory vegetation provides rhizodeposition important to soil microbes (Donegan et al., 2001; Gallardo and Schlesinger, 1994; Högberg et al., 2001). Inorganic fertilizers do not replenish organic matter essential as microbial substrates and may exacerbate soil microbial biomass and activity declines caused by organic matter removal (Blazier et al., 2005). In contrast, fertilization with poultry litter can increase soil microbial biomass and activity (Canali et al., 2004; Plaza et al., 2004). Soil organic matter also in part determines soil water availability and temperature (Attiwill & Adams, 1993), and poultry litter has been shown to increase soil water content and available water holding capacity and reduce soil temperature (Agbede et al., 2010; Warren & Fonteno, 1993).

Due to the potential influences of poultry litter on soil and tree nutrition, soil microbes, and tree growth, a series of experiments were conducted in the mid-South region of the U.S.A. This chapter will provide a review of the key results of these trials from 1996 through 2011. The focus of this chapter will be on the changes in soil nutrition, physical properties and microbes, tree nutrition and growth, and water nutrient contents in loblolly pine plantations and silvopastures in response to fertilization with conventional fertilizer and poultry litter.

STUDY DESCRIPTIONS

Results of five studies conducted in the mid-South U.S.A are described in this chapter. At least one treatment in each study received surface application of broiler litter as fertilizer, and loblolly pine was the tree component of each study. All studies occurred in the Western Gulf Coastal Region within areas identified by Friend et al. (2006) as having high occurrence of poultry production and southern pine forests. Two of the studies (SILVO, SWITCH) included poultry litter applied to silvopastures. The silvopasture in the SILVO study consisted as bahiagrass (Paspalum notatum Flüggé) established under thinned loblolly pine, and the silvopasture in the SWITCH study was comprised of switchgrass (Panicum virgatum L.) established under thinned loblolly pine. Two of the studies (AR-FORvsPAST, LA-FORvsPAST) included a comparison of broiler litter in loblolly pine and pastures. The STRAW study included poultry litter applied to a loblolly pine plantation in which straw was annually harvested.

The SILVO and SWITCH studies were conducted at the Louisiana State University Agricultural Center Hill Farm Research Station in Homer, Louisiana, U.S.A. The STRAW and LA-FORvsPAST trials were carried out at the Louisiana State University Agricultural Center Calhoun Research Station in Calhoun, Louisiana, U.S.A. The AR-FORvsPAST study was conducted at the University of Arkansas Southwest Research and Extension Center near Hope, Arkansas, U.S.A. Average annual precipitation of the region in which the studies were carried out is 120 cm, and average temperature is 18°C (Bailey, 1995). Primary study site characteristics are described in Table 2.

Table 2. Location, vegetation, tree and density at study initiation, and soil characteristics for studies of poultry litter fertilization of loblolly pine in the mid-South U.S.A.

Study	Geographical Coordinates	Vegetation	Tree Age (years)	Tree Density (trees ha^{-1})	Soil Classification
SILVO	32°44'N, 93°03'W	loblolly pine-bahiagrass silvopasture	12	247	Loamy, siliceous, thermic Arenic Paleudults
SWITCH	32°44'N, 93°03'W	loblolly pine-switchgrass silvopasture	17	124	Loamy, siliceous, thermic Arenic Paleudults
STRAW	32°31'N, 92°21'W	loblolly pine plantation	10	618	Fine-loamy siliceous thermic Typic Fragiudults
LA-FORvsPAST	32°31'N, 92°21'W	loblolly pine, bermudagrass as vegetation type treatments	5	1586	Fine-loamy siliceous thermic Typic Fragiudults
AR-FORvsPAST	33°42'N, 93°32'W	loblolly pine, bahiagrass as vegetation type treatments	26	201	Fine-loamy, siliceous thermic Typic Fraigudults; Clayey, mixed, thermic Aquic Hapludults

Table 3. Treatments conducted in studies of poultry litter fertilization of loblolly pine conducted in the mid-South U.S.A. [1]Italicized treatments were applied as sub-plot treatments, underlined treatments were applied in all possible combinations to whole plots, all other treatments were applied as whole-plot treatments

Study	Treatment[1]	Treatment Description
SILVO	CONTROL	No treatment
	IF	Inorganic fertilizer mixture (diammonium phosphate, ammonium nitrate, muriate of potash to annually supply 114 kg N ha^{-1}, 39 kg P ha^{-1}, 20 kg K ha^{-1})
	PL5	Poultry litter applied at 5 Mg ha^{-1} that supplied N, P, K, Ca, Mg, Fe, Mn, Cu, Zn, B at 112, 36, 78, 106, 23, 9, 2, 1, 1.5, 0.2 kg ha^{-1}, respectively
	PL10	Poultry litter applied at 10 Mg ha^{-1} that supplied N, P, K, Ca, Mg, Fe, Mn, Cu, Zn, B at 224, 73, 157, 211, 45, 18, 3, 2, 3, 0.3 kg ha^{-1}, respectively
SWITCH	CONTROL	No treatment
	IF80	Ammonium nitrate applied that supplied 80 kg N ha^{-1}
	IF160	Ammonium nitrate applied that supplied 160 kg N ha^{-1}
	PL1.5	Poultry litter applied at 1.5 Mg ha^{-1} to supply N, P, K, Ca, Mg, Fe, Mn, Cu, Zn, and B at 80, 42, 90, 54, 15, 2, 1.5, 0.15, 1, and 0.1 kg ha^{-1}, respectively.
	PL3	Poultry litter applied at 3 Mg ha^{-1} to supply N, P, K, Ca, Mg, Fe, Mn, Cu, Zn, and B at 160, 84, 180, 108, 30, 4, 3, 0.3, 2, and 0.2 kg ha^{-1}, respectively.
STRAW	CONTROL	No treatment
	RAKE	Straw harvesting
	RAKE-IF	Straw harvesting, diammonium phosphate and urea inorganic fertilizers that supplied N and P at 193 and 102 kg ha^{-1}, respectively
	RAKE-PL	Straw harvesting, poultry litter applied at 8 Mg ha^{-1} that supplied N and P at 193 and 102 kg ha^{-1}, respectively. Other nutrients added by poultry litter not tested due to budget constraints.

LA-FORvsPAST	CONTROL	No treatment
	PL5	Poultry litter applied at 5 Mg ha⁻¹ that supplied N, P, K, Ca, Mg, Fe, Mn, Cu, Zn, B at 112, 109, 92, 159, 34, 9, 2, 3, 2, 0.2 kg ha⁻¹, respectively
	PL10	Poultry litter applied at 10 Mg ha⁻¹ that supplied N, P, K, Ca, Mg, Fe, Mn, Cu, Zn, B at 224, 218, 184, 318, 68, 18, 3, 6, 3, 0.3 kg ha⁻¹, respectively
	PL20	Poultry litter applied at 20 Mg ha⁻¹ that supplied N, P, K, Ca, Mg, Fe, Mn, Cu, Zn, B at 448, 436, 368, 636, 136, 36, 6, 12, 6, 0.6 kg ha⁻¹, respectively
	FOREST	Loblolly pine plantation
	PASTURE	Bermudagrass pasture
AR-FORvsPAST	FOREST	Loblolly pine plantation
	PASTURE	Bahiagrass pasture
	CONTROL	No treatment
	PL9	Poultry litter at 9 Mg ha⁻¹ that supplied N, P, K at 30, 14, 22 kg ha⁻¹. Other nutrients added by poultry litter not tested due to budget constraints.

Treatments (Table 3) were replicated four times each in the STRAW, LA-FORvsPAST studies, three times in the SILVO and AR-FORvsPAST studies, and six times in the SWITCH study. Treatments were applied as a one-way treatment structure in the STRAW, LAFORvsPAST, SWITCH, and SILVO studies. Treatments were applied as a split-plot treatment structure in the AR-FORvsPAST study, with vegetation type (pasture, forest) as a whole-plot treatment and fertilization as a sub-plot treatment. The experimental design was a randomized complete block design for all studies. In statistical analyses of all variables assessed in these treatments, differences among treatments were determined by analysis of variance at $\alpha = 0.05$; correlation among variables was assessed at $\alpha = 0.05$ as well.

Soil and/or plant responses to treatments were observed in the studies (Table 4). In the SILVO, SWITCH, and LA-FORvsPAST studies, grass clippings were randomly collected within quadrats either at the end of growing seasons (in the SWITCH study) or multiple times during the season and averaged (in the SILVO and LA-FORvsPAST studies) to determine forage yields. In the SILVO, STRAW, and LA-FORvsPAST studies loblolly pine basal area was measured by converting diameter at breast height measurements into basal area for all trees in measurement plots; basal area measures were summed for each plot to estimate stand level basal area. In the SILVO study soil was sampled by a tractor-mounted auger to the bottom of the B_t horizon and separated into A, E, and B_t horizons, which had average depths of 0.15, 0.48, and 0.59 m, respectively. In the SWITCH study, soil was sampled with punch augers to a 15-cm depth for labile C determination and 30 cm for nutrient analyses. Soil in the STRAW study was sampled with punch augers to a 15-cm depth. In the LA-FORvsPAST study, soil was sampled to a 15-cm depth with punch augers pre-treatment and sampled to 0-15, 15-30, 30-45, 60-80, and 80-100 cm depths post-treatment. Soil was sampled to a 15 cm depth in the AR-FORvsPAST study using punch augers. In the SILVO and STRAW studies, loblolly pine foliage was sampled from the upper third of crowns. Organic matter in soil samples was quantified by the Walkley-Black method (Walkley, 1947) in the SILVO, LA-FORvsPAST, and SWITCH studies and by the loss on ignition method (Ball, 1964) in the STRAW and AR-FORvsPAST studies. Soil pH was determined by pH meters in a 2:1 mixture of deionized water to soil in the SILVO and SWITCH studies. Phosphorus in the samples was extracted by Bray 2 P (Bray & Kurtz, 1945) in the SILVO and LA-FORvsPAST studies and by Mehlich 3 (Mehlich, 1984) in the AR-FORvsPAST and SWITCH studies. Nutrients other than P were extracted by ammonium acetate (K, Ca, Mg, Na) and DTPA (Cu, Fe, Mn, Zn) in the SILVO and LA-FORvsPAST studies (Gambrell, 1996; Helmke & Sparks, 1996). Mehlich 3 was used to extract K, Ca, Mg, S, Cu, and Zn in the SWITCH study (Mehlich, 1984). All nutrients from soil samples were quantified via ICP spectrometry (Jones & Case, 1990) in all studies in which soil was analyzed for nutrient concentration. Exchangeable N (NH_4N, NO_3N) was extracted by KCl extraction (Mulvaney, 1996) in the SILVO and AR-FORvsPAST studies and measured colorimetrically on Bran Luebbe (Bran-Luebbe, Inc, Delavan, WI) and Lachat autoanalyzers (Lachat Instruments, Loveland, CO, U.S.A.) in the SILVO and AR-FORvsPAST studies, respectively. In the STRAW and SWITCH studies soil labile C was measured by sequential fumigation incubation (Zou et al., 2005). In the STRAW and SWITCH studies microbial biomass C was measured by fumigation incubation (Jenkinson and Powlson, 1976a,b) and microbial activity was measured by an assay of dehydrogenase activity (Lenhard, 1956; Alef, 1995). In the STRAW study N mineralization and nitrification was measured using the buried bag method (Eno, 1960). In the AR-FORvsPAST study, potential N mineralization and nitrification (Hart et al., 1994) were assessed in samples aerobically incubated in the laboratory for 28 days; NH_4N and NO_3N used to determine mineralization and nitrification in this procedure were measured by the cadmium reduction method (Mulvaney, 1996) using a Lachat autoanalyzer. Total N in soil samples from the ARFORvsPAST study was determined by dry combustion using an Elementar Vario MAX CN analyzer (Elementar Analysesysteme GmbH, Hanau, Germany).

All nutrients except N in foliage were analyzed by nitric acid digestion and ICP spectrometry; N in these samples was measured by Dumas combustion and thermal conductivity detection using a Leco N/protein analyzer (Leco Inc., St. Joseph, MI, U.S.A) (Helmke & Sparks, 1996; Tate, 1994; Zarcinas et al., 1987). In the SILVO study, N concentrations of bahiagrass samples were ascertained by Kjeldahl method; other nutrients in the samples were determined by Dumas combustion and nitric acid digestion and ICP spectrometry (Helmke & Sparks, 1996; Horneck & Miller, 1998; Tate, 1994; Zarcinas et al., 1987).

Table 4. Timeline of treatments and measurements in studies of fertilization of loblolly pine with poultry litter in the mid-South U.S.A. Shaded cells designate years in which treatments or measurements occurred. [1]OM = organic matter, STR = strength, BD = bulk density, WHC = water holding capacity, LABC = labile C, NMIN = N mineralization, CMIC = microbial biomass C, ACT = microbial dehydrogenase activity

Study	Treatment or Measurement[1]	1996	1997	1998	1999	2000	2001	2002	2003	2004	2005	2006	2007	2008	2009	2010	2011
SILVO	Fertilization			▓	▓	▓											
	Forage yield				▓	▓											
	Soil pH, OM, nutrients		▓				▓										
	Soil NH₄-N, NO₃-N		▓	▓	▓		▓										
	Pine foliage nutrients		▓			▓			▓								
	Pine basal area		▓		▓												
SWITCH	Fertilization																▓
	Forage yield																▓
	Soil pH, OM, nutrients																▓
	Soil LABC, CMIC, ACT																▓
STRAW	Straw raking						▓	▓	▓	▓	▓	▓	▓	▓			
	Fertilization							▓									
	Soil STR, P, BD, WHC										▓						
	Soil OM, LABC, NMIN, exchangeable N										▓	▓					
	Soil CMIC, ACT										▓	▓					
	Pine foliage nutrients										▓		▓				
	Pine basal area						▓				▓		▓				
LA-FORvsPAST	Fertilization																
	Forage yield			▓	▓												
	Soil pH, OM, nutrients			▓	▓												
	Pine basal area						▓										
	Runoff nutrients							▓									
AR-FORvsPAST	Fertilization							▓			▓						
	Forage yield										▓						
	Soil NH₄-N, NO₃-N										▓	▓					
	Soil potential NMIN										▓						
	Soil OM, N, P										▓						

Bulk density, porosity, soil moisture content, and air-filled porosity of soil samples collected in the STRAW study were analyzed using procedures of Blake & Hartge (1986) and Danielson & Sutherland (1986). Available water holding capacity was determined in the STRAW study using soil moisture retention curves (Brye, 2003; Gee et al., 1992). Soil strength in the STRAW study to 15- and 30-cm depths was measured with a Scout SCT compaction meter (Spectrum Technologies, Inc., Plainfield, IL, USA) (Bradford, 1986). Soil water was collected to a 30-cm depth using tension lysimeters in the AR-FORvsPAST study; NO_3-N and PO_4-P in the water samples was analyzed by ion chromatography and NH_4-N was measured with a Lachat autoanalyzer. Water samples were also digested using a Kjeldahl digestion procedure and analyzed for total Kjeldahl nitrogen (TKN) and total Kjeldahl phosphorus (TKP) using the Lachat spectrophotometer. In the LA-FORvsPAST study, water was collected from runoff troughs after every rain event. Water samples were analyzed for total P by acid persulfate digestion and ICP spectrometry and for dissolved P by ICP spectrometry (Clesceri et al., 1998; Pote & Daniel, 2000).

PLANT BIOMASS AND NUTRITION

Increases in forage yields were observed in response to poultry litter in the SILVO, LAFORvsPAST, AR-FORvsPAST, and SWITCH studies. In the SILVO study poultry litter increased bahiagrass yields, but the magnitude of response was rate-dependent. The PL10 treatment had greater bahiagrass yields than all other treatments, and the PL5 and IF treatments had greater bahiagrass yields than the CONTROL treatment (Evans, 2000). The PL10 treatment also led to greater P, Zn, and Cu concentrations in bahiagrass relative to the CONTROL and IF treatments (Evans, 2000). These results indicated that poultry litter increased yields and nutritional quality of bahiagrass. Gaston et al. (2003) similarly found in the LA-FORvsPAST study that bermudagrass yields increased with increasing litter application rate. In the AR-FORvsPAST study, bahiagrass yields of the PL9 treatment were ~1.5 times greater than those of the CONTROL treatment in the first two years of fertilization. Switchgrass yield response to poultry litter in the SWITCH study was not ratedependent as in the SILVO and LA-FORvsPAST studies, because both application rates led to comparable yields.

Loblolly pine growth was also improved by poultry litter in the SILVO and AR-FORvsPAST studies. In the SILVO study, tree- and stand-level basal area growth was increased by poultry litter at the 10 Mg ha^{-1} rate (Blazier et al., 2008a). As with forage yields, litter application rate affected the level of growth response. After four annual litter applications, the 10PL treatment had greater annual basal area growth per tree than that of all other treatments, and the 5PL treatment had greater annual basal area growth than the CONTROL treatment. Stand-level basal area growth of the 10PL treatment was greater than that of the CONTROL and IF treatments. All fertilizer treatments led to greater foliage N concentrations than the CONTROL treatment, and both poultry litter treatments had greater foliage P concentrations than the CONTROL treatment. These results, which are consistent with other studies (Dickens et al., 2004; Friend et al., 2006; Roberts et al., 2006), show that loblolly pine growth can be increased with poultry litter amendments. The levels of growth responses were somewhat surprising, because all foliage nutrient concentrations were above critical levels (Allen, 1987; Blazier et al., 2008a; Jokela, 2004). Due to the relatively low density of trees in silvopastures, trees may have more readily responded to fertilization by virtue of larger crown mass (which provides a larger nutrient sink per tree) and less competition for applied nutrients compared to that in typical pine plantations. The similarities in growth responses and N and P application rates of the 5LIT and INO treatments suggests that although the 5LIT treatment supplied more K and a wider array of nutrients than the INO treatment, N and P were likely the primary limiting nutrients in the stand (Blazier et al., 2008a). In the AR-FORvsPAST study, annual loblolly pine basal area growth in response to the PL9 treatment was 10.9% greater than that of the CONTROL treatment.

In the STRAW and LA-FORvsPAST studies, no significant loblolly pine growth responses to treatments were observed (Gaston et al., 2003). Before the studies were established, the land was intensively managed for forage production. As such, the decades of fertilization application at these locations had resulted in high nutrient availability. Foliage P and S concentrations were increased by the RAKE-PL treatment relative to the other treatments in the STRAW study, but these increases appeared to have been luxury consumption since these nutrient increases were not accompanied by increased loblolly pine growth.

SOIL PHYSICAL PROPERTIES AND ORGANIC MATTER

In the STRAW study, all treatments that included straw harvesting induced evidence of soil compaction by significantly increasing bulk densities (Table 5) to levels 0.6 to 3.3% greater than the 1.75 g cm^{-3} bulk density defined as a growth-limiting threshold for forests grown on loamy soils (Daddow and Warrington, 1983), whereas soil in the CONTROL treatment remained below this threshold. These bulk density increases were also associated with significant declines in porosity in all treatments that included straw harvesting (Table 5). These findings suggest that annual straw harvesting had potential to reduce tree growth through reduced rooting volume and aeration. Nevertheless, no decreases in loblolly pine growth were observed in response to raking, as described above. It is likely that equipment traffic and increased exposure of mineral soil to rainfall associated with straw harvesting led to these increases in bulk density. Similarities in bulk density and porosity among the RAKE treatment and treatments that included raking and fertilization suggest that the additional trafficking from fertilization equipment each season did not appreciably compact the soil and that straw harvesting was the predominant cause of soil compaction (Blazier et al, 2008b).

Table 5. Soil physical properties and organic matter in response to pine straw harvesting and fertilization with inorganic fertilizer and poultry litter in a loblolly pine plantation in north central Louisiana, U.S.A. Means within columns followed by different letters differ at P < 0.05. Adapted from Blazier et al. (2008b)

Treatment	Bulk Density (g cm^{-3})	Porosity (g kg^{-1})	Air-filled Porosity	Moisture (g kg^{-1})	Soil Strength (MPa)	Organic Matter (g kg^{-1})	Available Water Holding Capacity (g kg^{-1})
CONTROL	1.67 b	369 a	99 a	270 a	1.25 b	27.8 a	427 a
RAKE	1.81 a	318 b	51 b	268 ab	2.31 a	25.8 ab	367 b
RAKE-IF	1.76 a	334 b	86 a	248 b	2.45 a	19.0 b	353 b
RAKE-PL	1.78 a	329 b	48 b	281 a	0.99 b	25.8 ab	384 ab

The RAKE-PL treatment appeared to have ameliorated some of the soil physical impacts of the raking since soil strength, organic matter, and moisture in the RAKE-PL treatment were similar to that in the CONTROL treatment (Table 5). However, the RAKE-PL treatment was characterized by lower air-filled porosity than the CONTROL treatment (Table 5), so there may have been a compaction potential associated with the application of the poultry litter (Tekeste et al., 2007). Poultry litter did not alter soil physical properties in a manner similar to inorganic fertilizers. The RAKE-PL treatment was characterized by soil moisture content, strength, organic matter concentrations, and available water holding capacity similar to the CONTROL treatment (Table 5). The RAKE-PL treatment may have replenished some organic matter lost through straw harvesting and accelerated decomposition associated with increased nutrient levels, because broiler poultry litter typically consists of 44% organic matter (Adeli et al., 2006; Dick et al., 1998). These results suggest that use of poultry litter as a fertilizer source in an annual straw harvest regime was superior to inorganic fertilizers in sustaining soil physical quality.

In contrast with the RAKE-PL treatment, the RAKE and RAKE-IF treatments had detrimental effects on some soil physical properties. The RAKE and RAKE-IF treatments both had soil strengths 46% greater than the CONTROL treatment. Soil strengths of the RAKE and RAKE-IF treatments also exceeded the 2 MPa soil strength threshold defined as highly compacted because of demonstrated root growth restrictions (Taylor & Gardner, 1963; Tiarks & Haywood, 1996). Available water holding capacity was also reduced by the RAKE and RAKE-IF treatments relative to the CONTROL treatment. These findings suggest that the RAKE and RAKE-IF treatments made soil less amenable for root growth in the uppermost 5 cm of soil, which is the predominant zone in which tree roots, particularly fine roots, grow (Gilman, 1987). Relative to the CONTROL treatment, only the RAKE-IF treatment had greater soil strength, reduced moisture content, and reduced soil organic matter concentrations (Table 5). Repeated fertilization with inorganic nitrogen has been shown to reduce soil organic matter concentrations by increasing decomposition rates (Khan et al., 2007). Increased soil strength in response to the RAKE-IF treatment may have been due to the reductions in soil organic matter concentrations caused by this treatment. Soil strength tends to increase with decreasing soil organic matter concentrations because soil organic matter serves as an organic aggregate binding and bonding material (Munkholm et al., 2002). The relatively lower moisture content and available water holding capacity of the RAKE-IF treatment is consistent with its lower soil organic matter content because organic matter fosters soil moisture retention (Plaza et al., 2004; Powers et al., 2005).

There were no differences in soil organic matter among treatments in the SILVO, SWITCH, LA-FORvsPAST, and AR-FORvsPAST studies (Blazier et al., 2008a; Liechty et al., 2009), in which litter was not removed. As such, increases in forage and/or tree yields from fertilization in these studies were not associated with concomitant increases in soil organic matter. In the SILVO and SWITCH studies, the lack of declines in organic matter in response to inorganic fertilizer application as seen in the STRAW study was likely due to the straw raking done in tandem with fertilization in the STRAW study. As organic matter supply was drastically reduced by annual straw harvesting, stimulating decomposition with inorganic fertilizer led to significant declines in soil organic matter. Additionally, the increases in forage understory biomass of the SILVO and SWITCH studies may have been less prone to lead to increases in organic matter, as evidenced in the LA-FORvsPAST study. In that study no differences in organic matter were found among treatments in the pasture despite the increases in bermudagrass yields described above, whereas organic matter in the loblolly pine plantation differed among treatments as PL20 > PL10, PL5 > CONTROL.

SOIL LABILE C, MICROBIAL BIOMASS C, AND MICRO-BIAL ACTIVITY

Annual application of inorganic fertilizer had a profound effect on microbial biomass and activity in the STRAW study (Table 6). Microbial biomass C of the RAKE-IF treatment was lower than that of the CONTROL and RAKE treatments, and dehydrogenase activity of the RAKE-IF treatment was lower than all other treatments. The reductions in microbial biomass C and activity were apparently not a result of lower substrate supply, because labile C was similar among treatments (Table 6). Consequently, the higher potential turnover rate of the RAKE-IF treatment relative to all others is likely a result of reduced microbial biomass and activity rather than relatively high recalcitrance of organic matter.

The reductions in microbial biomass C and activity in the RAKE-IF treatment were likely associated with the lower pH of this treatment relative to all others. It has been welldemonstrated that intensive fertilization with inorganic N reduces soil pH and that declining pH is associated with reductions in soil microbial biomass and activity (Anderson and Domsch, 1993; Baath et al., 1995; Blazier et al., 2005). These results thus showed that microbial biomass and activity were reduced by declines in pH from inorganic fertilizer, whereas annual raking and fertilization with poultry litter had no such effects. In contrast to inorganic fertilizer, poultry litter tends to increase soil pH because litter contains calcium carbonate originating from poultry rations (Hue, 1992; Kingery et al., 1993). Although litter did not significantly increase pH in the STRAW study, litter sustained pH at levels comparable to the CONTROL treatment, which fostered microbial biomass C and activity levels comparable to the CONTROL treatment as well.

As in the STRAW study, inorganic fertilizer led to declines in microbial biomass C relative to the CONTROL treatment (Table 6) in the SWITCH study. Fertilization has been shown to reduce soil microbial biomass C in forest soils (Rifai et al., 2010; Wallenstein et al., 2006). Rifai et al. (2010) identified several possible mechanisms for soil microbial biomass declines in response to fertilization, including (1) pH reduction caused by nitrate leaching induced by application of high rates of NH_4NO_3, and (2) inhibition of organic compound decomposition from excess N that reduces organic matter available to soil microbes. In the SWITCH study there were no declines in pH among treatments consistent with declines in soil microbial biomass C, although pH of the inorganic fertilizer treatments were lower than those of the poultry litter treatments. Dehydrogenase activity decreased as fertilizer application rates increased for both fertilizer types. Since N was the sole nutrient added by inorganic fertilizer treatments in this study, these dehydrogenase activity trends suggest that excess N perturbed microbial decomposition of organic matter in this loblolly pine and switchgrass system. However, potential C turnover rate was shorter for the lower rate of inorganic fertilizer (IF80) relative to the lower rate of poultry litter (PL1.5) despite the equivalent N rate of the two treatments. Since labile C supply, microbial biomass C, and dehydrogenase activity were similar for the IF80 and PL1.5 treatments, the reason for the higher potential C turnover rate of the PL1.5 treatment was unclear and merited further study.

Table 6. Soil labile C, microbial, and pH responses to fertilization in an annually raked loblolly pine plantation (STRAW) and a loblolly pine and switchgrass silvopasture (SILVO) in the mid-South U.S.A. For each study, means within columns followed by different letters differ at P < 0.05. Adapted in part from Blazier et al. (2008b)

			Treatment		
STRAW	**CONTROL**	**RAKE**	**RAKE-IF**	**RAKE-PL**	
Labile C (mg kg^{-1})	475.1 a	522.3 a	457.0 a	582.5 a	
Potential C turnover rate (days)	46.0 b	53.2 b	92.9 a	62.8 ab	
Microbial biomass C (mg kg^{-1})	169.2 a	157.2 a	75.3 b	143.5 ab	
Dehydrogenase activity (µg g^{-1})	50.6 a	71.0 a	25.8 b	44.5 a	
pH	4.9 a	4.9 a	4.3 b	5.1 a	
	CONTROL	**IF80**	**IF160**	**PL1.5**	**PL3**
SWITCH					
Labile C (mg kg^{-1})	835.6 a	585.2 a	718.1 a	878.7 a	836.7 a
Potential C turnover rate (days)	29.7 ab	24.7 b	30.4 ab	43.8 a	37.3 ab
Microbial biomass C (mg kg^{-1})	410.9 a	341.6 b	348.4 b	320.4 ab	377.4 ab
Dehydrogenase activity (µg g^{-1})	11.0 ab	24.2 a	9.9 b	24.3 a	5.9 b
pH	5.5 bc	5.4 c	5.4 c	5.6 ab	5.7 a

SOIL NUTRIENTS

Nitrogen

In all studies in which exchangeable soil N was measured, NO_3N amounts or proportions of in soil increased in response to poultry litter application (Table 7). In the AR-FORvsPAST study, NO_3N significantly increased in the loblolly pine plantation and in pasture relative to the CONTROL treatment following two years of poultry litter application. The proportion of NO_3N to total exchangeable N was also greater in response to poultry litter than without litter application (Liechty et al., 2009). There was no difference in NO_3N concentrations among treatments in the SILVO study, but as in the AR-FORvsPAST study the ratio of NO_3N to total exchangeable N increased in response to poultry litter additions. This increase in the proportion of NO_3N in the SILVO study occurred in response to both rates of broiler litter tested; no such increase was observed in response to the inorganic fertilizer mixture (Blazier et al., 2008a). Results similar to the SILVO study were also found in the STRAW study; NO_3N increased in response to the treatment regime that included poultry litter, whereas no such increase was observed in response to non-fertilized treatments and the treatment regime that included a mixture of inorganic fertilizers (Liechty et al., 2009).

Increases in soil NO_3N in response to poultry litter were attributable to greater nitrification rates (Table 7). Soil in plots treated with broiler litter had greater N mineralization rates in the AR-FORvsPAST study, and a greater proportion of mineralized N was nitrified. There was also a significant positive correlation between NO_3-N in soil and nitrification rates (Liechty et al., 2009). Similar results were observed in the STRAW study, in which both rates of poultry litter had greater N mineralization and nitrification than CONTROL and IF treatments (Blazier et al, 2008b). The greater nitrification and NO_3-N of poultry litter treatments relative to CONTROL treatments in both studies was likely predominately a function of the addition of N to soil by litter. Relatively high NO_3-N in soil after fertilization is in part indicative of low plant sequestration of applied N (Adeli et al., 2006), so consecutive applications of litter at the rates in these studies likely exceeded loblolly pine, bermudagrass, and bahiagrass N demand. The higher nitrification rates seen in response to poultry litter in these studies relative to inorganic fertilizer, even when both fertilizer sources were applied to provide the same N rates, was likely due to the differences in the effects of the fertilizer sources on soil pH. In the SILVO and STRAW studies, soil pH declined in response to inorganic fertilization applications relative to all other treatments (Tables 6 and 7). Likewise, soil pH of the poultry litter treatments in the SWITCH study was greater in response to broiler litter than to CONTROL and inorganic fertilizer treatments (data not shown). Ellum (2010) found in the STRAW study that nitrification was significantly and positively correlated with pH. Nitrification rates have been shown to decline with decreasing pH due to reductions in populations and activity of nitrifying bacteria (Aune & Lal, 1997).

Table 7. Soil exchangeable N, mineralization, nitrification, and pH in response to fertilization in loblolly pine plantations, silvopasture, and bahiagrass pasture in a series of trials conducted in the mid-South U.S.A. Means within rows followed by different letters differ at P < 0.05. Adapted in part from Liechty et al. (2009)

	Treatment			
SILVO – 5 years post-treatment	CONTROL	IF	PL5	PL10
NO_3-N (mg kg^{-1})	7.2 a	0.1 a	4.8 a	16.2 a
Total exchangeable N (mg kg^{-1})	34.8 a	18.6 a	29.6 a	71.0 a
% NO_3-N	31.0 b	11.1 c	51.7 a	55.6 a
pH	4.9 a	4.5 a	5.0 a	5.0 a
STRAW – 5 years post-treatment	CONTROL	RAKE	RAKE-IF	RAKE-PL
NO_3-N (mg kg^{-1})	0.6 c	0.8 c	1.4 b	10.0 a
Total exchangeable N (mg kg^{-1})	6.5 b	5.1 c	6.3 b	14.4 a
% NO_3-N	15.9 c	17.8 c	25.2 b	65.6 a
N mineralization (mg kg^{-1})	23.6 b	18.5 b	13.7 b	51.2 a
N nitrification (mg kg^{-1})	23.2 b	17.4 b	17.3 b	48.2 a
% N nitrified	98.3 a	94.1 a	126.3 a	94.1 a
AR-FORvsPAST – 2 years post-treatment	PASTURE-CONTROL	PASTURE-PL9	FOREST-CONTROL	FOREST-PL9
NO_3-N (mg kg^{-1})	2.1 b	4.1 a	0.1 b	15.3 a
Total exchangeable N (mg kg^{-1})	8.9 b	11.0 b	6.2 b	24.4 a
% NO_3-N	21.4 bc	37.6 ab	1.1 c	57.1 a
N mineralization (mg kg^{-1})	14.3 b	23.6 a	7.3 c	14.3 b
N nitrification (mg kg^{-1})	15.6 b	26.3 a	3.4 d	13.3 c
% N nitrified	109.0 b	111.4 a	46.5 d	93.3 c

Although differences in soil NO_3N and nitrification between loblolly pine and bahiagrass pasture in the AR-FORvsPAST study in part reflected the differences in pH and C:N ratios of the soils in these two land uses (Richardson 2006), they also reflected the differences in uptake and use of available N forms by the loblolly pine and pastures. Although N mineralization and nitrification was greater in pasture when fertilized with poultry litter, the increase in NO_3N remaining in soil per unit increase in potential net nitrification was greater in loblolly pine plantation than in pasture by the second application of poultry litter (Liechty et al., 2009). Conifer tree roots have been shown to preferentially absorb NH_4N rather than NO_3N (Kronzucker et al., 1997), whereas NO_3N is preferentially taken up by forage (Blevins and Barker, 2007). Thus, loblolly pine plantation had a greater propensity to retain a higher proportion of NO_3N than pasture. Given this tendency of loblolly pine to retain proportionally greater NO_3N, it is likely that less poultry litter should be applied to such plantations than to pastures to minimize NO_3N pollution in surface and subsurface water.

Annual raking in the STRAW study reduced total exchangeable N, and fertilization, regardless of source, replaced at least a portion of the lost N and increased total exchangeable N (Table 7). Interestingly, although both fertilizers increased exchangeable N, poultry litter increased exchangeable N to a greater extent than the inorganic fertilizer, although both fertilizers were applied at the same N rate (Ellum, 2010). The higher exchangeable N concentrations in the RAKE-PL reflected the increases in NO_3N levels in the RAKE-PL treatment. The NO_3N concentrations were nearly 7 and 17 times greater in this treatment than those in the RAKE-IF and CONTROL treatments, respectively. This result provides evidence of the propensity of loblolly pine plantations to accumulate NO_3N in response to annual applications of broiler litter, even when exchangeable N is reduced by annual straw raking. To safeguard against such NO_3N accumulation, it is likely necessary to fertilize a raked loblolly pine plantation with broiler litter less frequently and at lower rates than in the STRAW study.

Phosphorus

Soil test P accumulation, determined as the annual difference in soil test P concentrations from pre-treatment concentrations, increased in the uppermost soil horizon in all studies in which soil test P was measured (Table 8). In the SILVO study, both litter treatments had significantly greater soil test P accumulation in the uppermost soil horizon than the CONTROL and IF treatments. After the first application, soil test P accumulation was similarly increased by both litter rates. After four annual applications, the PL10 treatment had greater soil test P accumulation than all other treatments. The

IF treatment did not result in a significant accumulation of soil test P at any point in the study (Liechty et al, 2009). Soil test P accumulation also increased in the SWITCH study in response to a single application of litter at both rates. In the LA-FORvsPAST and AR-FORvsPAST studies, soil test P accumulation increased in response to broiler litter in loblolly pine plantation and in pasture (Liechty et al., 2009). Increases in soil test P in surface soil in response to litter application have been similarly found in agricultural (Mitchell & Tu, 2006; Sharpley et al., 1993; Sistani et al., 2004) and forest (Friend et al., 2006) soils. In addition to these increases in upper soil horizons, soil test P accumulation was increased to the B_t horizon (an average depth of 0.59 m) by the 10PL treatment after four applications in the SILVO study (Blazier et al., 2008a). Additional evidence of increasing soil test P in lower soil profile was found in the LA-FORvsPAST study, in which soil test P concentrations of the 20PL treatment exceeded that of all others in the 30 to 45 cm depth in loblolly pine and bermudagrass soil in the seventh year of the study (data not shown). These increases in soil test P in surface and subsurface soil in response to annual litter applications suggest that vegetation P demands and soil P sorption capacity were exceeded at all sites irrespective of vegetation type and stand conditions.

Land use type and rate affected soil test P trends in response to broiler litter in the LAFORvsPAST study. Soil test P accumulation in the loblolly pine plantation averaged over all treatments exceeded that of the pasture for six years of the study (Figure 3). Initial soil test P concentrations of the pasture were 1.5 times greater than that of the loblolly pine plantation (data not shown), but in the first three years of treatment soil test P accumulation of the pasture was negative whereas soil test P accumulation of the loblolly pine plantation ranged from 51 to 76 mg kg[-1] year[-1] over the same period. Until the final fertilization, soil test P increased more markedly in the loblolly pine plantation than in the pasture. These differences in soil test P accumulation trends between land use types may have been indicative of lower P demand by loblolly pine than bermudagrass, which led to a greater P accumulation in the soils of the loblolly pine plantations than pastures. Litter application rate also influenced soil P accumulation in both land use types in the LA-FORvsPAST study (Table 8). Annual applications of litter at 5 Mg ha[-1] did not significantly increase soil test P relative to the CONTROL treatment during the study. Soil test P accumulation was greater in response to the 20 Mg ha[-1] litter application rate relative to the CONTROL and PL5 treatments throughout the study and greater relative to the 10 Mg ha[-1] rate by the fourth annual fertilization.

Table 8. Soil test P accumulation (mg kg⁻¹) in response to fertilization with poultry litter and inorganic fertilizer in the mid-South U.S.A. For each study site and soil depth, means within columns followed by different letters differ at P < 0.05. [a]Average depth of soil samples subdivided into the A horizon, [b]average depth of soil samples subdivided into the E horizon, [c]average depth of soil samples subdivided into the B_t horizon, [d]soil test P accumulation reported for study is an average of loblolly pine plantation and pasture soils because analyses did not reveal a treatment x land use type interaction. Adapted in part from Blazier et al. (2008a) and Liechty et al. (2009)

Study	Treatment	Depth (cm)	Year after treatment						
			1	2	3	4	5	6	7
SILVO	CONTROL	0–15[a]	13.0 b	------	16.9 b	10.5 c	------	------	------
	IF		19.0 b	------	22.1 b	24.1 c	------	------	------
	PL5		36.7 a	------	68.5 a	87.2 b	------	------	------
	PL10		48.8 a	------	84.2 a	146.5 a	------	------	------
	CONTROL	15–48[b]	3.8 a	------	-0.4 a	-5.2 b	------	------	------
	IF		0.9 a	------	-2.0 a	-5.8 b	------	------	------
	PL5		3.3 a	------	1.1 a	0.9 b	------	------	------
	PL10		2.8 a	------	4.4 a	80.2 a	------	------	------
	CONTROL	48–59[c]	2.2 a	------	-1.4 a	-4.0 b	------	------	------
	IF		-0.4 a	------	-4.5 a	-6.6 b	------	------	------
	PL5		-0.5 a	------	-3.5 a	-5.7 b	------	------	------
	PL10		-1.8 a	------	-3.5 a	44.6 a	------	------	------
SWITCH	CONTROL	0-15	0.2 c	------	------	------	------	------	------
	IF80		0.1 b	------	------	------	------	------	------
	IF160		0.1 b	------	------	------	------	------	------
	PL90		0.5 a	------	------	------	------	------	------
	PL180		0.5 a	------	------	------	------	------	
LA-FORvsPAST[d]	CONTROL	0-15	-19.9 b	-8.67 a	-24.1 b	-11.9 c	------	-11.0 c	-19.6 c
	PL5		3.2 b	9.3 a	-14.2 b	26.4 bc	------	162.5 bc	82.9 bc
	PL10		32.2 ab	29.7 a	65.3 a	103.2 b	------	328.2 b	210.0 b
	PL20		71.2 a	43.5 a	80.8 a	243.4 a	------	760.0 a	447.6 a
AR-FORvsPAST[d]	CONTROL	0-15	------	10.1 b	------	------	------	------	------
	PL9		------	47.2 a	------	------	------	------	------

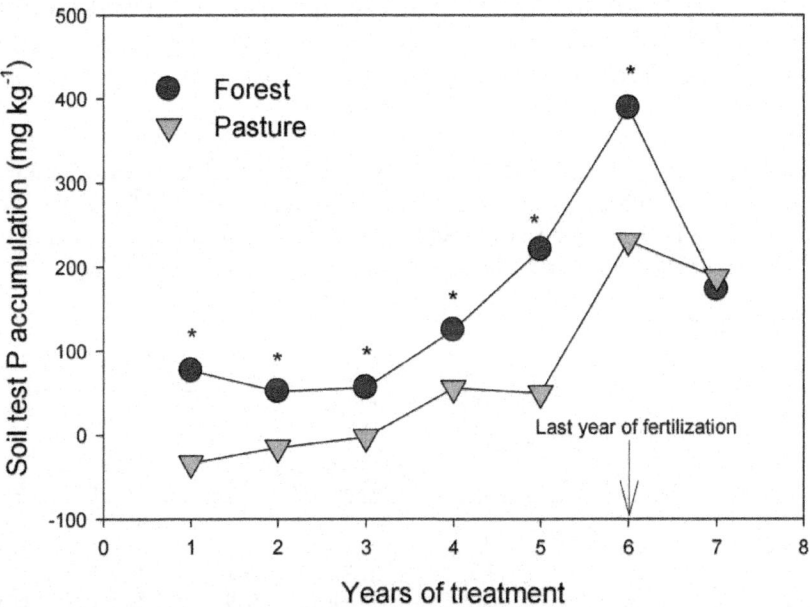

Figure 3. Soil test P accumulation (0 to 15 cm) as affected by annual fertilization with poultry litter in a loblolly pine plantation and a bermudagrass pasture in the mid-South U.S.A. Asterisks denote years in which soil test P accumulation differed among land use types at $P < 0.05$.

Other Nutrients

Soil K concentrations were increased by broiler litter in the SILVO study (Table 9). A single application of the 10PL treatment increased K concentrations in the A horizon, and subsequent applications led to increases in K concentrations in the E horizon. Increases in K concentrations in lower soil depths have also been observed in response to annual litter fertilization of pastures and agricultural crops on sandy soils (Kingery et al., 1994; Mitchell & Tu, 2006). A similar increase in soil K concentrations in the uppermost 15 cm of soil in response to a single application of broiler litter was found in the SWITCH study (data not shown). In that study soil K increased more in response to the PL3 treatment than all others, and K concentrations of all other fertilizer treatments exceeded that of the CONTROL treatment. Results of both studies indicate that poultry litter can lead to increases in soil K concentrations in these silvopastures, even after a single application.

Although soil K concentrations increased in both the SILVO and SWITCH studies, the amount of poultry litter required to increase the concentrations differed between the two types of silvopastures. An application of only 1.5

Mg ha^{-1} of litter was needed to increase K concentrations in the loblolly pine-switchgrass silvopasture while in the loblolly pinebahiagrass silvopasture K concentrations were observed only after two annual applications of 10 Mg ha^{-1} of poultry litter.

Table 9. Effects of annually fertilizing a loblolly pine and bahiagrass silvopasture with poultry litter and inorganic fertilizer on soil K and Mg in the A and E soil horizons and on Ca in the A horizon. For each nutrient and horizon, means within a column followed by a different letter differ at $P < 0.05$. Adapted from Blazier et al. (2008a)

Nutrient	Horizon	Treatment	Date 1997	1998	2001	2002
K	A	CONTROL	42.1 a	30.1 b	23.3 a	30.7 a
		IF	33.9 a	42.9 b	23.7 a	26.0 a
		PL5	33.2 a	44.3 b	31.4 a	30.1 a
		PL10	39.8 a	62.6 a	34.9 a	36.1 a
	E	CONTROL	22.8 a	22.2 a	21.3 c	34.1 b
		IF	22.7 a	27.2 a	28.8 bc	37.8 b
		PL5	21.7 a	34.2 a	39.7 ab	43.8 b
		PL10	30.8 a	36.8 a	51.5 a	60.0 a
Mg	A	CONTROL	30.5 b	33.8 b	107.7 bc	96.2 c
		IF	32.0 a	26.9 bc	103.7 c	89.3 c
		PL5	31.8 a	38.4 ab	113.5 ab	100.9 b
		PL10	35.2 a	44.4 a	120.6 a	114.8 a
	E	CONTROL	26.6 a	25.8 b	107.8 b	100.4 b
		IF	29.8 a	25.8 b	112.4 b	104.4 b
		PL5	34.8 a	28.2 ab	126.6 ab	114.3 ab
		PL10	57.6 a	34.9 a	145.1 a	137.6 a
Ca	A	CONTROL	184.6 a	194.6 a	134.4 c	70.2 c
		IF	177.2 a	157.2 a	89.6 c	20.8 c
		PL5	171.2 a	196.0 a	162.9 b	95.5 b
		PL10	186.0 a	226.0 a	229.9 a	174..3 a

Since the soil type was identical for these two studies, these results suggest that loblolly pine-bahiagrass silvopasture had a greater K demand than the loblolly pine-switchgrass silvopasture. The higher demand of the loblolly pine and bahiagrass pasture was likely due in part to loblolly pine density that was nearly double that in the loblolly pine and switchgrass silvopasture. The switchgrass also likely had a lower K demand than bahaiagrass , because switchgrass is characterized by relatively low nutrient demand despite its relatively high biomass growth potential (Tilman et al., 2006). Nevertheless, annual broiler application at 10 Mg ha^{-1} apparently exceeded vegetation K demand and sorption capacity of the A horizon in the loblolly pine and bahiagrass silvopasture as indicated by increased in K concentrations in the E horizon after four annual applications.

As with K, soil Mg concentrations were increased in the A and E horizons by repeated applications of litter in the SILVO study (Table 9; Blazier et al., 2008a). After two applications soil Mg in the A and E horizons was increased by the 10 Mg ha^{-1} rate relative to the CONTROL and IF treatments, and after four applications the 5 Mg ha^{-1} rate led to greater soil K concentrations in the A horizon than in the CONTROL and IF treatments. However, the 5 Mg ha^{-1} did not increase soil K concentrations in the E horizon and did not increase soil K concentrations to levels in the A horizon comparable to that of the 10 Mg ha^{-1} rate after the fourth applications. By the fourth application, soil Ca concentrations in the A horizon were also increased by the poultry litter treatments, with that of the PL10 treatment exceeding all other treatments and that of the PL5 treatment greater than the CONTROL and IF treatments.

WATER NUTRIENTS

Poultry litter applications led to increases in NO$_3$N in soil water in the AR-FORvsPAST study. Total N concentrations in soil water were greater for pastures than the loblolly pine plantation and greater for the PL9 treatment than the CONTROL treatment; differences in NO$_3$N accounted for the majority of the total N differences between land use types and treatments. In both pasture and loblolly pine plantation, NO$_3$N concentrations increased in response to poultry litter application (Figure 4). Soil water NO$_3$N concentrations were significantly positively correlated with potential nitrification rates.

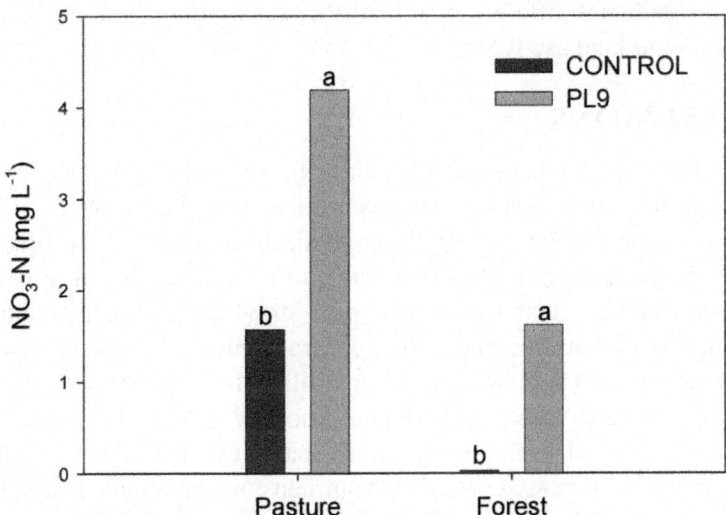

Figure 4. Mean bi-weekly soil water (30 cm) NO$_3$N concentrations in bermudagrass pasture and loblolly pine plantation treated with poultry litter. For each land use type,

means headed by different letters differ at P < 0.05. Adapted from Liechty et al. (2009).

Although bi-weekly NO_3-N concentrations in soil water never exceeded the 10 mg L^{-1} drinking water standard of the U.S. Environmental Protection Agency in the loblolly pine plantation, this standard was exceeded in two or more sampling periods in pasture plots fertilized with poultry litter. Soil water N increased 51% more in pastures than in loblolly pine plantation, which suggests the potential for N pollution of water is greater for pastures fertilized with poultry litter than for loblolly pine plantations fertilized with poultry litter. However, because forest soils have an apparently greater propensity than pastures to retain proportionally greater NO_3-N in soil (described above), with long-term litter applications N losses in soil water from forests could be greater than in pastures (Liechty et al., 2009).

Repeated fertilization with poultry litter led to increases in total and dissolved P concentrations in runoff in pasture and loblolly pine plantation in the LA-FORvsPAST study. Total and dissolved P concentrations increased with increasing litter application rate in both land use types, although the P concentrations increased more markedly to 10 and 20 Mg ha^{-1} rates in pasture than in loblolly pine plantation. Total and dissolved P concentrations in runoff were positively correlated with Bray P concentrations in soil. These results indicate potential for losses of P in runoff in response to litter application in pasture and loblolly pine plantation, with modest evidence that P loss potential in loblolly pine plantation was lower. In the AR-FORvsPAST study, there were no significant differences in total P concentrations in soil water among treatments and land use types.

CONCLUSIONS

Poultry litter was a beneficial fertilizer for loblolly pine plantations and silvopastures in this series of studies. Unlike with inorganic fertilizer, soil pH did not decrease with poultry litter application, which sustained microbial biomass and activity at levels comparable to non-fertilized soil. Poultry litter application to soils that had annual pine straw harvesting maintained soil strength, organic matter, and soil moisture similar to those without straw harvesting, whereas applying inorganic fertilizer to soils with straw harvesting negatively impacted these soil attributes. Loblolly pine trees in plantations and silvopastures, as well as the grasses in silvopastures, responded to poultry litter fertilization with increased growth and nutrient concentrations. These increases in plant growth and nutrition provided some buffering against increasing soil nutrient concentrations when these plantations and silvopastures were annually fertilized with poultry litter. Nevertheless, poultry litter was more prone to lead to accumulation of NO3- N and P in soil than inorganic fertilizer. Loblolly pine plantations were also more prone to increases in soil NO_3-N and P than pastures. Accumulations in soil NO_3-N and P were also associated with increased NO_3-N and P concentrations in soil water and runoff, respectively. As such, poultry litter fertilization of these loblolly pine plantations and silvopastures had the potential to contaminate soil water with N and P. Any poultry litter fertilization regimes for loblolly pine plantations and silvopastures must account for the greater tendencies of N and P accumulation in soil and water of these ecosytems; lower rates and/or frequencies than those used in these trials will likely be necessary for ecologically sustainable fertilization with poultry litter.

REFERENCES

1.   Adams, P.L., Daniel, T.C., Edwards, D.R., Nichols, D.J., Pote, D.H. & Scott, H.D. (1994). Poultry litter and manure contributions to nitrate leaching through the vadose zone. Soil Science Society of America Journal, Vol. 58 (No. 4): 1206-1211.

2.   Adeli, A., Rowe, D.E. & Read, J.J. (2006). Effects of soil type on bermudagrass response to broiler litter application. Agronomy Journal, Vol. 98 (No. 1): 148-155.

3.   Agbede, T.M., Oladitan, T.O., Alagha, S.A., Ojomo, A.O. & Ale, M.O. (2010). Comparative evaluation of poultry manure and NPK fertilizer on soil physical and chemical properties, leaf nutrient concentrations, growth and yield of yam (Dioscorea rotundata Poir) in southwestern Nigeria. World Journal of Agricultural Sciences, Vol. 6 (No. 5): 540-546.

4.   Alef, K. (1995). Dehydrogenase activity, In: Methods in applied soil microbiology and biochemistry, Alef, K. & Nannipieri, P. (eds.). pp. 228-231. Academic Press, San Diego, CA.

5.   Allen, H.L. (1987). Forest fertilizers: nutrient amendment, stand productivity, and environmental impact. Journal of Forestry, Vol. 85 (No. 2): 37-46.

6.   Allen, H.L., Fox, T.R. & Campbell, R.G. (2005). What is ahead for intensive pine plantation silviculture in the South? Southern Journal of Applied Forestry, Vol. 29 (No. 2): 62-69.

7.   Alikhani, H.A., Saleh-Rastin, N. & Antoun, H. (2006). Phosphate solubilization activity of rhizobia native to Iranian soils. Plant and Soil, Vol. 287 (No. 1-2): 35-41.

8.   Anderson, T.H. & Domsch, K.H. (1993). The metabolic quotient for CO_2 ($qCO2$) as a specific activity parameter to assess the effects of environmental conditions, such as pH, on the microbial biomass of forest soils. Soil Biology and Biochemistry, Vol. 25 (No. 3): 393-395.

9.   Attiwill, P.M. & Adams, M.A. (1993). Tansley review no. 50: Nutrient cycling in forests. New Phytologist, Vol. 124: 561-582.

10. Aune, J.B. & Lal, R. (1997). Agricultural productivity in the tropics and critical limits of properties of oxisols, ultisols, alfisols. Tropical Agriculture, Vol.74 (No. 2): 96-103.

11. Baath, E., Frostegard, A., Pennanen, T., & Fritze, H. (1995). Microbial community structure and pH response in relation to soil organic matter quality in wood-ash fertilized, clear-cut or burned coniferous forest soils. Soil Biology and Biochemistry, Vol. 27 (No. 2): 229-240.

12. Bailey, R.G. (1995). Description of the ecoregions of the United States (2nd edition), USDA Forest Service Miscellaneous Publication No. 1391, USDA Forest Service, Washington, D.C.

13. Ball, D.F. (1964). Loss on ignition as an estimate of organic matter and organic carbon in non-calcareous soils. Journal of Soil Science, Vol. 15 (No. 1): 84.92.

14. Beem, M., Turton, D.J., Barden, C.J. & Anderson, S. (1998). Application of poultry litter to pine forests, OSU Extension Factsheet F-5037, Oklahoma Cooperative Extension Service, Oklahoma State University, Stillwater, OK.

15. Binkley, D., Burnham, H. & Allen, H.L. (1999). Water quality impacts of forest fertilization with nitrogen and phosphorus. Forest Ecology and Management, Vol. 121 (No. 3): 191- 213.

16. Blake, G.R. & Hartge, K.H. (1986). Bulk density, In: Methods of Soil Analysis, Part 1: Physical and Mineralogical Methods. (2nd Edition), Klute, A. (ed.). pp. 363-375. Soil Science Society of America, Madison, WI.

17. Blazier, M.A., Hennessey, T.C. & Deng, S.P. (2005). Effects of fertilization and vegetation control on microbial biomass carbon and dehydrogenase activity in a juvenile loblolly pine plantation. Forest Science, Vol. 51 (No. 5): 449-459.

18. Blazier, M.A., Hennessey, T.C., Dougherty, P.M. & Campbell, R. (2006). Nitrogen accumulation and use by a young loblolly pine plantation in southeast Oklahoma: Effects of fertilizer formulation and date of application. Southern Journal of Applied Forestry, Vol. 30 (No. 2): 66-78.

19. Blazier, M.A., Gaston, L.A., Clason, T.R., Farrish, K.W., Oswald, B.P. & Evans, H.A. (2008a). Nutrient dynamics and tree growth of silvopastoral systems: impact of poultry litter. Journal of Environmental Quality, Vol. 37 (No. 4): 1546-1558.

20. Blazier, M.A., Hotard, S.L. & Patterson, W.B. (2008b). Straw harvesting, fertilization, and fertilizer type alter soil microbiological and physical properties in a loblolly pine plantation in the mid-South U.S.A. Biology and Fertility of Soils, Vol. 45 (No. 2): 145- 153.

21. Blevins, D.G. & Barker, D.J. (2007). Nutrients and water in forage crops, In: Forages: The Science of Grassland Agriculture. Volume II (6th edition), Barnes, R.R., Nelson, C.J., Moore, K.J. & Collins, M. (Eds.). pp 67-80. Blackwell Publishing, Ames, IA.

22. Bradford, J.M. (1986). Penetrability, In: Methods of Soil Analysis, Part 1: Physical and Mineralogical Methods. (2nd edition), Klute A. (Ed.). pp. 463-478.

23. Soil Science Society of America, Madison, WI. Bray, R.H. & Kurtz, L.T. (1945). Determination of total, organic, and available forms of phosphorus in soils. Soil Science, Vol. 59 (No. 1): 39-45.

24. Breeuwsma, A. & Silva, S. (1992). Phosphorus fertilization and environmental effects in the Netherlands and the Po region (Italy), Report 57, Winand Staring Centre for Integrated Land, Soil and Water Research, Wageningen, The Netherlands.

25. Brye, K.R. (2003). Long-term effects of cultivation on particle size and water-retention characteristics determined using wetting curves. Soil Science, Vol. 168 (No. 7): 459- 468.

26. Cabrera, M.L. & Sims, J.T. (2000). Beneficial use of poultry by-products: challenges and opportunities, In: Land application of agricultural, industrial, and municipal by-products (1st edition), Power, J.F. & Dick, W.A. (Eds.). pp. 425-450. Soil Science Society of America, Madison, WI.

27. Canali, S., Tinchera, A., Intrigliolo, F., Pompili, L., Nisini, L., Mocali, S. & Torrisi, B. (2004). Effect of long term addition of composts and poultry manure on soil quality of citrus orchards in Southern Italy. Biology & Fertility of Soils, Vol. 40 (No. 3): 206-210.

28. Clason, T.R. (1995). Economic implications of silvipastures on southern pine plantations. Agroforestry Systems, Vol. 29: 227-238.

29. Clason, T.R. & Robinson, J.L. (2000). From a pine forest to a silvopasture system, Agroforestry Note 18, USDA Forest Service, USDA Natural Resource Conservation Service, Washington, D.C.

30. Clason, T.R. & Sharrow, S.H. (2000). Silvopastoral practices, In: North American Agroforestry: An Integrated Science and Practice, Garrett, H.E., Rietveld, W.J. & Fisher, R.F. (Eds.). pp. 119-147. Agronomy Society of America, Madison, WI.

31. Clesceri, L.S., Greenberg, A.E. & Eaton, A.D. (Eds.). (1998). Standard methods for the examination of water and wastewater (20th edition), American Public Health Association, Washington, DC.

32. Colbert, S.R., Jokela, E.J. & Neary, D.G. (1990). Effects of annual fertilization and sustained weed control on dry matter partitioning, leaf area, and growth efficiency of juvenile loblolly and slash pine. Forest Science, Vol. 36 (No. 4): 995-1014.

33. Daddow R.L. & Warrington, G.E. (1983). Growth-limiting soil bulk densities as influenced by soil texture, Watershed Systems Development Group Report WSDG-TN-00005, USDA Forest Service, Fort Collins, CO.

34. Daniel, T.C., Sharpley, A.N. & Lemunyon, J.L. (1998). Agricultural phosphorus and eutrophication: a symposium overview. Journal of Environmental Quality, Vol. 27 (No. 2): 251-257.

35. Danielson, R.E. & Sutherland, P.L. (1986). Porosity, In: Methods of Soil Analysis, Part 1: Physical and Mineralogical Methods (2nd edition). Klute, A. (Ed.). pp. 443-462. Soil Science Society of America, Madison, WI.

36. Dick, W.A., Eckert, D.J. & Johnson, J,W. (1998). Land application of poultry litter, Ohio State University Cooperative Extension Fact Sheet ANR-4-98, Ohio State University, Columbus, OH.

37. Dickens, E.D., Bush, P.B., & Morris, L.A. (2003). Poultry litter application recommendations in pine plantations. Warnell School of Forestry and Natural Resources, College of Agricultural and Environmental Sciences, University of Georgia, Athens, GA, Retrieved from http://www.bugwood.org/fertilization/PLARPP.html

38. Dickens, E.D., Richardson, B.W. & McElvany, B.C. (2004). Old-field thinned loblolly pine plantation fertilization with diammonium phosphate plus urea and poultry litter: 4 year growth and product class distribution results, In: Proceedings of the 12th Biennial Southern Silvicultural Research Conference, General Technical Report SRS-48, Outcalt, K. (Ed.), pp. 395-397. USDA Forest Service, Southern Research Station, Asheville, NC.

39. Donegan, K.K., Watrud, L.S., Seidler, R.J., Maggard, S.P., Shiroyama, T., Porteous, L.A. & DiGiovanni, G. (2001). Soil and litter organisms in Pacific Northwest forests under different management practices. Applied Soil Ecology, Vol. 18 (No. 2): 159-175.

40. Duryea, M.L. & Edwards, J.C. (1989). Pine-straw management in Florida's forest, Florida Cooperative Extension Service Institute of Food

and Agricultural Science Circular 831, University of Florida, Gainsville, FL.

41. Edmeades, D.C. (2003). The long-term effects of manures and fertilizers on soil productivity and quality: a review. Nutrient Cycling in Agroecosystems, Vol. 66 (No. 2): 165-180.

42. Edwards, D.R. & Daniel, T.C. (1992). Environmental impacts of on-farm poultry waste disposal: a review. Bioresource Technology, Vol. 41 (No. 1): 9-33.

43. Eichhorn, M.M. (2001). Impact of best management practices and organic wastes on water quality and crop production: Poultry litter application demonstration project, Louisiana Department of Environmental Quality Projects CFMS514283 and CFMS554784 Final Report, Hill Farm Research Station, Louisiana Agricultural Experiment Station, Baton Rouge, LA.

44. Ekinci, K., Keener, H.M. & Elwell. D.L. (2000). Compositing short paper fiber with broiler litter and additives Part 1: Effects of initial pH and carbon/nitrogen ratio on ammonia emissions. Compost Science & Utilization, Vol. 8 (No. 2): 160-172.

45. Ellum, K. (2010). Pine straw raking and fertilizer source impacts on nitrogen mineralization, pine needle gas exchange, and tree water stress in a loblolly pine plantation, M.S. thesis, School of Forest Resources, University of Arkansas, Monticello, AR.

46. Elser, J. J., Bracken, M.E.S., Cleland, E.E., Gruner, D.S., Harpole, W.S., Hillebrand, H., Ngai, J.T., Seabloom, E.W., Shurin, J.B. & Smith, J.E. (2007). Global analysis of nitrogen and phosphorus limitation of primary producers in freshwater, marine, and terrestrial ecosystems. Ecology Letters, Vol. 10 (No. 12): 1135–1142.

47. Eno, C.F. (1960). Nitrate production in the field by incubating the soil in polyethylene bags. Soil Science Society of America Proceedings, Vol. 24: 277–279.

48. Evans, H.A. (2000). Application of poultry litter and commercial fertilizer in a loblolly pinebahiagrass silvopasture, M.S. thesis, College of Forestry, Stephen F. Austin University, Nacogdoches, TX.

49. Falkiner, R.A. & Polglase, P.J. (1997). Transport of phosphorus through soil in an effluentirrigated tree plantation. Australian Journal of Soil Research, Vol. 35: 385-398.

50. Farley, K.A., Jobbágy, E.G. & Jackson, R.B. (2005). Effects of afforestation on water yield: a global synthesis with implications for policy. Global Change Biology, Vol. 11 (No. 10): 1565-1576.

51.   Fauci, F. & Dick, R.P. (1994). Microbial biomass as an indicator of soil quality: effects of long-term management and recent soil amendments, In: Defining soil quality for a sustainable environment (1st edition). Doran, J.W., Coleman, D.C., Bezdicek, D.F. & Stewart, B.A. (Eds). pp. 229-234. Soil Science Society of America, Madison, WI.

52.   Fisher, D.S., Steiner, J.L., Endale, D.M., Stuedemann, J.A., Schomberg, H.H., Franzluebbers, A.J. & Wilkinson, S.R. (2000). The relationship of land use practices to surface water quality in the Upper Oconee Watershed in Georgia. Forest Ecology and Management, Vol. 128 (No. 1-2): 39-48.

53.   Friend, A.L., Roberts, S.D., Schoenholtz, S.H., Mobley, J.A. & Gerard, P.D. (2006). Poultry litter application to loblolly pine forests: growth and nutrient containment. Journal of Environmental Quality, Vol. 35 (No. 3): 837-848.

54.   Funderberg, E. (2009). Poultry litter as fertilizer. In: Ag News and Views, January 2009, Soil & Crops. The Samuel Roberts Noble Foundation. Available from: http://www.noble.org/ag/Soils/PoultryLitter/index.html.

55.   Gallardo, A. & Schlesinger, W.H. (1994). Factors limiting microbial biomass in the mineral soil and forest floor of a warm-temperate forest. Soil Biology & Biochemistry. Vol. 26 (No 10): 1409-1415.

56.   Gallimore, L.E., Basta, N.T., Storm, D.E., Payton, M.E., Huhnke, R.H. & Smolen, M.D. (1999). Water treatment residual to reduce nutrients in surface runoff from agricultural land. Journal of Environmental Quality, Vol. 28 (No. 5): 1474-1478.

57.   Gambrell, R.P. (1996). Manganese, In: Methods of soil analysis, part 3: Chemical methods (3rd edition). Bartels, J.M. (Ed.) pp. 665-682. Soil Science Society of America, Madison, WI.

58.   Gaston, L.A., Clason, T.R. & Cooper, D. (2003). Poultry litter fertilizer on pasture, silvopasture, and forest soils. Louisiana Agriculture, Vol. 46 (No. 3): 22-23.

59.   Gaston, L.A., Drapcho, C.M., Tapadar, S. & Kovar, J.L. (2003). Phosphorus runoff relationships for Louisiana Coastal Plain soils amended with poultry litter. Journal of Environmental Quality, Vol. 32 (No. 4): 1422-1429.

60.   Gee, G.W., Campbell, M.D., Campbell, G.S. & Campbell, J.H. (1992). Rapid measurement of low soil water potentials using a water activity meter. Soil Science Society of America Journal, Vol. 56 (No. 4): 1068-1070.

61.   Gilman, E.F. (1987). Where are tree roots? Florida Cooperative Extension Service, Institute of Food and Agricultural Sciences, University of

Florida, Extension Bulletin ENH137, Retrieved from http://edis.ifas.ufl. edu/pdffiles/WO/WO01700.pdf

62. Gupta, G., Borowiec, J. & Okoh, J. (1997). Toxicity identification of poultry litter aqueous leachate. Poultry Science, Vol. 76 (No. 10): 1364-1367.

63. Harmel, R.D., Torbet, H.A., Haggard, B.E., Haney, R. & Dozier, M. (2004). Water quality impacts of converting to a poultry litter fertilization strategy. Journal of Environmental Quality, Vol. 33 (No. 6): 2229-2242.

64. Harris, J.A. (2003). Measurements of the soil microbial community for estimating the success of restoration. European Journal of Soil Science, Vol. 54 (No. 4): 801-808.

65. Hart, S.C., Stark, J.M., Davidson, E.A. & Firestone, M.K. (1994). Nitrogen mineralization, immobilization, and nitrification, In: Methods of Soil Analysis. Part 2. Microbiological and Biochemical Properties. Weaver, R.W., Angele, S., Bottomly, P. (Eds.), pp. 985- 1018. Soil Science Society of America. Madison, WI.

66. Haywood, J.D., Tiarks, A.E. & Sword, M.A. (1997). Fertilization, weed control, and pine litter influence loblolly pine stem productivity and root development. New Forests, Vol. 14 (No. 3): 233-249.

67. Haywood, J.D., Tiarks, A.E., Elliott-Smith, M.L. & Pearson, H.A. (1998). Response of direct seeded Pinus palustris and herbaceous vegetation to fertilization, burning, and pine straw harvesting. Biomass Bioenergy, Vol. 14 (No. 2): 157-167.

68. Helmke, P.A. & Sparks, D.L. (1996). Lithium, sodium, potassium, rubidium, and cesium, In: Methods of soil analysis, part 3. Chemical methods, Bartels, J.M. (Ed.), pp. 551-574. Soil Science Society of America, Madison, WI.

69. Henry, C.L., Cole, D.W., Hinckley, T.M. & Harrison, R.B. (1993). The use of municipal and pulp and paper sludges to increase production in forestry. Journal of Sustainable Forestry, Vol. 1 (No. 3): 41-55.

70. Högberg, P., Nordgren, A., Buchmann, N., Taylor, A.F.S., Ekblad, A., Högberg, M.N., Nyberg, G., Ottoson-Löfvenius, M. & Read, D.J. (2001). Large-scale forest girdling shows that current photosynthesis drives soil respiration. Nature, Vol. 411: 789-792.

71. Horneck, D.A. & Miller, R.O. (1998). Determination of total nitrogen in plant tissue, In: Handbook of reference methods for plant analysis, Yask, P. (Ed.), pp. 75-83. CRC Press, Boca Raton, FL.

72. Hue, N.V. (1992). Correcting soil acidity of a highly weathered ultisol with chicken manure and sewage sludge. Communications in Soil Science and Plant Analysis, Vol. 23 (No.3- 4): 241-264.

73. Jackson, M.J., Line, M.A., Wilson, S. & Hetherington. S.J. (2000). Application of composted pulp and paper mill sludge to a young pine plantation. Journal of Environmental Quality, Vol. 29 (No. 2): 407-414.

74. Jenkinson, D.S. & Powlson, D.S. (1976a). The effects of biocidal treatments on metabolism in soil-I. Fumigation with chloroform. Soil Biology & Biochemistry, Vol. 8 (No. 3): 167- 177.

75. Jenkinson, D.S. & Powlson, D.S. (1976b). The effects of biocidal treatments on metabolism in soil-V: A method for measuring soil biomass. Soil Biology & Biochemistry, Vol. 8 (No. 3): 209-213.

76. Jokela, E.J. (2004). Nutrient management for southern pines, In: Slash pine: still growing and growing! Proceedings of the slash pine symposium, General Technical Report SRS-76, Dickens, E.D., Barnett, J.P., Hubbard, W.G. & Jokela, E.J. (Eds.), pp. 27-35. U.S. Department of Agriculture, Forest Service, Southern Research Station, Asheville, NC.

77. Jones, J.B. & Case, V.W. (1990). Sampling, handling, and analyzing plant tissue samples. In: Soil testing and plant analysis (3rd edition), Westerman, R.L. (Ed.), pp. 389-447. Soil Science Society of America, Madison, WI.

78. Jorgensen, J.R. & Wells, C.G. (1986). Foresters' primer in nutrient cycling: a loblolly pine management guide, General Technical Report SE-37, USDA Forest Service, Southeastern Forest Experiment Station, Asheville, NC.

79. Kellogg, R.L., Lander, C.H., Moffitt, D.C. & Gollehon, N. (Eds.). (2000). Manure nutrients relative to the capacity of cropland and pastureland to assimilate nutrients: Spatial and temporal trends for the United States. USDA Natural Resources Conservation Service Publication NPS 00-0579, GSA National Forms and Publication Center, Fort Worth, TX.

80. Khan, S.A., Mulvaney, R.L., Ellsworth, T.R. & Boast, C.W. (2007). The myth of nitrogen fertilization for soil carbon sequestration. Journal of Environmental Quality, Vol. 36 (No. 6): 1821-1832.

81. Kingery, W.L., Wood, C.W., Delaney, D.P., Williams, J.C. & Mullins, G.L. (1993). Implications of long-term application of poultry litter on tall fescue pastures. Journal of Production Agriculture, Vol. 6 (No. 3): 315-395.

82. Kingery, W.L., Wood, C.W., Delaney, D.P., Williams, J.C. & Mullins, G.L. (1994). Impact of long-term land application of broiler litter on

environmentally related soil properties. Journal of Environmental Quality, Vol. 23 (No. 1): 139-147.

83. Kronzucker, H.J., Yaeesh Siddiqi, M. & Glass, A.D.M. (1997). Conifer root discrimination against soil nitrate and the ecology of forest succession. Nature, Vol. 385: 59-61.

84. Lemunyon, J. & Gilbert, R. (1993). The concept and need for a phosphorus assessment tool. Journal of Production Agriculture, Vol.6 (No. 4): 483-486.

85. Lenhard, G. (1956). The dehydrogenase activity in soil as a measure of the activity of soil microorganisms. Z. Pflanzenernäh Düng Bodenkd, Vol. 73:1-11.

86. Liechty, H.O., Blazier, M.A., Wight, J.P., Gaston, L.A., Richardson, J.D. & Ficklin, R.L. (2009). Assessment of repeated application of poultry litter on phosphorus and nitrogen dynamics in loblolly pine: implications for water quality. Forest Ecology and Management, Vol. 258 (No. 10): 2294-2303.

87. Lopez-Zamora, I., Duryea, M.L., McCormac, W.C., Comerford, N.B. & Neary, D.G. (2001). Effect of pine needle removal and fertilization on tree growth and soil P availability in a Pinus elliottii Engelm. var. elliottti stand. Forest Ecology and Management, Vol. 148 (No 1-3): 125-134.

88. Lynch, L. & Tjaden, R. (2004). Would forest landowners use poultry manure as fertilizer? Journal of Forestry, Vol. 102 (No. 5): 40-45.

89. McLeod, R.V. & Hegg, R.O. (1984). Pasture runoff quality from application of inorganic and organic nitrogen sources. Journal of Environmental Quality, Vol. 13 (No. 1): 122-126.

90. Mehlich, A. (1984). Mehlich 3 soil test extractant: A modification of the Mehlich 2 extractant. Communications in Soil Science and Plant Analysis, Vol. 15 (No. 12): 1409-1416.

91. Mills, R. & Robertson, D.R. (1991). Production and marketing of Louisiana pine straw, Louisiana Cooperative Extension Service Publication 2340, Louisiana State University Agricultural Center, Baton Rouge, LA.

92. Minogue, P.J., Ober, H.K. & Rosenthal, S. (2007). Overview of Pine Straw Production in North Florida: Potential Revenues, Fertilization Practices, and Vegetation Management Recommendations, School of Forest Resources and Conservation Department, Florida Cooperative Extension Service, Institute of Food and Agricultural Sciences Publication 125. University of Florida, Gainesville, FL.

93. Mitchell, C.C. & Donald, J.O. (1995). The value and use of poultry manures as fertilizers, Alabama Cooperative Extension System Circular ANR-224, Auburn University, Auburn, AL.

94. Mitchell, C.C. & Tu, S. (2006). Nutrient accumulation and movement from poultry litter. Soil Science Society of America Journal, Vol. 70 (No. 6): 2146-2153.

95. Mulvaney, R.L. (1996). Nitrogen-inorganic forms, In: Methods of Soil Analysis. Part 3. Chemical methods, Sparks, D.L. (Ed.), pp. 1123-1184. Soil Science Society of America, Madison, WI.

96. Munkholm, L.J., Schjønning, P., Debosz, K., Jensen, H.E. & Christensen, B.T. (2002). Aggregate strength and mechanical behavior of a sandy loam under long-term fertilization treatments. European Journal of Soil Science, Vol. 53 (No. 1): 129-137.

97. Murthy, R., Zarnoch, S.J. & Dougherty, P.M. (1997). Seasonal trends of light-saturated net photosynthesis and stomatal conductance of loblolly pine trees grown in contrasting environments of nutrition, water, and carbon dioxide. Plant, Cell, and Environment, Vol. 20 (No. 5): 558-568.

98. Nair, V.D. & Graetz, D.A. (2004). Agroforestry as an approach to minimizing nutrient loss from heavily fertilized soils. the Florida experience. Agroforestry Systems, Vol. 60: 269-279.

99. National Agricultural Statistics Service. (2008). Poultry-Production and Value 2007 Summary, United States Department of Agriculture National Agricultural Statistics Service, Washington, D.C.

100. O'Neill, G.J. & Gordon. A.M. (1994). The nitrogen filtering capacity of Carolina poplar in an artificial riparian zone. Journal of Environmental Quality, Vol. 23 (No. 6): 1218-1223.

101. Plaza, C., Hernández, D., García-Gil, J.C. & Polo, A. (2004). Microbial activity in pig slurryamended soils under semiarid conditions. Soil Biology & Biochemistry, Vol. 36 (No. 10): 1577-1585

102. Polglase, P.J., Tompkins, D. & Falkiner, R.A. (1995). Mineralization and leaching of nitrogen in an effluent-irrigated pine plantation. Journal of Environmental Quality, Vol. 24 (No. 5): 911-922.

103. Pote, D.H. & Daniel, T.C. (2000). Analyzing for total phosphorus and total dissolved phosphorus in water samples. In: Methods of Phosphorus Analysis for Soils, Sediments, Residuals, and Waters, Southern Cooperative Series Bulletin No. 396., Pierzynski, G.M. (Ed.), pp. 91-93. Retrieved from: http://www.sera17.ext.vt.edu/Documents/Methods_of_P_Analysis_2000.pdf

104. Pote, D.H. & Daniel, T.C. (2008). Managing pine straw harvests to minimize soil and water losses. Journal of Soil and Water Conservation, Vol. 63 (No.1): 27-28.

105. Pote, D.H., Kingery, W.L., Aiken, G.E., Han, F.X., Moore Jr., P.A. & Buddington, K. (2003). Water-quality effects of incorporating poultry litter in perennial grassland soils. Journal of Environmental Quality, Vol. 32 (No. 6): 2392-2398.

106. Powers, R.F., Scott, D.A., Sanchez, F.G., Voldseth, R.A., Page-Dumroese, D., Elioff, J.D. & Stone, D.M. (2005). The North American long-term soil productivity experiment: findings from the first decade of research. Forest Ecology and Management, Vol. 220 (No. 1-3): 31-50.

107. Powlson, D.S. & Brookes, P.C. (1987). Measurement of soil microbial biomass provides an early indication of changes in total soil organic matter due to straw incorporation. Soil Biology & Biochemistry, Vol. 19 (No. 2): 159-164.

108. Prestemon, J.P. & Abt, R.C. (2002). The southern timber market to 2040. Journal of Forestry, Vol. 100 (No. 7): 16-22.

109. Pritchett, W.L. & Fisher, R.F. (1987). Properties and management of forest soils (2nd edition), John Wiley & Sons, Inc., New York, NY.

110. Richardson, J. (2006). Effects of poultry litter applied to pine plantations and pastures on water quality and soil nitrogen mineralization. M.S. thesis, University of Arkansas at Monticello.

111. Rifai, S.W., Markewitz, D. & Borders, B. (2010). Twenty years of intensive fertilization and competing vegetation suppression in loblolly pine plantations: impacts on soil C, N, and microbial biomass. Soil Biology & Biochemistry, Vol. 42 (No. 5): 713-723.

112. Roberts, S.D., Friend, A.L. & Gerard, P.D. (2004). The effect of large applications of nutrients from organic waste on biomass allocation and allometric relations in loblolly pine, In: Proceedings of the 12th Biennial Southern Silvicultural Research Conference, General Technical Report SRS-48, Outcalt, K. (Ed.), pp. 398-402. USDA Forest Service, Southern Research Station, Asheville, NC.

113. Roberts, S.D., Friend, A.L. & Schoenholtz, S.H. (2006). Growth of precommercially thinned loblolly pine four years following application of poultry litter, In: Proceedings of the 12th Biennial Southern Silvicultural Research Conference, General Technical Report SRS-92, Outcalt, K. (Ed). pp. 139-142. USDA Forest Service, Southern Research Station, Asheville, NC.

114. Roise, J.P., Chung, J. & Lancia, R. (1991). Red-cockaded woodpecker habitat management and longleaf pine straw production: an economic analysis. Southern Journal of Applied Forestry, Vol. 15 (No. 2): 88-92.

115. Samuelson, L.J., Wilhoit, J. Stokes, T. & Johnson, J. (1999). Influence of poultry litter fertilization on 18-year-old loblolly pine stand. Communications in Soil Science and Plant Analysis, Vol. 30 (No. 3-4): 509-518.

116. Sanchez, F.G., Scott, D.A. & Ludovici, K.H. (2006). Negligible effects of severe organic matter removal and soil compaction on loblolly pine growth over 10 years. Forest Ecology and Management, Vol. 227 (No. 1): 145-154.

117. Sauer, T.J., Daniel, T.C., Nichols, D.J., West, C.P., Moore, P.A. & Wheeler, G.L. (2000). Runoff water quality from poultry litter-treated pasture and forest sites. Journal of Environmental Quality, Vol. 29 (No. 2): 515-521.

118. Schindler, D. (1978). Factors regulating phytoplankton production and standing crop in the word's freshwaters. Limnology and Oceanography, Vol. 23 (No. 3): 478-486.

119. Sharpley, A.N. (1999). Agricultural phosphorus, water quality, and poultry production: are they compatible. Poultry Science, Vol. 78 (No. 5): 660-673.

120. Sharpley, A.N. & Menzel, R.G. (1987). The impact of soil and fertilizer P on the environment. Advances in Agronomy, Vol. 41: 297-324.

121. Sharpley, A.N., Smith, S.J. & Bain, W.R. (1993). Nitrogen and phosphorus fate from longterm poultry litter applications to Oklahoma soils. Soil Science Society of America Journal, Vol. 57 (No. 4): 1131-1137.

122. Shober, A.L. & Sims, J.T. (2003). Phosphorus restrictions and land application of biosolids: current status and future trends. Journal of Environmental Quality, Vol. 32 (No. 6): 1955-1964.

123. Sims, J.T. (1986). Nitrogen transformations in a poultry manure amended soil: Temperature and moisture effects. Journal of Environmental Quality, Vol. 15 (No. 1): 59-63.

124. Sims, J.T. & Wolf, D.C. (1994). Poultry waste management: Agricultural and environmental issues. Advances in Agronomy, Vol. 52: 1-83.

125. Sistani, K.R., Adeli, A., McGowen, S.L., Tewolde, H. & Brink, G.E. (2008). Laboratory and field evaluation of broiler litter nitrogen mineralization. Bioresource Technology, Vol. 99 (No. 7): 2603-2611.

126. Sistani, K.R., Brink, G.E., Adeli, A., Tewolde, H. & Rowe, D.E. (2004). Year-round soil nutrient dynamics from broiler litter application to three bermudagrass cultivars. Agronomy Journal, Vol. 96 (No. 2): 525-530.

127. Tan, X., Chang, S.X. & Kabzems, R. (2005). Effects of soil compaction and forest floor removal on soil microbial properties and N transformations in a boreal forest longterm soil productivity study. Forest Ecology and Management, Vol. 217 (No. 2-3): 158- 170.

128. Tate, D.F. (1994). Determination of nitrogen in fertilizer by combustion: Collaborative study. Journal of AOAC International, Vol. 77 (No. 4): 829-839.

129. Tekeste, M., Hatzhghi, D.H. & Stroonsnijder, L. (2007). Soil strength assessment using threshold probability approach on soils from three agro-ecological zones in Eritrea. Biosystems Engineering, Vol. 98 (No. 4): 470-478.

130. Tilman, D., Hill, J. & Lehman, C. (2006). Carbon-negative biofuels from low-input highdiversity grassland biomass. Science, Vol. 314 (No. 5805): 1598-1600.

131. United States Department of Agriculture (USDA) Economic Research Service. (2009). Poultry and Eggs: Background. Retrieved from: http://www.ers.usda.gov/Briefing/Poultry/Background.htm

132. United States Department of Agriculture Soil Conservation Service (USDA SCS). (1989). Soil Survey of Claiborne Parish, Louisiana. USDA SCS, Washington, D.C.

133. Vose, J.M. & Allen, H.L. (1988). Leaf area, stemwood growth, and nutrition relationships in loblolly pine. Forest Science, Vol. 34: 547-563.

134. Wagner, G.H. & Wolf, D.C. (1999). Carbon transformations and soil organic matter formation, In: Principles and Applications of Soil Microbiology, Sylvia, D.M., Fuhrmann, J.J., Hartel, P.G. & Zuberer, D.A. (Eds.). pp. 218-258. Prentice Hall Inc., Upper Saddle River, NJ.

135. Walkley, A. (1947). A critical examination of a rapid method for determining organic carbon in soils: Effect of variations in digestion conditions and of inorganic soil constituents. Soil Science, Vol. 63: 251-263.

136. Wallenstein, M.D., McNulty, S., Fernandez, I.J., Boggs, J. & Schlesinger, W.H. (2006). Nitrogen fertilization decreases forest soil fungal and bacterial biomass in three long-term experiments. Forest Ecology and Management, Vol. 222 (No. 1-3): 459-468.

137. Warren, S.L. & Fonteno, W.C. (1993). Changes in physical and chemical properties of a loamy sand soil when amended with poultry litter. Journal of Environmental Horticulture, Vol. 11 (No. 4): 186-190.

138. Weaver, T. (1998). Managing poultry manure nutrients. Agricultural Research, Vol. 46: 12-13.

139. Will, R.E., Markewitz, D., Hendrick, R.L., Meason, D.F., Crocker, T.R. & Borders, B.E. (2006). Nitrogen and phosphorus dynamics for 13-year-old loblolly pine stands receiving complete competition control and annual N fertilizer. Forest Ecology and Management, Vol. 227 (No. 1-2): 155-168.

140. Williams, C.M, Barker, J.C & Sims, J.T. (1999). Management and utilization of poultry wastes. Reviews of Environmental Contamination & Toxicology, Vol. 162: 105-157.

141. Zarcinas, B.A., Cartwright, B. & Spouncer, L.R. (1987). Nitric acid digestion and multinutrient analysis of plant material by inductively coupled plasma spectrometry. Communications in Soil Science and Plant Analysis, Vol. 18 (No. 1): 131-146.

142. Zimmermann, B., Elsenbeer, H. & DeMoraes, J.M. (2006). The influence of land-use changes on soil hydraulic properties: Implications for runoff generation. Forest Ecology and Management, Vol. 222 (No. 1-3): 29-38.

143. Zinkhan, F.C. & Mercer, D.E. (1996). An assessment of agroforestry systems in the southern U.S.A. Agroforestry Systems, Vol. 35: 303-321.

144. Zou, X.M., Ruan, H.H., Fu, Y., Yang, X.D. & Sha, L.Q. (2005). Estimating soil labile organic carbon and potential turnover rates using a sequential fumigation-incubation procedure. Soil Biology & Biochemistry, Vol. 37 (No. 10): 1923-1928.

CITATION

CHAPTER 1

Suzanne Simard, and Mary Austin (2010). The Role of Mycorrhizas in Forest Soil Stability with Climate Change, Climate Change and Variability, Suzanne Simard (Ed.), ISBN: 978-953-307-144-2, InTech, DOI: 10.5772/9813.

CHAPTER 2

Shih-Hao Jien, Chung-Chi Wang, Chia-Hsing Lee, and Tsung-Yu Lee. Stabilization of organic matter by biochar application in compost-amended soils with contrasting pH values and textures, Sustainability 2015, 7(10), 13317-13333; doi:10.3390/su71013317.

CHAPTER 3

Wang C, He N, Zhang J, Lv Y, Wang L (2015) Long-Term Grazing Exclusion Improves the Composition and Stability of Soil Organic Matter in Inner Mongolian Grasslands. PLoS ONE 10(6): e0128837. doi:10.1371/journal.pone.0128837

CHAPTER 4

John J. Sloan, Peter A.Y. Ampim, Raul I. Cabrera, Wayne A. Mackay and Steve W. George (2011). Moisture and Nutrient Storage Capacity of Calcined Expanded Shale, Principles, Application and Assessment in Soil Science, Dr. Burcu E. Ozkaraova Gungor (Ed.), ISBN: 978-953-307-740-6, InTech, DOI: 10.5772/30141.

CHAPTER 5

B. Suneel Kumar and T. V. Preethi, Behavior of clayey soil stabilized with rice husk ash & lime, International Journal of Engineering Trends and Technology (IJETT) – Volume 11 Number 1 - May 2014, ISSN: 2231-5381

CHAPTER 6

T. Watanabe, M. S. H. Khan, I. M. Rao, J. Wasaki, T. Shinano, M. Ishitani, H. Koyama, S. Ishikawa, K. Tawaraya, M. Nanamori, N. Ueki and T. Wagatsuma (2011). Physiological and Biochemical Mechanisms of Plant Adaptation to Low-Fertility Acid Soils of the Tropics: The Case of Brachiariagrasses, Principles, Application and Assessment in Soil Science, Dr. Burcu E. Ozkaraova Gungor (Ed.), ISBN: 978-953-307-740-6, InTech, DOI: 10.5772/30334.

CHAPTER 7

O.M. Nieto, J. Castro and E. Fernández (2011). Long-Term Effects of Residue Management on Soil Fertility in Mediterranean Olive Grove: Simulating Carbon Sequestration with RothC Model, Principles, Application and Assessment in Soil Science, Dr. Burcu E. Ozkaraova Gungor (Ed.), ISBN: 978-953-307-740-6, InTech, DOI: 10.5772/31064.

CHAPTER 8

Magdalena Borzecka-Walker, Antoni Faber, Katarzyna Mizak, Rafal Pudelko and Alina Syp (2011). Soil Carbon Sequestration Under Bioenergy Crops in Poland, Principles, Application and Assessment in Soil Science, Dr. Burcu E. Ozkaraova Gungor (Ed.), ISBN: 978-953-307-740-6, InTech, DOI: 10.5772/29678.

CHAPTER 9

Mareike Ließ, Bruno Glaser and Bernd Huwe (2011). Soil-Landscape Modelling – Reference Soil Group Probability Prediction in Southern Ecuador, Principles, Application and Assessment in Soil Science, Dr. Burcu E. Ozkaraova Gungor (Ed.), ISBN: 978-953-307-740-6, InTech, DOI: 10.5772/29468.

CHAPTER 10

J. Shamshuddin and Noordin Wan Daud (2011). Classification and Management of Highly Weathered Soils in Malaysia for Production of Plantation Crops, Principles, Application and Assessment in Soil Science, Dr. Burcu E. Ozkaraova Gungor (Ed.), ISBN: 978-953-307-740-6, InTech, DOI: 10.5772/29490.

CHAPTER 11

Michael A. Blazier, Hal O. Liechty, Lewis A. Gaston and Keith Ellum (2011). Poultry Litter Fertilization Impacts on Soil, Plant, and Water Characteristics in Loblolly Pine (Pinus taeda L.) Plantations and Silvopastures in the Mid-South USA, Principles, Application and Assessment in Soil Science, Dr. Burcu E. Ozkaraova Gungor (Ed.), ISBN: 978-953-307-740-6, InTech, DOI: 10.5772/29356.

INDEX